PREPARING FOR HYBRID THREATS TO SECURITY

This book examines hybrid threats within the broader context of a security crisis in Europe.

As geopolitical tensions increase and great power rivalries intensify, can states protect their communities? While conventional wars are fought, parallel battles take place through more subtle and non-violent means. This multidisciplinary book examines how hybrid threats undermine political governance and social stability in liberal democracies, covering aggressors, targeted states, and victimized communities. It seeks to address how aggressor states undermine liberal democracies under the threshold of conflict and the role played by hybrid threats as aggressor states prepare for full-scale war. The chapters also explore how liberal democracies organize and interact to detect hybrid threats, arguing that, in order to increase resilience, politicians and government agencies must involve the private sector and citizens in threat-reduction policies. The analysis builds upon the latest research in the international crisis management literature.

This book will be of interest to students of security studies, hybrid warfare, defence studies, and international relations, as well as professional practitioners.

Odd Jarl Borch is a professor of Strategy at the Nord University Business School, Bodo, Norway. He is the author of more than 150 scientific publications. He has served on several government committees within industry development, safety, and preparedness, including the Norwegian Total Preparedness Commission.

Tormod Heier is a professor of Military Strategy and Operations at the Norwegian Defence University College, Norway. He has edited or authored 14 books on crisis management, security and defence policy, and military conflicts. He was awarded the Norwegian PEN's Ossietzky Price for Freedom of Speech in 2017 and the Norwegian Army's Medal of Merit in 2023.

PREPARING FOR HYBRID THREATS TO SECURITY

Collaborative Preparedness and Response

Edited by Odd Jarl Borch and Tormod Heier

Routledge
Taylor & Francis Group

LONDON AND NEW YORK

Designed cover image: © KanawatTH – Getty Images

First published 2025
by Routledge
4 Park Square, Milton Park, Abingdon, Oxon OX14 4RN

and by Routledge
605 Third Avenue, New York, NY 10158

Routledge is an imprint of the Taylor & Francis Group, an informa business

Open Access funding was provided by the Inland Norway University
of Applied Sciences, the Nord University, the NTNU Social Research
and the Norwegian Defense University College.

British Library Cataloguing-in-Publication Data
A catalogue record for this book is available from the British Library

Library of Congress Cataloging-in-Publication Data
Names: Borch, Odd Jarl, editor. | Heier, Tormod, editor.
Title: Preparing for hybrid threats to security : collaborative
    preparedness and response / edited by Odd Jarl Borch, Tormod Heier.
Description: Abingdon, Oxon ; New York : Routledge, 2025. |
    Includes bibliographical references and index.
Identifiers: LCCN 2024022471 (print) | LCCN 2024022472 (ebook) |
    ISBN 9781032617923 (hardback) | ISBN 9781032627014 (paperback) |
    ISBN 9781032617916 (ebook)
Subjects: LCSH: Hybrid warfare—Europe. | National security—Europe.
Classification: LCC U167.5.I8 P74 2025 (print) | LCC U167.5.I8
    (ebook) | DDC 355/.03304—dc23/eng/20240708
LC record available at https://lccn.loc.gov/2024022471
LC ebook record available at https://lccn.loc.gov/2024022472

ISBN: 978-1-032-61792-3 (hbk)
ISBN: 978-1-032-62701-4 (pbk)
ISBN: 978-1-032-61791-6 (ebk)

DOI: 10.4324/9781032617916

Typeset in Sabon LT Pro
by Apex CoVantage, LLC

CONTENTS

TABLES

FIGURES

CONTRIBUTORS

Gordan Akrap is the president of the Hybrid Warfare Research Institute, Zagreb, Croatia. He graduated from Zagreb Faculty of electronics and computing in 1994 and received his Ph.D. at the University of Zagreb in the field of Information and Communication sciences. He is active in research of national and regional security, information and communication sciences, intelligence, and the history of Homeland War. He has published several books and articles in journals and proceedings. Akrap is founder of Zagreb Security Forum, Editor-in-chief of *National Security and The Future Journal*, and Assistant Rector at "Dr. Franjo Tuđman" Defense and Security University in Zagreb, Croatia.

Nina Andriianova is a leading researcher of the Center for Military and Strategic Studies of the National Defence University of Ukraine. She received her Ph.D. in political sciences in 2019, in area of the transformation of the armed forces and intelligence services during Euro-Atlantic integration of Poland Republic and comparison with Ukraine. Andriianova's research field of interest covers among other things countermeasures in the Russian-Ukrainian war, foreign military assistance to Ukraine during the Russian-Ukrainian War, Ukrainian problems of Euro-Atlantic integration, and the foundations of the Russian Federation's imperial policy. Nina participates in the joint Ukrainian-Norwegian research project "Total Defence" within the context of the National Defence University of Ukraine and the Norwegian Defence University College. Here, her latest publication is "Theoretical and Applied Aspects of the Russian-Ukrainian War: Hybrid Aggression and National Resilience", edited by M. Koval, 2023.

Bjørn T. Bakken is Associate Professor of Crisis Management at the Inland Norway University of Applied Sciences (INN University). Bakken is currently Head of the BA Program and MA Specialization in Preparedness and Crisis Management. He holds master's and bachelor's degrees from the Norwegian School of Economics and Business Administration (NHH) in Bergen, Norway, and a Ph.D. in Business and Economics from the BI Norwegian Business School in Oslo, Norway. His doctoral work is on the psychology of decision-making in crisis management. Bakken has published his research in international scientific books and journals, such as the *Journal of Occupational and Organizational Psychology*, *Journal of Behavioral Decision Making*, and *Journal of Contingencies and Crisis Management*. He has managed international research and development projects on preparedness and crisis management, and is currently researching decision-making in crisis, and how technology can improve the effectiveness of crisis management training—for example, by employing VR, AR, and AI to support simulations in training and exercises.

Joakim Berndtsson is Associate Professor in the School of Global Studies, University of Gothenburg, Sweden. Berndtsson finished his Ph.D. in 2009 on *The Privatisation of Security and State Control of Force: Changes, Challenges and the Case of Iraq*. Berndtsson's research focuses on the privatisation of war and security, security studies, civil-military relations, the transformation of war, as well as Swedish defence and security policies. He is currently leading a research project on Civilian-Military Collaboration in Scandinavian total defence organisations, funded by the Swedish Research Council. He also works on projects on military professionalism and on public opinion on defence, funded by the Swedish Armed forces. He is also PI for the project "Outsourcing migration control: Externalizing EU borders to Africa", funded by the Swedish Research Council.

Nina M. Bjørge is an assistant professor at the Royal Norwegian Air Force Academy. She holds a master's degree in political science from NTNU specialized in Norwegian foreign policy and international political economy. Her research interests are in hybrid threats and warfare, technological developments and their implications for international politics, and the mediation of war. She has previously researched narratives about Russia in Norwegian newspapers and different perspectives on societal resilience and cybercrime. Her current research projects are about hybrid threats and the mediation of the current Russian-Ukrainian War through TikTok.

Ole Boe is a full professor of organization and leadership at the Norwegian Police University College and the University of South-Eastern Norway and a professor II of crisis management and preparedness at the Inland Norway University of Applied Sciences (INN University). He holds a Ph.D. in cognitive psychology and previously served as an associate professor at the Norwegian Military Academy and the Norwegian Defence University College, where he taught military leadership and conducted leadership development for officers. He has published over 370 scientific works on stress management, character strengths, decision-making, military leadership, and leadership development. He is an editor and author/coauthor of eight books. Boe is a graduate of the Norwegian Defence Command and Staff College and served almost 20 years in the military, including several years as an operations officer (Captain, Retd.) in a Norwegian military special unit.

Odd Jarl Borch is a professor of Strategy at the Nord University Business School, Bodo, Norway. He received his M.Sc. from The Norwegian School of Economics and Business Administration in Bergen, Norway in 1979, and his Ph.D. from Umea University in Sweden in 1990. Borch has a Master Mariner education from Bodin Maritime Academy. He conducts research within the field of organization and strategic management, focusing on crisis management. Dr. Borch has authored more than 150 scientific publications. He has served at several government committees within industry development, safety, and preparedness, including the Norwegian Total Preparedness Commission. He was the founder and first leader of the University of the Arctic (UArctic) Thematic Network on Arctic Safety and Security, including 21 universities and research institutions in nine countries, and the first UArctic chair within safety and security.

Patrick Cullen is a senior researcher at the Fridtjof Nansen Institute. He was the program director for the Multinational Capability Development Campaign's Countering Hybrid Warfare Project from 2015 to 2019. He has been a common contributor to various projects related to countering Hybrid Threats, such as the European Centre for Excellence for Countering Hybrid Threats, as well as the European Commission. He has been invited to speak on the topic at various defense schools, international organizations, and multinational agencies worldwide, including the UN Security Council, the Pentagon, OSCE, and NATO. He received his Ph.D. from the London School of Economics in 2009 with an international security specialization and focus on violent non-state actors and the evolution of warfare.

Tanja Ellingsen is an associate professor of Political Science at the Faculty of Social Science at Nord University, Norway. Ellingsen holds a Ph.D. in Political Science from UiO on "Towards a Clash of Civilizations?" and has published in various journals, including *Journal of Conflict Resolution, Journal of Peace Research*, and *American Political Science Review*. Ellingsen has a particular interest in peace and conflict studies and was an associated member of Center of Excellence: Study of Civil War in Oslo, PRIO until 2014. Her research has particularly focused on the role of ethnic and religious differences for armed conflict and terrorism/extremism. She is currently part of large transdisciplinary research project Words and Violence funded by the Norwegian Research Council.

Jannicke Thinn Fiskvik is a researcher at NTNU Social Research. She holds a Ph.D. from the Norwegian University of Science and Technology (NTNU) on "Expeditionary Warfare and Changing Patterns of Civil-Military Relations: The Politics of War in the Nordic Countries, 2001–2014". Fiskvik has been a visiting scholar to the Center of Security Studies at ETH Zürich, the Center for War Studies at Southern Denmark University, the Finnish Institute of International Affairs, the Norwegian Institute for Defence Studies, and the Swedish Defence University. Her research includes examinations of potential foreign influence campaigns in Norwegian local elections, trends and dynamics of social media, and societal resilience. She is currently part of a large interdisciplinary project on new geopolitics and security governance of petroleum infrastructures, funded by the Research Council of Norway.

Tormod Heier is a professor of Military Strategy and Operations at the Norwegian Defence University College. He is also Professor-II at Innland Norway University of Applied Science and Associate Professor at the Swedish Defence University in Stockholm. As author, co-author, editor, and co-editor, Heier has published numerous books on crisis management, security and defence policy, and military conflicts in Afghanistan, Libya, and Ukraine. Heier served 30 years as army officer. He holds a Ph.D. in Political Science from the University of Oslo and a Masters' Degree in War Studies at King's College, University of London. As an officer, he has previously served in Brigade North, Norway's National Intelligence Service, the Norwegian Ministry of Defence, and NATO's mission in Afghanistan. Heier was awarded the Norwegian Army's Medal of Merit in 2023 and Norwegian PEN's Ozzietzky Price in 2017.

Thorvald Hærem earned his Ph.D. in Department of Informatics at Copenhagen Business School and is a professor of organizational psychology at BI Norwegian Business School, where he also serves as Head of the Department of Leadership and Organizational Behavior. His research interests include organizational and individual routines, behavioral decision-making, and information processing in general. He has published his research in journals, such as the *Journal of Applied Psychology*, *Journal of Behavioral Decision Making*, *MISQ*, *Organizational Studies*, *Academy of Management Review*, and *Organization Science*.

Valerii Hordiichuk is Chief of research department at the Centre for Military and Strategic Studies, National Defence University of Ukraine. Hordiichuk graduated from the Military Institute of Telecommunication and Informatization, Kyiv, in 2009. In 2020, he received his Ph.D. at the Odesa National Academy of Telecommunication named after O. Popov in the field of Information and Communication sciences. In 2020, Hordiichuk also received a Master of Science degree in operational naval services. His research fields of interest cover national security and defence, the art of war, military transformation and integration processes, and information and communication sciences.

Marte Høiby is a senior research scientist at Sintef Digital with research interests in hybrid threats, cyber threats, propaganda, and information war. She has background from academia as a former assistant professor in journalism and media studies. Høiby holds a Ph.D. in information sciences. She is also a course instructor in journalist safety training and has given lectures on safety to journalists in countries across Asia, Africa, and Latin America. She has been a visiting scholar at the United Nations University for peace and coordinated several cross-country research projects on topics related to media, war, and conflict.

Andrii Ivashchenko is a leading researcher at the Center for Military Strategic Studies of the National Defense University of Ukraine since 2008. Prior to that, he worked at the National Center for Defense Technologies and Military Security of Ukraine, the National Institute for Strategic Studies, and as an external advisor to the Committee on National Security and Defense of the Verkhovna Rada (Parliament) of Ukraine. Ivashchenko holds the rank of colonel. He served in various positions in the Air Defense Forces and the Ministry of Defense of Ukraine. He participated in the NATO SFOR operation in Bosnia and Herzegovina. He holds a Ph.D. and the rank of associate professor. Ivashchenko's research interests are modern wars and military conflicts, military strategies, and strategic planning.

Maksym Kamenetskyi is a candidate of Historical Sciences (Ph.D.), an associate professor, a foreign policy expert, and a Western Balkan region expert. Kamenetskyi graduated in 1986 from the Faculty of History at Taras Shevchenko National University of Kyiv, Ukraine. He has served as an associate professor in the International Relations and Foreign Policy Department at the Institute of International Relations at Taras Shevchenko National University of Kyiv. Kamenetskyi lectures on several courses, including foreign policy analysis, international terrorism in world policy, and special tools of foreign policy. From 1988 to 1996, he worked as an interpreter for the highest state leaders of Ukraine. He is an expert in international relations in the Balkan-Adriatic region, the theory of international relations, and foreign policy analysis, and he is one of the first researchers of the Soviet-Yugoslav conflict from 1948 to 1955 in the USSR.

Ørjan Nordhus Karlsson is a Ph.D. candidate at Nord University Business School, Bodø, Norway. His research interest is related to crisis management on the strategic level, including inter-organizational collaboration, trans-boundary threats, and total defense. Karlsson has an M.Sc. in Sociology from the University of Oslo (UiO). He holds a position as the specialist director at the Norwegian directorate for civil protection. He is also a national assigned EU, UN, and NATO expert in the fields of civil-military cooperation, crisis management, and hybrid threats.

Kairi Kasearu is a professor of empirical sociology at the University of Tartu, Estonia. Her primary research areas are military-civilian relations, European conscription systems, Estonian conscription, reserve force, the will to defend, public opinion, and social exclusion and inclusion. She is the principal investigator for the project "Development of Resource Management in the Defense Sector." She has recently published in *Armed Forces and Society*, *Journal of Baltic Studies*, *Current Sociology*, and *Journal of Political and Military Sociology*.

Mass Soldal Lund is Professor of Cybersecurity and Societal Security and Assistant Head of Department at Inland Norway University of Applied Sciences. He also holds a position as Adjunct Professor of Cyberspace Operations at the Norwegian Defence University College. Lund co-authored the book *Model-Driven Risk Analysis: The CORAS Approach* (2011, with B. Solhaug and K. Stølen). Lund holds an M.Sc. and a Ph.D. in informatics from the University of Oslo, Norway, and has 20 years of research experience from SINTEF, the Norwegian Defence University College, and Inland Norway University of Applied Sciences.

Inger Lund-Kordahl holds a medical degree from the University of Oslo and a Ph.D. from Arctic university of Norway, Tromsø. Her professional specialization relates to clinical specialization in emergency, pediatric, and intensive care medicine. Lund-Kordahl works in clinical medicine. She is also Associate Professor in emergency medicine at the Norwegian University of Science and Technology and in crisis management and preparedness at Inland Norway University of Applied Sciences. Her doctoral work is on prehospital emergency medicine in the Norwegian armed forces. Lund-Kordahl teaches courses in operative leadership, law enforcement medicine, emergency medicine, and complex decision-making in emergency medicine and critical incidents. Her research fields are mainly tactical emergency medicine and critical incident training and exercises with VR and AI support.

Leif Inge Magnussen is Professor Dr. of Education at the University of South-Eastern Norway (USN). For the past 20–25 years, he has held positions at the Norwegian Police Academy, Norwegian School of Sports Sciences, Norwegian Military Academy, and at USN doing research and teaching subjects such as education, psychology, leadership, and innovation. He is also a certified mountain guide and has been the president of the Norwegian Mountain Guides Association and the leader of Center of Emergency Preparedness at USN. Currently, he is a focus area leader of the Center of Excellence in advanced simulator training. He has edited books on nature guiding (*Friluftsliv og guiding i natur*, 2018), emergency preparedness collaboration (*Samvirke—en lærebok i beredskap*, 2017), both at Universitetsforlaget, and a Nato science for Peace publication: *Disaster, Diversity and Emergency Preparation* (2019). His research interests are related to learning processes, vocational education, risky situations, emergency preparedness, and outdoor adventures. Magnussen is currently involved in multiple research projects related to play, learning, and the unforeseen in a variety of contexts.

Yevhen Mahda holds a Ph.D. in Political Sciences and is Associate Professor in Publishing and Printing Institute of NTUU "Ihor Sikorsky Kiev Polytechnic Institute", Ukraine. He has previously been the executive director of the Institute of World Policy. Mahda graduated from the Kyiv National Taras Shevchenko University in 1996, History Department. He is author of the following: *Hybrid War: Survive and Win* (2015), *Russia's Hybrid Aggression: Lessons for the Europe* (2017), *Games of Images: How Europe Perceives Ukraine* (2016, co-author Tetyana Vodotyka), and *The Sixth: Memories of the Future*, which is the study of Ukrainian presidents (2017). Mahda's research fields of interest are as follows: political consulting; election campaign technologies; political advertising; media activities in a hybrid warfare context.

Herner Saeverot is Professor of Education at Western Norway University. For the last 20–30 years, he has conducted research on forms of teaching, existential education (a concept he introduced in a paper in 2011), and unforeseeable education. Since 2015, he has been the editor-in-chief of the international journal *Nordic Studies in Education*, and he has been awarded a lifelong membership of the Royal Norwegian Society of Sciences and Letters. Recently, Saeverot has been project manager for several international projects (funded by EU, Erasmus+, and the Norwegian Directorate for Education and Training). With Routledge, he has recently published three books: the research monograph *Education and the Limits of Reason: Reading Dostoevsky, Tolstoy, and Nabokov* (2018, in collaboration with Professor Dr. Peter Roberts); the research monograph *Indirect Education: Exploring Indirectness in Teaching and Research* (2022, Open Access); and the edited volume *Meeting the Challenges of Existential Threats through Educational Innovation: A Proposal for an Expanded Curriculum* (2022). He has recently published two other books: the edited volume *Continental Philosophy of Education* (2024, with Professor John Baldacchino), and the research monograph *Educational Theory of the Unforeseen: Educating for an Unpredictable Future* (October 2024, with Professor Glenn-Egil Torgersen).

Line Djernæs Sandbakken is a Ph.D. student at Nord University Business School, Bodø, Norway. Her research interests are related to instrumental and institutional perspectives on inter-organizational collaboration in a crisis management context. Sandbakken has an M.Sc. in Globalization—Global Politics and Culture from the Norwegian University of Technology and Science (NTNU), as well as an MBA in Management from Nord University. Sandbakken previously worked for the Norwegian Red Cross with topics like organization development and emergency preparedness. She has also worked at the Nord University Emergency Preparedness Laboratory (NORDLAB) before starting her Ph.D.

Viacheslav Semenenko is the colonel of the Armed Forces of Ukraine. Semenenko holds a Ph.D. in technical sciences and has since 2015 been Associate Professor at the National Defence University of Ukraine. He has military education from the Kyiv Institute of the Army to the strategic level at the National Defence University of Ukraine. Semenenko's fields of interest and research lie within the Russian-Ukrainian War, hybrid warfare, national resilience, military policy and military strategy, logistics, and armament. Semenenko is the co-chair of the International Project/Military Aspects of Countering Hybrid Warfare: Experiences, Lessons, Best Practices.

Liina-Mai Tooding is an associate professor emeritus, employed at the Institute of Social Studies, University of Tartu in Estonia. Her research has focused on the life-course sociology and quantitative research methods in social sciences. She has authored several books of statistical analysis. Currently, she conducts research that focuses on conscription and reserve force.

Glenn-Egil Torgersen is Professor of Education at the Institute for Educational Science, University of South-Eastern Norway (USN). He holds a Ph.D. in Psychology (NTNU, Trondheim), M.Sc. in Pedagogy (University of Oslo), and Docent scientific competence in Organization and Management, and he is also educated as a teacher educator. He has over 30 years experiences in higher education and has worked for over 20 years at the Norwegian Defence University College (NDUC), and 15 years as a senior scientist II at Institute for Energy Technology Norway, the nuclear reactor project (HRP/OECD). For the past 20 years, he has worked with basic research on the unforeseen, social interaction under risk, and educational theory development. He is the leader and editor of several major basic research projects in this field, including *The Norwegian Armed Forces' Pedagogical Basic View* (2006), the scientific anthologies *Interaction: 'Samhandling' Under Risk—A Step Ahead of the Unforeseen* (2018), and *Pedagogy for the Unforeseen* (2015). He is USN responsible for the multi-institutional basic research project *Education for the Unforeseen and Innovation*, funded by the Norwegian Research Council. Prof. Dr. Torgersen has been awarded a lifelong membership of the Royal Norwegian Society of Sciences and Letters.

Tiia-Triin Truusa currently works at the Baltic Defense College, Tartu, Estonia, as the Manager of Academic and Outreach Activities and as a Military Sociology Research Fellow at the University of Tartu, Estonia. She has been engaged in research on defense human resources for the past nine years. Her main areas of research include human resources in defense (military sociology, veteran's issues, conscription, and leadership) and sociological approaches to civil-military relations. Her Ph.D. thesis "The entangled gap: the male Estonian citizen and the interconnections between civilian and military spheres in society" (2022, University of Tartu) proposed a holistic analytical framework for tracking and understanding the intricate lacework of civil-military interaction that takes place in societies. She has been actively contributing to the voluntary defense of Estonia as a member of the Women's Voluntary Defense Organization since 2002.

PREFACE

This book is directed towards students, scholars, and leaders with an interest in the complicated field of hybrid threats.

The aim is to illuminate how hybrid threats may challenge political governance and social stability in Western societies. We reflect upon aggressors, their motivations, and their instruments used against a target state in peace time and full-scale war. We describe and explore the development over time, where both aggressors and target states adjust to a context of increased interdependency in crisis, and as a prelude to war. Empirical examples are derived from Australia, Estonia, Ukraine, Croatia, Sweden, and Norway.

The main purpose is to gain more knowledge. Not least about how Western liberal states may enhance resilience. The method we use is to encourage our fellow coauthors to reflect and discuss, partly on how preparedness and response tools unfold in a security context characterized by hybrid threats and partly on how cooperation may proceed to detect and respond effectively to this kind of malign activity. In this book, we show how a range of institutions must interact closely facing hybrid threats. This includes politicians and government officials at all levels, crises response institutions, companies, and organizations, as well as the single inhabitant. For early-warning and continuous resilience building, we are dependent on the agility and joint efforts of the inhabitants of every community. The study illuminates how society may build a total defense or comprehensive security approach that may deter the aggressor from attacking. This also includes measures for counterattacks if needed.

This study is based on empirical studies over several years including several countries. We thank our respective institutions, especially the Norwegian Defence University College and Nord University for their support. We are grateful to the leaders of numerous government institutions, organizations, and companies that have contributed with their knowledge.

Open Access funding was provided by the Inland Norway University of Applied Sciences, the Nord University, the NTNU Social Research, and the Norwegian Defense University College.

Odd Jarl Borch and Tormod Heier
Bodø and Oslo

1
UNDERSTANDING HYBRID THREATS

An introduction

Odd Jarl Borch and Tormod Heier

This book seeks answers to questions such as: How do aggressors undermine liberal democracies under the threshold of war? And what role has hybrid threats as aggressor states prepares for full-scale war? Can targeted states in a world of interconnectivity protect their transparent communities against such malign activities? If so, how may liberal and transparent democracies respond and increase their resilience as civic vulnerabilities are targeted by external aggressors?

These questions have become increasingly important. As Europe's post–Cold War security architecture unravels, a timeless characteristic in international politics emerges: While conventional wars are fought, numerous parallel battles are also waged; not with brute force but with other, more subtle, and non-violent means. The threat comes from political instruments of a non-kinetic character. Whether you live in metropolis like London, Berlin, or Paris, or in tiny communities in the Lapland or the Orkney Islands, decision-makers are constantly targeted by subversion, cyberattacks, and various kinds of information operations. Their purpose is, as always, to exploit the adversary's critical vulnerabilities. Most often, this is to achieve a political outcome. Among the most exposed targets are national cohesion, political trust, and societal resilience. Together, these targets constitute key ingredients in national models: the confidence-based contract between the citizens and the state; the social glue that keeps governed and governance together; the cohesive unit that allows democracies to be a beacon for rule by law and individual liberty.

The non-military target list illustrates a paradox in contemporary politics. While many democracies possess some of the most sophisticated military forces the world has witnessed, the same countries are also "the type of states

DOI: 10.4324/9781032617916-1

most frequently targeted by hybrid measures"; they are even some of the most vulnerable entities in the global community of states (Nilsson et al., 2021, p. 2). Constrained by the same values that most citizens adhere to, like rule of law and individual liberty—democracies are nevertheless more likely to be confronted by non-kinetic means. For any inferior rival, an indirect approach towards a liberal democracy is more rational than risking a full-scale conventional or nuclear war. Allied nations' collaborative preparedness and response options are thereby tied to a paradigm consisting of opposing societies, not opposing combatants (Treverton, 2021, p. 38).

This again allows us to explore more comprehensively how security and safety are amalgamated in an era of global political competition but also interdependency and dense interconnectivity. Aggressors' courses of action seem increasingly to manifest themselves in the internal space: a malign threat operating from the inside rather than from the outside of the state border. Bridging the gap between external and internal security means that a broader range of actors must be involved—in response, in research, as well as in training and exercises. Not least, there have to be a focus on how democracies organise themselves to prepare for, and counter, so-called hybrid threats. Crisis management below the threshold of war therefore demands a research field that builds upon a comprehensive theoretical platform—an analytical framework enabling modern states to organise and interpret a broad variety of complex empirical variations systematically and coherently.

Aim and purpose

The aim of this book is to provide students, researchers, politicians, and civil servants with different perspectives on hybrid threats and adjoining response measures. We illuminate the wide range of malign activities, such as fake news, harassment, election tampering, cyberattacks, and sabotage, directed towards critical infrastructure. These actions are often synchronised and coordinated, partly to undermine the social contract that ties citizens to their political constituencies, but also, in a narrower operational context, to prepare for large-scale wars.

The book's aim also serves a broader purpose: to enhance the readers' competence, skills, and awareness as more societies are faced with a more unpredictable security environment. Deeper insights contribute to mitigate political, societal, and institutional vulnerabilities flourishing within transparent communities. Stakeholders from public or private sectors may thereby more coherently take steps to mobilise adequate response measures.

The hybrid threat concept

The empirical complexity of this research field rests on contemporary threat perceptions. Throughout this volume, therefore, 'hybrid threat' is used as a concept of analysis. Using the term, however, is not without problems. Being a contested expression for malign activities with security implications, often below the threshold of war, hybrid threats have since its 2006 inception (Hoffman, 2007) flourished within academic and policy-oriented circles. Clearly, the concept has its advantages. Hybrid threat is a catchy phrase encapsulating numerous large and small unwanted operations. Ranging from synchronised manipulation and dissemination of information to coordinated subversion and coercion, with violent and non-violent methods for the purpose of policy—hybrid threats are a relevant and comprehensive expression of a 21st-century security context.

Hybrid threats are also associated with "grey zones". These are blurred arenas that are hard to delineate, largely because global digitalization permeates the way modern societies, and their inhabitants, chose to organise their lives. Clearly defined actors, intentions, and capabilities, the key ingredients in any threat assessment is therefore difficult to accurately identify or measure objectively. The same goes to what is, and what is not, a clearly defined battlefield. This again opens a new Pandora's box. Not least when it comes to define who is in charge when huge state bureaucracies and numerous local municipals are soaked into complex crisis management operations. These entities are states; political and administrative agencies trying to cooperate within an operational framework where roles, responsibilities, and authorities have been separated and delegated throughout a hierarchical and parliamentary chain-of-command.

This is a code of conduct that for centuries have thrived inside fragmented and sector-oriented state apparatuses; it has thus fostered numerous small and large subcultures that ". . . form a backdrop for action" (Smircich, 1983, p. 58). The numerous grey zones even rise epistemological questions of whether nations live in a time of peace or in a time of perpetual war (Galeotti, 2022). Maybe we live in a time where war and peace are surpassed by a perennial struggle, or a constant shift, in and out of war (Kennan, 1948)? In a digital age of interdependence and interconnectivity between global, national, regional, and local levels, the hybrid threat concept is regarded useful and has thus thrived enormously.

But as more and more meaning is put into the concept, hybrid threat also loses some of its analytical clout. As pointed out by Rob Johnson (2018), the concept's versality makes the concept useful inside political and strategic circles. But as guidance for operational and tactical planning and executing responses, utility is limited. For scholarly purposes, the implications are grave. This is because hybrid threats easily end up as a highly politicised concept; a trendy and politically correct label used to stamp opposing states or non-state actors one personally disguises. This leaves the entire field of research on hybrid threats highly biased. Driven by emotions and prejudice towards authoritarian rivals that undermine democracy, human rights, and rule by law, the liberal belief and value system preferred by Western scholars may easily challenge scientific criteria like balance, objectivity, and critical thinking towards own practice. Using Alexander Wendt's constructivist dualism *Self* and *Others* (Wendt, 2012), hybrid threats are something allotted to Others, while Self belongs to a more legitimate liberal security community reigning the moral high ground. Scrutinising the related concept of hybrid warfare, Friedman's (2018, p. 1) phenomenological study finds the following:

> A concept that had initially been intended to offer a better understanding of the nature of the contemporary conflicts has been weaponised, becoming a tool in internal manoeuvring for finance, public opinion and political power in Russia and the West, as well as a means of intimidation in relations between the two.

Friedman's argument visualises a deeper problem related to hybrid threats: different regions across the globe tend to develop different meanings and perceptions of what hybrid threats are. Different regions also develop different strategies or courses of action regarding how hybrid threats should be addressed. As hybrid threats are analysed, the primary frame of analytical reference has been inside hostile environments where military rather than civilian authorities have been in charge. This leaves the field of research with a slightly combatant and militarised varnish.

Hybrid threats are thereby more easily confined to "different modes of warfare including conventional capabilities, irregular tactics, terrorism, and criminal behaviour in the battlespace to obtain its political objectives" (Hoffman, 2010, p. 443). This approach is useful for military commanders operating in a hostile environment where civilian authorities are subordinated "wing-mates" rather than leading the crisis management operation. The chapters in this book, however, scrutinises hybrid threats inside civic societies; these are operational arenas governed by civilian rather than military agencies. Hybrid threats are as such seen as a broader societal and judicial concern; a challenge addressed under the auspices of civilian institutions. The military role is to be the *supporting* rather than the *supported* element. However, as the Ukrainian and Croatian experiences illustrate in this book, hybrid threats below the threshold of war are also used as a prelude to something larger—a full-scale war for national existence. In Europe's contemporary security environment, more research and discussion on hybrid threats have become even more pertinent.

Hybrid threats and strategy

Despite its vagueness, the hybrid threat concept nevertheless entails the essence of strategy. As pointed out by Edwards M. Earle in the classical anthology *Makers of Modern Strategy* from 1971: strategy is not confined to wartime issues; on the contrary, war is an "inherent part of society", defined as the coordination of all instruments of power towards the attainment of a specific political objective (Earle, 1971, p. viii). This logic resembles warnings voiced by the American diplomat and strategist, George Kennan. Writing *The Long Telegram* in 1948–post-war Moscow, on the eve of the Cold War, Kennan warned his fellow Americans in the state department: inter-state relations are always characterised by "the perpetual rhythm of struggle, in and out of war" (Kennan, 1948). Hybrid threats go to the core of Earls' 1971 description and Kennan's warning from the late 1940s: The synchronised influencing of political decision-making processes—preferably so by non-violent means that are hard to attribute. Retaliation may thereby be avoided, and this again may challenge a population's perception of war.

Hybrid threats towards civic communities, at local, regional, and national levels, therefore, demand us to scrutinise more thoroughly, more systematically, and more analytically, the many grey zones embodied inside modern states. Modern states are often seen as more effective and legitimate than less advanced states, among other things due to increased specialisation. But specialisation also leads to more delegation of roles, responsibilities, and authorities, which again makes seamless coordination and effective responses more difficult (Cohen et al., 1972). In a crisis short of war, these institutional grey zones may easily become a strategic impediment. Organizational ambiguities within our own state structure thereby contribute to blur the boundaries between us and them, between peace and war, between friends and foes, or between internal and external security, or even between security and safety, the absence of a clearly defined chain-of-command also means absence of control.

This vulnerability may again undermine public support and legitimacy, because trust and confidence are dwindling. Particularly so if public agencies display a tardy, inefficient, or indecisive response to crisis where public expectations are high. As pointed out by former U.S. Secretary of Defense, James Mattis, the hybrid ambiguity constitutes one of the most "demanding operational environments" (Mattis, 2009). This is probably true beyond U.S. battlefields in Central Asia or in the Middle East. No matter where targeted victims are forced to respond, on behalf of a governmental agency, a military unit, or a small local community: organising and synchronising resources at the right time, at the right place, is an operational art as relevant to mayors, police chiefs, state administrators and ministers as it is to generals.

Addressing hybrid threats therefore requires us to think in terms of *grand strategy*. This thinking builds on essential ideas codified by the British strategist, Sir Basil Liddell Hart, almost 60 years ago (1967, pp. 222–223). Any response requires a tight coordination of all instruments of power towards the attainment of a specific political objective. In this anthology, the specific objective is not to enhance the state's survival against an existential military threat. The scope is on building resilience, as well as responding effectively towards the more likely threats: challenges stemming from non-violent, non-kinetic, or non-military-means, but which may undermine public safety and national security. Particularly so within the numerous smaller and larger civic communities that all strive for political and social order as heads of state prepare for a more unpredictable future.

Focusing on most likely scenarios rather than worst-case scenarios is important. It allows us to scrutinize the concerted and often destructive employment of political, economic, civil, and information activities. In sum, this orchestration may easily undermine the social stability and the political order in which any liberal state needs for effective governance. The book's argument is that democracies' first line of defence is not the military but the resilience located inside the individual nations' civic communities.

A multi-disciplinary field of research

We interpret hybrid threats as part of a broader and more complex phenomenon. This is a phenomenon where states and non-state actors are pitted against each other, preferably below the threshold of conventional war. Competition between contenders, large and small, is therefore a key characteristic: a quest where opposing parties seek mutually incompatible objectives in a continuous struggle for more power, more influence, and ultimately survival. The complexity has increased even more as a myriad of public and private entities vigilantly are tied to the phenomenon, most notably diverse civic societies that all seek adequate responses to hybrid threats.

Understanding the threat picture, and the social and political context in which the threats derive from, is important. This understanding is crucial to comprehend the aggressors' intentions and capacities. But also, for the target state's precise estimation of the risk potential. Often, it is opposite value-systems with rivalling ideologies and interests that fill the context; a context operationalized by courses of action containing judicial, economic, ethnic, territorial, and military components. This context drives hybrid threat assessments into multi-disciplinary processes. Knowledge-based insight, that is, from political scientists, social anthropologists, lawyers, economists, military strategists, and tacticians, is therefore needed. Not least to protect vulnerable municipals by means of research- and experience-based knowledge; deeper insight that allows mayors, police chiefs, and citizens to comprehend and discuss what is a malign actor's "most dangerous" and "most likely" course of action in our community.

The classical intelligence approach still represents a necessary platform, especially so as situation awareness and decision-making processes inside a state apparatus takes place. Intelligence failures, that is, in sensing critical societal currents, nevertheless calls for a broader set of analytical tools.

This may include "strange bed fellows": new partners with new perspectives; multi-disciplined collaborators with a broader register of analytical ideas; private entities, that is, from international finance or emerging markets. They are all needed to fuel a deeper, more nuanced, and diverse understanding of aggressors, their intentions, and capabilities (Treverton, 2021).

The multi-disciplinary approach reflects a 21st-century context where aggressors are weaponized by a broad range of instruments. This is particularly so in the cyber domain, where millions of citizens can be easily reached with a minimum of costs, that is, through social media and cell phones. This calls for a broad understanding of technological trends. Partly so from a user perspective, but also from a technical perspective so that future arenas of potential aggression can be prepared. The "internet of things", artificial intelligence, and machine learning technologies empower aggressors with an even more potent toolbox. These digital instruments have become increasingly user friendly, cheap to purchase, and sophisticated in their technological output. For intelligence services worldwide, there is a need to significantly broaden the information and knowledge domain into a truly multi-disciplinary profession.

Of particular concern is states that deliberately seek ambiguity through a broad range of destructive instruments (Cullen, 2018). The blurred characteristics and uncertainty about target groups call for a broad mobilization to create situational awareness. "Intelligence crowd sourcing" (CROSINT) has, among others, been launched as a domain expanding and innovative tool (Treverton, 2021). A widening of the intelligence and situational awareness response set may imply an inclusion of the whole society. To understand the tradecraft of this toolset, inter-disciplinary research extracts new insight from a range of social science disciplines as well as the humanistic disciplines.

Hence, as response measures are discussed, all the political and administrative levels inside the civic community must be engaged. Encapsulating local, regional, and national levels simultaneously allow response measures to thrive inside a truly multi-disciplinary framework. For any targeted community, this is of strategic importance. Particularly so when it comes to comprehend and exploit the potential trust, goodwill, and positive attitude that lies inside millions of citizens that want to support its own local community. Public administration, management theory, and an even broader institutional theory perspective gives us the necessary understanding of how to deal with a wicked, transboundary phenomenon like hybrid threats in a coordinated, resilient way. By this, we mean a broad range of civilian and military, public and private, as well as informal community volunteers' institutions; together, they form resilient building blocks in local municipals. How they organize themselves, how they are activated and intertwined, is key to understand how small, transparent, and often vulnerable communities seek to mitigate and respond to hybrid threats.

Finally, there is also a need to scrutinize hybrid threats from a humanitarian perspective. The Ukraine war clearly illustrates how aggressor motivation and choice of tools deliberately target the most vulnerable groups inside a civic society. Perspectives that allow us to comprehend hybrid threats through the lenses of historical, religious, ethical, and linguistic perspectives are prerequisites for a deeper inter-disciplinary acknowledgement of how hybrid threats unfold in the 21st century.

Book structure

The 17 chapters are organised into five parts. Following this introduction, *Part I: The Threat* describes the hybrid threat phenomenon. First, in Chapter 2, Nina Bjørge and Marte Høiby explore the status and future avenues of contemporary research. Thereafter, in Chapter 3, Tanja Ellingsen describes how hybrid threats undermine Western liberal democracies. Part I ends with Chapter 4, where Patrick Cullen concretizes the threat by exploring China's *modus operandi* towards Australian communities.

On this basis, *Part II: Ukrainian Experiences* delves deeper into Europe's security context. Chapter 5 by Mass Soldal Lund analyses the role that Russia's cyberoperations played in the Ukraine war. Thereafter, in Chapter 6, Yevhen Mahda and Viacheslav Semenenko explore how Russia transformed its hybrid *modus operandi* into a full-scale war during the invasion phase. Part II ends with Chapter 7, where Valerii Hordiichuk, Andrii Ivashchenko and Nina Andriianova describe how Ukraine's state structure improved its resilience as Russia's war unfolded.

The empirical descriptions from contemporary Europe serve as a basis for *Part III: Response Strategies*. In Chapter 8, Tormod Heier first outlines why European civilian communities rather than NATO's military forces are a top priority for Russia. Jannicke Thinn Fiskvik and Tormod Heier thereafter explore how European municipals more effectively may address such a situation by systematically exploiting military experiences to improve communal resilience in Chapter 9. Joakim Berndtsson follows up in Chapter 10 by exploring how Sweden prepares for hybrid threats with its "Total Defence Concept". The analytical focus thereafter changes to Norway. Partly in Chapter 11, as Line Sandbakken and Ørjan Nordhus Karlsson describe how governmental agencies have reorganised across sectors and domains for better information sharing, and partly in Chapter 12, where Odd Jarl Borch explores how network entrepreneurs in local communities play a significant role to mitigate hybrid threats.

Having analysed various response strategies, *Part IV: Knowledge and Resilience* dwells on targeted states. How do they address the problem and prepare themselves? In Chapter 13, Kairi Kasearu, Tiia-Triin Truusa, and Liina-Mai Tooding scrutinize Estonian experiences with its Russian diaspora.

In Chapter 14, Gordan Akrap and Maksym Kamenetskyi highlight the importance of a common national identity by comparing hybrid threats in Croatia and Ukraine. In Chapter 15, Bjørn Bakken, Thorvald Hærem, and Inger Lund-Kordahl analyse methods of training and exercise is a hybrid threat environment. Part IV ends with a pedagogical perspective; in Chapter 16, Leif Inge Magnussen, Glenn-Egil Torgersen, Ole Boe, and Herner Saeverot propose a development model that may improve strategic skills related to hybrid threat mitigation.

Part V: Conclusions ends with Chapter 17, where the volume editors Odd Jarl Borch and Tormod Heier compile the findings presented throughout the book and deduce a hybrid threat-response model.

References

Cohen, M. D., March, J. G., & Olsen, J. P. (1972). A garbage can model of organizational choice. *Administrative Science Quarterly, 17*(1), 1–25. https://doi.org/10.2307/2392088.

Cullen, P. (2018). *Hybrid threats as a new 'wicked Problem' for early warning.* Hybrid CoE Strategic Analysis 8. The European Centre of Excellence for Countering Hybrid Threats.

Earle, E. M. (1971). Makers of modern *strategy.* Princeton University Press.

Friedman, O. (2018). *Russian 'hybrid Warfare': Resurgence and politicisation.* Oxford University Press.

Galeotti, M. (2022). *The weaponisation of everything: A field guide to the new way of war.* Yale University Press.

Hoffman, F. G. (2007). *Conflict in the 21st century: The rise of hybrid wars* (p. 72). Potomac Institute for Policy Studies.

Hoffmann, F. (2010). "Hybrid treats": Neither omnipotent nor beatable. *Orbis, 54*(3), 441–451.

Johnson, R. (2018). Hybrid war and its countermeasures: A critique of the literature. *Small Wars & Insurgencies, 29*(1), 141–163. https://doi.org/10.1080/09592318.2018.1404770.

Kennan, G. (1948, May 4). *Policy planning staff memorandum.* 9964-kennan-memo-political-warfarepdf (upenn.edu).

Liddell Hart, B. (1967). *Strategy.* Faber & Faber.

Mattis, J. (2009, December 3). Joint warfare in the 21st Century. *Small Wars Journal,* Blog.

Nilsson, N., Weissmann, M., Palmertz, B., Thunholm, P., & Häggström, H. (2021). Security challenges in the grey zone: Hybrid threats and hybrid warfare. In M. Weissmann, N. Nilsson, B. Palmertz, & P. Thunholm (Eds.), *Hybrid warfare: Security and asymmetric conflict in international relations.* I.B. Tauris.

Smircich, L. (1983). Is the concept of culture a paradigm for understanding organizations and ourselves? In P. N. Frost et al. (Eds.), *Organizational culture* (pp. 55–72). Sage Publishers.

Treverton, G. F. (2021). An American view: Hybrid threats and intelligence. In M. Weissman, M. Nilsson, B. Palmertz, & P. Thunholm (Eds.), *Hybrid warfare* (pp. 36–45). I. B. Tauris.

Wendt, A. (2012). *Social theory of international politics.* Cambridge University Press.

PART I
The threat

2

CONTEMPORARY RESEARCH ON HYBRID THREATS

Status and future avenues

Nina M. Bjørge and Marte Høiby

Introduction

Hybrid threats (HT) and hybrid warfare (HW) are modern academic concepts referring to a combination of actions targeting adversary states, including actions below threshold of conventional warfare. The concepts were introduced to the academic literature by Hoffman in 2007, who emphasized the blurring effect of different levels of warfare (Hoffman, 2007), but became wider used in academia after NATO introduced a definition during the Wales Summit in 2014. The contribution of HT and HW academic literature may add value to policy development and recommendations to state officials and practitioners, and widen the general understanding of modern threats (Libiseller, 2023). It is useful to understand HT and HW through the social sciences because their malicious acts mainly target civil society and can easily exploit the openness and transparency of democratic governance. This means that the concepts should be investigated through several social science fields to be fully understood, such as political science and international relations, military and strategic studies, the societal security field, media studies, and the psychological field. All depending on which domain the malicious acts are targeting. The social science field might come short to explain all components of HT and HW, such as technological aspects of the cyber domain and critical infrastructure, and subjects of the economic domain.

NATO defines hybrid threats as

> combine[d] military and non-military as well as covert and overt means, including disinformation, cyber-attacks, economic pressure, deployment of irregular armed groups and use of regular forces. Hybrid methods are used

DOI: 10.4324/9781032617916-3

to blur the lines between war and peace and attempt to sow doubt in the minds of target populations. They aim to destabilise and undermine societies.

(NATO, n.d.)

The hybrid warfare concept also attempts to grasp the complex actions of modern warfare, involving several actors, blurring the traditional peace-war dichotomy, and mixing conventional and irregular means of conflict. The term hybrid warfare is in English academic literature often illustrated with the example of Russia's combined traditional and irregular techniques against Ukraine starting 2014. *Military Balance 2015*, as explained by Wither (2020), provided a complex definition of hybrid warfare and defines it as

> the use of military and nonmilitary tools in an integrated campaign, designed to achieve surprise, seize the initiative and gain psychological as well as physical advantages utilizing diplomatic means; sophisticated and rapid information, electronic and cyber operations; covert and occasionally overt military and intelligence action; and economic pressure.
>
> *(Wither, 2020, p. 8)*

The definitions of HT and HW demonstrate a difference between the two in which HT applies to contexts where malicious actors target populations to spread confusion and conflict level remains below threshold of war, while HW refers stricter to the use of military and non-military means in more specific and integrated campaigns or operations. A further difference between the two inevitably lies in the difference between *threats* and *war* given that *hybrid* remains the same. In this chapter, we assume that HW can indicate a combined use of different forces, such as ground forces and marine (see Hoffman, 2010), and such forces used in combination with non-conventional strategies such as weaponizing insurgency groups or other as proxies. The term hybrid interference was introduced by Wigell (2019) to distinguish hybrid threats from hybrid warfare, however, without significant impact.

To classify actions as warfare has judicial, political, cultural, and other implications. If favourable, actors may deliberately stay below threshold of war, in which terminology such as hybrid threats can alleviate pressure in political discourse. HT can apply to contexts in which war is not explicitly declared, accused of, or labelled by a considerable unit in international society and thus remains vague and open to interpretation. It may be exploited by those who direct the theatre—or aims to control the narrative. This chapter investigates both the concepts of HT and HW since they seem to be used somewhat interchangeably (i.e. HT can be part of HW) and possibly inconsistent despite the potentially serious implications of using the word *war*. The most important reason to include both in literature searches is that we wanted to understand their prominence and significance in the scholarship.

This chapter provides in-depth insight into the scholarship on hybrid threats and warfare within social science, presenting an overview of topics, perspectives, and approaches. The current hybrid security landscape calls for increased attention to scholars' understanding and interpretation of these concepts and lays the foundation for this chapter. This chapter therefore investigates the following question: How are hybrid threats and warfare presented in English language peer-reviewed research literature?

Meta-narratives and fuzzy problems

One of the challenges of conducting meta studies is that concepts and phenomena are understood differently by different fields of study (Greenhalgh et al., 2005). Traditional research disciplinary organization has divided thoughts, ideas, and findings of scientists into fields that do not necessarily intersect. These fields of study further apply and develop terms and definitions to describe such concepts and phenomena and reinforce the closed system surrounding them. Thomas Kuhn's *structure of scientific revolutions* pointed at this in 1962, suggesting that science often is based on a set of rules and standards that have developed uniquely within a field of study and that most scientists thus approach their subjects though a paradigmatic lens that they themselves are accustomed to and perhaps unaware of, but which may not be universally accepted (Kuhn, 2012). Kuhn claimed that while this may grant scholars the necessary freedom to approach fuzzy problems in most constructive ways, it at the same time limits our knowledge to segmented pieces of understanding even of complex phenomena. It is necessary in this respect to acknowledge that threats and wars are defined by the parties involved, and that narratives are constructed from their interests which usually differ. We assume that no scientific institution or individual can be entirely unattached the political, popular, legal, or religious tractions of their surrounding society.

A fuzzy problem can be described as a poorly defined problem without a clear goal or path to solution, stemming from "fuzzy logic", a mathematical term representing vagueness, imprecise information, or partial truth. The broad and all-encompassing nature of concepts like hybrid threats and warfare may contest traditional organization of science and unidisciplinary lenses and challenge what Kuhn describes as "conceptual boxes supplied by professional education" (2012, p. 5). It can however explain why defining the concepts and acknowledging their complexity appear to be a popular exercise in the literature. First, the word *hybrid* invites a plethora of possible ingredients to any attempted definition. Second, the concept of *war* is contested not only within its field of study, but as well by political, judiciary, cultural, religious, and historical perspectives—perspectives that are further obscured every day by cutting-edge military technology. It is suggested in more recent literature that hybrid threats and warfare demand a whole-of-government approach while pointing at the risks of seeing the issue through a military prism (Lawson, 2021). The question is perhaps if such strive for conceptual frames and attempts of trajectory "boxing" at all benefit our knowledge about hybrid threats and warfare, or if these concepts quest for more disciplinary freedom and dynamic interpretation altogether.

Methodology

This study is a systematic mapping study (SMS) that aims to map research trends in the HT and HW field and detect gaps in research literature. In this way, it differs from systematic literature reviews that aim to synthesize information from the research literature (Page et al., 2021). The literature was obtained through Boolean searches in four large databases for social science literature: SAGE Journals, Taylor Francis, Science Direct (Elsevier), and Wiley. These databases were chosen for their international relevancy and strong social science profile. The terms HT and HW became increasingly used in explaining the changing security situation in Europe after the Russian annexation of Crimea in 2014 (Lawson, 2021; Libiseller, 2023), and as a result, the focus is on research articles published during 2014–2022. The data was collected in September 2022.

The terms used in searches are "hybrid threats", "hybrid warfare", and "hybrid war" in title, abstract, or keywords. All articles were collected from each search. The first step of the data collection was conducting searches and download results into Zotero. After removing duplicates, our sample contained 193 articles (see Table 2.1).

TABLE 2.1 Overview of search engines, search words, and hits.

Overview of searches

Search engine	Time period	Search words	Specification	Number of hits
SAGE Journals	2014–2022	"hybrid threats" OR "hybrid warfare" OR "hybrid war"	Title	6
			Abstract	13
			Keyword	15
Wiley Online Library	2014–2022	"hybrid threats" OR "hybrid warfare" OR "hybrid war"	Title	6
			Abstract	10
			Keyword	7
Taylor and Francis Online	2014–2022	"hybrid threats" OR "hybrid warfare" OR "hybrid war"	Title	78
			Abstract	85
			Keyword	68
Science Direct	2014–2022	"hybrid threats" OR "hybrid warfare" OR "hybrid war"	Abstract, title, keyword (together)	9
Number of articles collected				217
After removing duplicates				193

Source: created by the authors

Table 2.1 shows an overview of search engines, search words, and hits. The combination "hybrid threats" OR "hybrid warfare" OR "hybrid war" has been used for searches in the title, abstract and keywords in the SAGE Journal, Wiley Online Library, Taylor and Francis Online, and Science Direct search engines. The searches were limited to articles published between 2014 and 2022. Two hundred and seventeen articles were collected in the initial searches, and after removing duplicates, the data collection consisted of 193 articles.

When conducting a qualitative assessment of each abstract to ensure relevance to the study, following exclusion criterions applied: missing content; absent abstracts, or other important information. Use of non-English language, prominence of hybrid threats or hybrid warfare. If the HT or HW was only cited, lack of the prominence of the theme was regarded insufficient.

All abstracts were reviewed by two researchers and categorised on the basis of the main topic of focus, research perspective, and approach (see Table 2.2). The categorization used is composed and piloted by the researchers before coding and largely based on NATO's concept (NATO, n.d.). Thematic domains included are military, disinformation, cyberattacks, economic pressure, CBRN, as well as those considered important to civil institutions such as democratic institutions (political) and civil society (civil). We also added critical infrastructure, as this is also commonly targeted by adversary states. Generic and other where added last, generic being articles that covers the concept in a general manner, while other meaning that article had different main topic than the aforementioned domains.

The perspectives were chosen based on the level of society each author sees HT/HW, and to cover all relevant perspectives, we included perspectives reaching from the international level, down to more human level. The article's perspectives were classified in seven different distinctions; extra state (i.e. involving two or more states), holistic (sees the concept holistically), human/individual (i.e. the cognitive), organizational (such as NATO or the EU), national and technical perspectives, or other.

Different scholarly approaches towards HT and HW were categorised based on the article's structure, and how data is collected. Analytical or empirical means that the article analyses empirical data. A conceptual approach discusses the HT and HW concepts, a policy-oriented/normative approach is developing guidelines or recommendations for policy makers, and theoretical approaches are assessing the concept in a mere theoretical matter. If disagreements on relevancy or categorizations occurred between researchers, the decisions on whether to keep or dismiss an article were based on plenary discussions.

TABLE 2.2 Overview of characterization

Main theme	Topic	Perspective	Approach
Hybrid threats	CBRN	extra-state	analytical/empirical
Hybrid war	civil	holistic	conceptual
	critical infrastructure	human/individual	policy-oriented/ normative
	cyber	organizational	other
	economic	national	theoretical
	generic (general article about the subject)	other	
	information	technical	
	military		
	other		
	political		

Source: created by the authors

Table 2.2 shows an overview of the characterization applied to all articles in the survey process. The articles are surveyed based on four parameters: main theme, topic, perspective, and approach. The subcategories of main themes are hybrid threats and hybrid war. The subcategories of the topic are CBRN, civil, critical infrastructure, cyber, economic, generic (general article about the subject), information, military, other, and political. The subcategories of the perspectives are extra-state, holistic, human/individual, organizational, national, other, and technical. The subcategories of approach are analytical/empirical, conceptual, policy-oriented/normative, other, and theoretical.

This study is limited to investigate research published in English language journals, which may cause an inherited bias towards Western perceptions and framing of the concepts. Much of the literature is centered around Europe and the USA, and we are lacking a broader perspective on the concepts, such as Asian or African perspectives.

Descriptive results

After completing the categorization and screening of articles, the sample encompassed 123 articles about HT and HW. Descriptive statistics of this sample show that 83 focus on HW, 17 on HT, and 23 on both. A timeline in Figure 2.1 shows an increase in publications in 2016, continuing a steady growth afterwards. Twenty-two of the articles were published in 2022, even though the sample was collected in September, meaning that this number is higher by the end of year.

Number of articles published per year

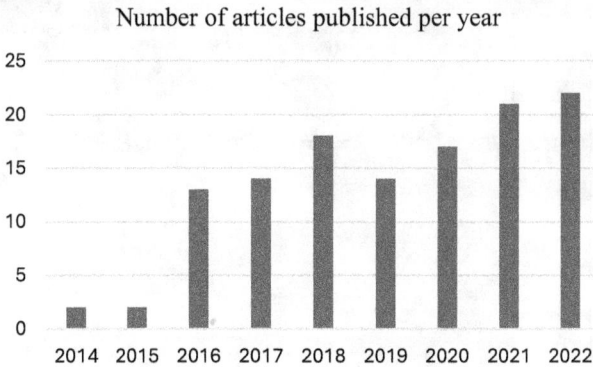

FIGURE 2.1 Year distribution of the articles.

Source: Created by the authors

The most prevalent topics are within the military (35), generic (31), information (19), and political (17) domain. Other categorizations are less mentioned, such as cyber (4), critical infrastructure (3), CBRN (2), civil issues (1), and economics (2). The other (9) section was a collective term for all articles that did not fit into the respective topics. The most used perspective is an extra-state point of view—discussing two countries or more (50), the second most used is a national perspective (27), and the third on the list is holistic (16). The less-mentioned perspectives are human/individual (7), organizational (11), technical (3), and other (9) domain.

Investigating topics and perspective combined, there are most articles with a Generic focus and an Extra-state perspective (17), while military (12), information (7), and politics (5) are the other topics prevalent in combination with an extra state perspective.

The most prevalent scientific approaches taken are the analytical/empirical (67), conceptual (25), and policy-oriented/normative (24), while very few use a theoretical approach (6). Most using analytical or empirical approach combined it with a thematic focus on military (18), generic (16), and information (14) (see Table 2.3).

Table 2.3 shows an overview of topics combined with perspectives and approaches applied by the authors while surveying abstracts. Each column has different shades of grey, and a number based on how often the combinations are applied. CBRN are combined once with the perspectives extra-state and other, as well as once with the approaches analytical/empirical and policy-oriented/normative. The civilian topic is applied once with the national perspective and once with analytical/empirical approach. The critical infrastructure topic is applied once with extra-state, holistic, and human/individual perspectives, and three times with the analytical/empirical approach. The cyber topic is applied once with extra-state, national, technical, and other

TABLE 2.3 Topics combined with research perspective and approach.

Topics combined with perspective and approach

Perspective	CBRN	Civil	Critical infrastructure	Cyber	Economic	Generic	Information	Military	Politics	Other	Total
Extra-state	1		1	1	1	17	7	12	5	5	50
Holistic			1			5	2	3	5		16
National		1		1		6	4	8	5	2	27
Human/individual			1		1		4	1			7
Organizational						1	2	5	2	1	11
Technical				1				1	1	1	3
Other	1			1		2		5			9
Total	2	1	3	4	2	31	19	35	17	9	123
Approaches											
Analytical/empirical	1	1	3	1	2	16	14	18	6	5	67
Conceptual				1		12		8	4		25
Policy-oriented/normative	1			2		3	2	7	6	3	24
Theoretical							2	2	1	1	6
Other							1				1
Total	2	1	3	4	2	31	19	35	17	9	123

Source: created by the authors

Key: 1 combination = white

2 combinations = light grey

Between 3 and 10 combinations = medium grey

Above 10 combinations = dark grey

perspectives; once with analytical/empirical and conceptual approaches; and twice with policy-oriented/normative approaches. The economic approach is applied once with extra-state and human/individual perspectives, and twice with an analytical/empirical approach.

The generic topic is combined with the extra-state 17 times, the holistic five times, the national six times, the organizational one times, and the other perspective two times. Generic is combined 16 times with an analytical/empirical approach, 12 times with conceptual, and three times with policy-oriented/normative. The information topic is applied with the perspectives extra-state seven times, holistic two times, national four times, human/individual four times, and organizational two times. The information topic is combined 14 times with analytical/empirical approaches, two times with policy-oriented/normative and theoretical, and one time with other as approach. The military topic is presented 12 times with an extra-state perspective, three times with a holistic, eight times with a national, one time with a human/individual, five times with an organizational, one time with technical, and five times with other as perspectives. Military is also combined 18 times with analytical/empirical approaches, eight times with conceptual, seven times with policy/normative, and two times with theoretical approaches. The political topic is combined five times with extra-state perspectives, holistic, and national, and two times with organizational. It is also combined six times with the analytical/empirical approaches, four times with conceptual, six times with policy-oriented/normative, and one time with theoretical approaches. The other topic is combined five times with extra-state, twice with national, and once with organizational and technical perspective. Regarding approaches, other is combined five times with analytical/empirical, three times with policy-oriented/normative, and one time with theoretical.

Extra searches through the abstracts show that emphasis is put on Russia, which is mentioned in 62 of the 123 articles. China is mentioned in eight, Iran and North Korea in two, and Turkey in one.

Due to the frequent use of some combinations, we decided to qualitatively assess each abstract with the specific combinations presented in "Summary of findings".

Quality

This study has a strong methodological setup, covering great amounts of social science research on hybrid threats and hybrid warfare, and offers a systematic mapping of the field, conveyed through tables and figures, and textual examples. The study presents a thorough discussion of Western concepts of HT and HW within English language literature. A limitation is the lack of non-western perspectives, such as from Russia, China, the Middle East, or other. Due to the nature of an SMS, this study also just scratches the surface, in which an SLR would go deeper into the literature and synthesize results.

Articles that explicitly use hybrid threats and hybrid warfare are covered; the study does not include the uses of cyber warfare or hybrid interference. This is a weakness and might exclude important aspects or perspectives from the field.

The studies investigated further are based on the combination of categories and topics largely covered, providing thorough insight into the field.

Summary of findings

This section presents findings from the articles that use the most regular combinations of characteristics. The subchapter titles are named after each topic and approach applied and are largely based on Table 2.3.

Generic domain

The field of study is itself engaged in the task of conceptualizing the term hybrid warfare. Articles within the generic domain are characterised by presenting general observations and discusses the nature of HT and HW. Thirty-one articles had a generic main topic, in which 12 took a conceptual approach, 16 an analytical/empirical approach, and 17 a extra-state perspective.

Generic topic and conceptual approach

Articles focusing on conceptualization outline contemporary understandings, often with comparative reference to a historical, traditional, or cultural manifest. Mumford (2020) outlines an ongoing discussion of the concepts HT and HW. Here, HW is conceptualised with five prevalent features: synergy, ambiguity, asymmetry, innovative disruption, and psychological features (Yan, 2020). The term "hybrid" in HW is problematised in this part of the sample and questioned whether it is the most appropriate term for studying the topic (Seely, 2017). Another issue discussed relates to the use of the concepts on different activities that are part of regular statecraft which are coercive in nature (Lawson, 2021). The Russian HW is presented with the aim of delimiting the foreign policy manoeuvres of the former Soviet republics (Muradov, 2022). HW and HT are repeatedly argued to be new terms deriving from old military tactics and war philosophy and linked to doctrines and actions from the Cold War (Muradov, 2022; Veljovski et al., 2017). It is also argued to be "self-defeating, normatively problematic, and strategically impractical" (Eberle & Daniel, 2022, p. 1), because the discourse portrays HW to be something too mysterious, hidden, and shifting to be solved (Eberle & Daniel, 2022). State rivalry and conflict are argued to be investigated through a broader lens than the traditional peace-war dichotomy (Eberle & Daniel, 2022). And measures to fight HW are argued to lay within strategic and political domains rather than in the operational or tactical spheres (Johnson, 2018).

Generic topic and analytic approach

In this section, contextual factors are emphasized as important to how the conflicting partners conceptualize occurring events (Almäng, 2019). It is also questioned whether HT and HW as analytical framework have any value at all (Bressan & Sulg, 2020).

The section on Russian HW ranges widely in topics. In an analysis of Kreml's strategy of brute force and HW, it is argued that Russia is experiencing defensive inferiority (Boulegue, 2017). How neighbouring states and former Soviet countries react to Russia's HW actions are analysed (Ambrosio, 2016; Ploumis, 2022; Polese et al., 2016), and scholars propose strategies for countering Russian aggression (Ploumis, 2022). Belarus' and Kazakhstan's responses are investigated by analysing their national security documents before and after Russia's annexation of Crimea (Ambrosio, 2016). Russian activities are argued to contribute to internal uprisings and insurgencies in the affected countries (Polese et al., 2016). With Ukraine as a case, strategies for countering Russian HW are proposed in relation to the country's specific context (Ploumis, 2022). Means used by Russia in Bulgaria are argued to prevent the strengthening of NATO's position in the Black Sea, sabotaging a new defence reform (Naydenov, 2018). The literature also covers Russian activities in the Middle East and is suggested to have negative effects on the vulnerable security structure, contributing to higher levels of conflict and insurgency (Zhou, 2019). Russian conceptual and theoretical publications and political documents suggest that there are conceptual differences between the Western and Russian understandings of hybrid warfare (Fridman, 2017). The HW "strategy" used by Russia is put forward as a Western myth rather than a formal Russian strategic concept (Fabian, 2019).

The literature focuses on other countries' use of HW, for instance Iran's use of military and paramilitary tools, like proxy forces, cyber tools, maritime force, and information operations, and how the U.S. military is countering these threats (Dalton, 2017). The literature also covers how HW is used by China in the Asia Pacific region (Ong, 2018).

Generic and extra-state perspective

Seventeen articles used a generic topic, and the perspective on extra-state relations, in which 14 focused directly or indirectly on Russia as a threat towards other countries. HT and HW are referred to as cultural statecraft, urban warfare, and grey zone conflicts (Forsberg & Smith, 2016; Ploumis, 2022), and are conceptualised into two different concepts linked to each of the most prominent countries: China's offensive hybridism and Russia's hybridism in retreat (Belo, 2020). China's offensive hybridism refers to their approach on achieving power and influence through a combination of economic, military, and diplomatic measures. Russian hybridism in retreat refers to the Russian way of HW has lost effectiveness (Belo, 2020). It is emphasised that hybrid conflicts should not be treated as homogenous phenomena, and that they differ based on contextual factors (Belo, 2020).

Russia is presented as a threat towards NATO and EU countries (Ambrosio, 2022; Boulegue, 2017), Ukraine, Belarus, Bulgaria, and Kazakhstan (Ambrosio, 2022; Ploumis, 2022), and Syria in the Middle East (Zhou, 2019). Due to escalation and outbreak of war in Eastern Ukraine in 2014, many articles focus on hybrid measures used by Russia in this conflict (Freudenstein, 2014; Kormych & Malyarenko, 2022; Mastriano, 2017; Muradov, 2022; Zhou, 2019). China's threat towards other countries is main area of study in 4 out of 17 articles in this category. China was presented as a threat towards their geographic interest sphere such as the Asian Pacific areas (Ong, 2018), but also towards the United States (Mittelmark, 2021). The threat from North Korea, Turkey, and Iran was also brought up in the literature, although they were less prominent (Dalton, 2017; Ong, 2018; Ploumis, 2022).

Military domain

Articles within the military domain are characterised by having their focus on different perspectives of military, defence, and armed forces. This was a large category in the material, counting 35 articles in total. Hereby, the combination of military and extra-state was used in 12 articles, and military and analytical approach 18 times.

Military topic and extra-state perspective

Articles in this section focus on military elements, invasions and statecraft (Halas, 2019; Lanoszka, 2016; Schroefl & Kaufman, 2014), armed forces (Englund, 2019), and the use of conventional force (Fox, 2022). Military thinking, doctrines (Thomas, 2016), and discourses (Janičatová & Mlejnková, 2021) are also covered.

Regular and irregular warfare methods by Russian proxy forces (Fox, 2022) and ISIS are often referred to in this section (Batyuk, 2017; Beccaro, 2018, 2022). The deployment of U.S. Special Operation Forces in Russia's neighbouring countries is also drawn upon as a response to Russian hybrid measures (Thompson, 2020). It is also argued that Russia's hostile activities does not reflect a specific definition or concept of HW (Janičatová & Mlejnková, 2021). The threat relations between Russia and NATO (Halas, 2019), Russia and the United States (Thompson, 2020), and Russia and Ukraine are frequently mentioned in the literature (Batyuk, 2017; Fox, 2022; Janičatová & Mlejnková, 2021; Lanoszka, 2016; Thomas, 2016). The articles in this category also focus on extra state relations, such as actions by Iran in Iraq and Syria, the U.S. counter-terrorism actions in Afghanistan (Englund, 2019), and ISIS warfare techniques in Syria, Libya Iraq, and the Levant (Batyuk, 2017; Beccaro, 2022).

Military topic and analytical approach

The content in this section is wide and covers general articles about HT and HW, Russian HW, insurgency and terrorist actions and on military force itself. Articles covering general perspectives on HT and HW propose several measures to address the issue. The wagers of HW are often discussed as revisionist powers, more powerful than their targeted states. The targeted states also often have cultural or ethnic ties towards the subversive power, which is easy to exploit (Lanoszka, 2016). HW is also put forward as measures for state and non-state actors that can be used to bypass deterrent threats and change the status quo into the best for their situation (Wirtz, 2017).

Deterrence itself is argued to be an efficient tool to tackle HT and HW. It is argued that states should deter hybrid actions that are below the threshold of war (Leimbach & Levine, 2021). Military countermeasures are also put forward as measures to counter HT and HW but should be applied in combination with civilian measures (Lanoszka, 2016). Military air power is also regarded as a countermeasure and is required to be affordable and accessible for such use (Blount, 2018). NATO has invested in its Response Force and Special Operation Force to support the European collective defence towards HW (Oren, 2016). The UK's political and military understanding and discourse of HW has direct implications for policymaking (Janičatová & Mlejnková, 2021).

This section also covers information on Russian HW and HT covering the use of proxy actors (Fox, 2022; Østensen & Bukkvoll, 2022), and how proxy forces have forced Kyiv into negotiations with Moscow (Fox, 2022). Analysis of Russian military literature compared to actual tactics used in Ukraine states that Russia has developed military theories on HW (Berzina, 2020). HW has also through the Russian strategic discourse become embedded in elite thinking and is visible in policy making (Suchkov, 2021). Other articles show insight into soldiers' motivation for serving in Ukraine's National Guard, and the most common motives revolve around the preservation of professional military dignity and enhance the states military abilities because of professional responsibilities (Prykhodko et al., 2019).

How insurgency groups take part of HW is also covered by a civil war perspective, which argues that a certain structure of organization contributes to enable the use of HW and strengthen rebel groups. The structures needed for a safe space are a centralized command and unity across social and political issues (Hartwig, 2020). The use of HW tactics by non-state actors such as ISIS, Boko Haram, and Houthi Movement is covered in the literature (Stoddard, 2020), and their activities in different regions, in which North Africa is often struck (Beccaro, 2018, 2022). How to develop countermeasures for these groups is also covered, in which it is brought up that Poland has a key focus in their defence strategy on developing anti-terror instruments (Gasztold & Gasztold, 2022), and that the United States has practiced counter-terrorist measures in Afghanistan (Englund, 2019).

Information domain

The articles within the information domain primarily cover topics such as influence operations and disinformation campaigns, and the more traditional field of study on propaganda, psychological operations, and the role of media in war and conflict. The information domain is combined with an analytical approach 14 times.

Information and analytical approach

The focus on disinformation campaigns has evolved in Western countries, and one article analyses the EU's resilience towards hybrid threats, considering the Russian influence operation towards the U.S. presidential elections (Kalniete & Pildegovičs, 2021). Anti-feminist movements with roots in the USA, Russia, and Europe are brought up, and discourses are argued to be shaped by strong, global forces (Slavova, 2022). Fake news and disinformation as threats towards democracy are studied in the United States, Germany, and Czechia. With perspectives of media literacy and hybrid warfare in mind, one author argues that attention needs to be drawn to why people share fake news (Monsees, 2021). Another study on Russian disinformation in the Czech Republic explains how embedding narratives in a broad cultural context makes them more efficient (Eberle & Daniel, 2019).

Responses and countermeasures towards influence and information campaigns are also a topic covered (Dorosh et al., 2022), including Canadian response to influence and disinformation regarding security and foreign policy issues. Findings suggest that understanding and defining disinformation are among main issues (Jackson, 2021). Media literacy and education receive growing attention in the Czech Republic and are linked to the risk of HT information campaigns, disinformation, and fake news (Supa et al., 2020).

Russian domestic disinformation is addressed, in which authorities target its own populations through different narratives and campaigns in state-owned media channels (Ambrosio, 2016; Lupion, 2018; Watanabe, 2017). From 2008 to 2014, Russian state-owned media increased influence due to unrestricted publication space in the Russian public sphere, and the content is argued to have become more persuasive since 2014 (Lupion, 2018). The Kremlin's influence campaigns are argued to set the agenda for domestic debate and promote foreign policy narratives to legitimate the war in Eastern Ukraine (Ambrosio, 2016; Pasitselska, 2017). Influence operations are spread not only through national television (Pasitselska, 2017) but also through international media to spread propaganda abroad (Watanabe, 2017). Digital activities of certain individuals close to the Russian state apparatus (i.e. Yevgeny Prigozhin, Konstantin Malofeev, Alexander Yonov, Alexander Malkevich, and Luc Michel) are analysed to see if they promote Russian state ideologies. Their activities are steered by their agenda and interest, rather than Russian state ideology (Laruelle & Limonier, 2021). Russia is also argued to wage a strategic information war against Ukraine (Sopilko et al., 2022), and evidence from Donetsk, Luhansk, and Crimea shows that levels of political and social trust were lower after the outbreak of war in 2014 than before. Inhabitants perceived a higher degree of economic instability in these regions compared to the rest of Ukraine (Hoyle et al., 2022). In-depth analyses of regulatory instruments for countering information warfare are presented with action recommendations for the case of Ukraine (Sopilko et al., 2022).

Discussion

The heritage of military doctrines

Results from our sample suggest that the peer-reviewed articles on HT and HW focus heavily on military issues while civil perspectives are nearly absent. Without paying much attention to neither the cyber domain nor critical infrastructure, the scholarship is occupied with entangling concepts and terminology in light of military and political issues. The information domain receives some but limited attention, and within this part of the scholarship, we find attempts to address impacts of HT and HW on civil society—while merely from a state and media political point of view. Civil and societal resilience to HT and HW strategies, and how cyberattacks, espionage, and critical infrastructure sabotage affect civil society and ideas of civil-military cooperation, are issues that receive little to no attention at all. A missing economic point of view suggests that the literature does not fully reflect recent developments in international relations below the threshold of war such as sanctions in the U.S.-China relationship or in the context of international response to Russia's invasion of Ukraine. Global infrastructure developments (such as China's One Belt One Road) and international business initiatives are other accelerating trends that affect international relations, which could have been discussed within the influence and economic soft power perspectives but were not.

The lack of research on CBRN, civil society, critical infrastructure, cyber, and economics within the HT/HW frame can be explained by the fact that these fields of study traditionally belong to other subject areas, and that researchers may not use (or be familiar with) the HT and HW terms. The data was collected from social science publishers, which may have contributed to the fact that some of these topics are left out. The sample of literature that we have studied was not selected with a premise of HT and HW focus, given that the terms were only instructed to be *included* in the sample. The research gaps suggest that the investigated HT/HW scholarship may not be entirely up to date or in line with the most recent development.

What the body of literature analysed in this study *does* pay considerable attention to is the general nature and conceptualization of hybrid threats and warfare and how these materialise in today's military and geopolitical context. Argued to be new terms deriving from old military tactics and war philosophy, HT and HW are largely linked to doctrines and actions from the Cold War. Whether the terminology is suited to cover today's wars, to what extent new warfare is any different from earlier wars, and discussions about understandings of hybrid measures and their included tactics are main issues.

With new technology and fast-growing, far-reaching digitalization, we argue that hybrid strategies and tactics enable power projection below the threshold of war to a far greater extent than before. Critical infrastructure disruption, energy supply cuts, and cyberattacks on private companies or public institutions such as media, police, or government networks are examples of such tactics. These acts can be carried out by state or proxy actors or criminal groups, but attribution is usually difficult because of the asynchronicity in time and space that is particular to cyberspace. Moreover, examples of grey zone activity, such as those mentioned earlier, mainly target civil society. It is thus concerning that civil perspectives receive so little attention in this field of war and conflict scholarship. Cyber tactics also enable the intelligence tradecraft to gain new opportunities (Stenslie et al., 2019) in a way that nations can gain substantial information about critical parts of modern societies.

Scholars focus largely on extra-state relations and national perspectives when investigating HT and HW, by either focusing on foreign policy threats, coercion and force or preparing national or military policies for countering these threats.

Russia as the main wager of hybrid war

Russia is dominantly presented as a wager of HT and HW, mentioned in 62 of 123 articles to different degrees. The focus on Russia is prominent in all combinations of topic, perspective, and approach, and particularly in the articles with an extra-state perspective. Here, Russia is presented as a threat to multiple countries (i.e., Ambrosio, 2022; Freudenstein, 2014; Kormych & Malyarenko, 2022; Mastriano, 2017; Muradov, 2022; Ploumis, 2022; Veljovski et al., 2017; Zhou, 2019). There may be several reasons for this, such as a reference to the Gerasimov doctrine (Bartles, 2016) and U.S.-led and -oriented discourse in Western military and political studies.

Scholarship on Russia as a wager of HT and HW may as well have increased because their hybrid approach *in fact* has increased. After the dissolution of the Soviet Union and privatization of property in Russia in the 1990s, the country experienced a period of insecurity and loss of influence in the international system. In the 2010s and 2020s, this changed, and the Kremlin showed force in Georgia in 2008 and in Crimea in 2014 (Matlary & Heier, 2016). Along with these military operations came the increased use of hybrid measures to destabilise and undermine regions using psychological warfare and decreasing political trust (Hoyle et al., 2022), influence operations in the United States (Mueller, 2019) and economic coercion in Ukraine (Balcaen et al., 2022).

Scholars' interest in Russian HT and HW is also centered on the understanding or misunderstanding of the Gerasimov Doctrine, interpreted by Western actors as a recipe to how Russia would wage future wars. The document was launched one year before the Maidan revolution, eventually leading to Russian annexation of Crimea (Bartles, 2016). Many Western military thinkers and scholars related the document to Russian aggression and thus interpreted it as a military doctrine. But the so-called Gerasimov Doctrine was instead an analysis of the Russian security landscape, with a focus on the United States (Bartles, 2016).

The geopolitical rivalry tradition between USSR and the United States dating back to the Cold War (McGlinchey, 2022) appears to still affect Western military discourse and research. Literature much cited in this research is founded in Western military thought (Fridman, 2017; Hoffman, 2007). Historically, the USSR has been perceived as a threat and enemy of the West/U.S., based on ideological differences and rivalry for global hegemony (McGlinchey, 2022). Similar notions of rivalry may be drawn to the current literature on HT and HW, in which conflicting interests between Western countries/NATO allies and Russia are quite prominent. In some cases, Russia is presented as an enemy with malicious intentions (Thompson, 2020). Despite of an asymmetric competition, with the Western countries/NATO allies having the largest military, economic, and diplomatic weight, the competition with Russia is yet of large interest for scholars in this field.

The second most frequently discussed country is China, which is mentioned in eight articles. It was put relatively little focus on China as a threat towards Western countries/NATO allies, compared to the Russian threat. Instead,

China was presented as a regional hegemon in the Southeast Asian Sea, and a potential regional threat (Ong, 2018; Patalano, 2018; Takahashi, 2018). That China receives such modest attention in the HT and HW literature is unexpected, due to the country's dominant role in geopolitics and vast span of soft power advantages, particularly within economic and diplomatic relations. China has purchased ports and telecommunication infrastructure in the Faroe Islands (Poulsen, 2020), engaged in building rail and roads infrastructure in Africa and Asia through the One Belt One Road Initiative (Huang, 2016), and are developing the Polar Silk Road along the Arctic coast, which will have considerable impact on the Arctic region (Lim, 2018). All these actions combined, China is a dominant force in large parts of the world, and researchers should indeed increase their focus on China as an executor of HT and HW. In addition to infrastructure and trade initiatives and activities across the world, China has large cyber capabilities and ranges second after the United States at Belfers Centre's Cyber Power Index (Voo et al., 2022). This itself argues for increased scholarly attention to China's use of hybrid tactics beyond conceptual discussions and the origin of military philosophy.

Conclusion

The willingness to wage or participate in war would by logic have an impact on the extent to which a nation or cross-national alliance resort to covert methods and tactics below threshold. This is an aspect that does not appear to be clearly addressed either in the literature or in definitions of the concepts. Given their concealing capacities, our findings suggest that the extent to which hybrid measures are used will depend on whether it is gainful to openly project power or not and whether attribution is desirable. A considerable novelty of HT and HW lays in the opportunity for grey zone activity brought by with spread and reach of cheaper, more sophisticated, and user-friendly information and communication technology. The notion held by some scholars that HT/HW is merely "old wine in new bottles" thereby falls short. Based on these findings, we suggest future research on HT and HW to further consider a broader aspect of civil society issues and civil-military cooperation, total defence, critical infrastructure, and societal consequences of digitalization, automatization, IoT, and cybertechnology. These suggestions serve as a rationale for the remainder of this volume.

It appears that knowledge of both HT and HW suffers from academic disciplinary encapsulation and benefits from efforts to grapple with them. But repeated attempts to settle an understanding of what they may or may not contain seem deemed to fail. This, we suggest, is because the concepts are ambiguous, subtle, and compound in nature, and their advantage lays in the changing landscape surrounding them and what is yet unknown, unexpected, or extraordinary. Scholarly effort that may instead contribute to expand our knowledge of these concepts should rather investigate other fields of study and move around the prism. Acknowledging that the focal point of our study in fact is limited to investigate concepts within this prism, we also believe it contributes with an important empirical argument for recommending new strains of future research. Future mapping studies should also include a wider span of search words, such as cyber warfare and hybrid interference or irregular tactics. More focus on China as a dominant actor and Chinese capacities to project power below threshold would add insight to the general HT/HW field of study, while cultural structural issues linked to hybrid threats and warfare, the global economic system, and trade infrastructure and diplomacy are all aspects that would add value to Western scholarship in this field.

References

Almäng, J. (2019). War, vagueness and hybrid war. *Defence Studies*, *19*(2), 189–204. https://doi.org/10.1080/14702436.2019.1597631.

Ambrosio, T. (2016). The rhetoric of irredentism: The Russian Federation's perception management campaign and the annexation of Crimea. *Small Wars & Insurgencies*, *27*(3), 467–490. https://doi.org/10.1080/09592318.2016.1151653.

Ambrosio, T. (2022). Belarus, Kazakhstan and alliance security dilemmas in the former Soviet Union: Intra-alliance threat and entrapment after the Ukraine crisis. *Europe-Asia Studies*, *0*(0), 1–29. https://doi.org/10.1080/09668136.2022.2061425.

Balcaen, P., Bois, C. D., & Buts, C. (2022). A game-theoretic analysis of hybrid threats. *Defence and Peace Economics*, *33*(1), 26–41. https://doi.org/10.1080/10242694.2021.1875289.

Bartles, C. (2016). Getting Gerasimov right. *Military Review*, *96*, 30–38.

Batyuk, V. I. (2017). The US concept and practice of hybrid warfare. *Strategic Analysis*, *41*(5), 464–477. https://doi.org/10.1080/09700161.2017.1343235.

Beccaro, A. (2018). Modern irregular warfare: The ISIS case study. *Small Wars & Insurgencies*, *29*(2), 207–228. https://doi.org/10.1080/09592318.2018.1433469.

Beccaro, A. (2022). ISIS in Libya and beyond, 2014–2016. *The Journal of North African Studies*, *27*(1), 160–179. https://doi.org/10.1080/13629387.2020.1747445.

Belo, D. (2020). Conflict in the absence of war: A comparative analysis of China and Russia engagement in gray zone conflicts. *Canadian Foreign Policy Journal*, *26*(1), 73–91. https://doi.org/10.1080/11926422.2019.1644358.

Berzina, I. (2020). From 'total' to 'comprehensive' national defence: The development of the concept in Europe. *Journal on Baltic Security*, *6*(2), 7–15.

Blount, C. (2018). Useful for the next hundred years? Maintaining the future utility of airpower. *The RUSI Journal*, *163*(3), 44–51. https://doi.org/10.1080/03071847.2018.1494348.

Boulegue, M. (2017). The Russia-NATO relationship between a rock and a hard place: How the 'defensive inferiority syndrome' is increasing the potential for error. *The Journal of Slavic Military Studies*, *30*(3), 361–380. https://doi.org/10.1080/13518046.2017.1341769.

Bressan, S., & Sulg, M.-L. (2020). Welcome to the grey zone: Future war and peace. *New Perspectives*, *28*(3), 379–397. https://doi.org/10.1177/2336825X20935244.

Dalton, M. G. (2017). How Iran's hybrid-war tactics help and hurt it. *Bulletin of the Atomic Scientists*, *73*(5), 312–315. https://doi.org/10.1080/00963402.2017.1362904.

Dorosh, L., Astramowicz-Leyk, T., & Turchyn, Y. (2022). The impact of post-truth politics as a hybrid information influence on the status of international and national security: The attributes of interpretation and the search for counteraction mechanisms. *European Politics and Society*, 23(3), 340–363. https://doi.org/10. 1080/23745118.2021.1873041.

Eberle, J., & Daniel, J. (2019). "Putin, you suck": Affective sticking points in the Czech narrative on "Russian hybrid warfare". *Political Psychology*, 40(6), 1267–1281. https://doi.org/10.1111/pops.12609.

Eberle, J., & Daniel, J. (2022). Anxiety geopolitics: Hybrid warfare, civilisational geopolitics, and the Janus-faced politics of anxiety. *Political Geography*, 92, 102502. https://doi.org/10.1016/j.polgeo.2021.102502.

Englund, S. H. (2019). A dangerous middle-ground: Terrorists, counter-terrorists, and gray-zone conflict. *Global Affairs*, 5(4–5), 389–404. https://doi.org/10.1080/2334 0460.2019.1711438.

Fabian, S. (2019). The Russian hybrid warfare strategy—neither Russian nor strategy. *Defense & Security Analysis*, 35(3), 308–325. https://doi.org/10.1080/14751 798.2019.1640424.

Forsberg, T., & Smith, H. (2016). Russian cultural statecraft in the Eurasian space. *Problems of Post-Communism*, 63(3), 129–134. https://doi.org/10.1080/1075821 6.2016.1174023.

Fox, A. C. (2022). The Donbas in flames: An operational level analysis of Russia's 2014–2015 Donbas campaign. *Small Wars & Insurgencies*, 0(0), 1–33. https://doi. org/10.1080/09592318.2022.2111496.

Freudenstein, R. (2014). Facing up to the bear: Confronting Putin's Russia. *European View*, 13(2), 225–232. https://doi.org/10.1007/s12290-014-0330-6.

Fridman, O. (2017). Hybrid warfare or Gibridnaya Voyna? *The RUSI Journal*, 162(1), 42–49. https://doi.org/10.1080/03071847.2016.1253370.

Gasztold, A., & Gasztold, P. (2022). The Polish counterterrorism system and hybrid warfare threats. *Terrorism and Political Violence*, 34(6), 1259–1276. https://doi. org/10.1080/09546553.2020.1777110.

Greenhalgh, T., Robert, G., Macfarlane, F., Bate, P., Kyriakidou, O., & Peacock, R. (2005). Storylines of research in diffusion of innovation: A meta-narrative approach to systematic review. *Social Science & Medicine*, 61(2), 417–430. https:// doi.org/10.1016/j.socscimed.2004.12.001.

Halas, M. (2019). Proving a negative: Why deterrence does not work in the Baltics. *European Security*, 28(4), 431–448. https://doi.org/10.1080/09662839.201 9.1637855.

Hartwig, J. (2020). Composite warfare and civil war outcome. *Terrorism and Political Violence*, 32(6), 1268–1290. https://doi.org/10.1080/09546553.2018.1464444.

Hoffman, F. G. (2007). *Conflict in the 21st century: The rise of hybrid wars* (p. 72). Potomac Institute for Policy Studies.

Hoffman, F. G. (2010). 'Hybrid Threats': Neither omnipotent nor unbeatable. *Orbis*, 54(3), 441–455. https://doi.org/10.1016/j.orbis.2010.04.009.

Hoyle, A., van den Berg, H., Doosje, B., & Kitzen, M. (2022). On the brink: Identifying psychological indicators of societal destabilization in Donetsk, Luhansk and Crimea. *Dynamics of Asymmetric Conflict*, 15(1), 40–54. https://doi.org/10.1080 /17467586.2021.1895262.

Huang, Y. (2016). Understanding China's belt & road initiative: Motivation, framework and assessment. *China Economic Review*, 40, 314–321. https://doi. org/10.1016/j.chieco.2016.07.007.

Jackson, N. J. (2021). The Canadian government's response to foreign disinformation: Rhetoric, stated policy intentions, and practices. *International Journal*, 76(4), 544–563. https://doi.org/10.1177/00207020221076402.

Janičatová, S., & Mlejnková, P. (2021). The ambiguity of hybrid warfare: A qualitative content analysis of the United Kingdom's political–military discourse on Russia's hostile activities. *Contemporary Security Policy*, 42(3), 312–344. https://doi.org/10.1080/13523260.2021.1885921.

Johnson, R. (2018). Hybrid war and its countermeasures: A critique of the literature. *Small Wars & Insurgencies*, 29(1), 141–163. https://doi.org/10.1080/09592318.2018.1404770.

Kalniete, S., & Pildegovičs, T. (2021). Strengthening the EU's resilience to hybrid threats. *European View*, 20(1), 23–33. https://doi.org/10.1177/17816858211004648.

Kormych, B., & Malyarenko, T. (2022). From gray zone to conventional warfare: The Russia-Ukraine conflict in the Black Sea. *Small Wars & Insurgencies*, 0(0), 1–36. https://doi.org/10.1080/09592318.2022.2122278.

Kuhn, T. (2012). *The structure of scientific revolutions* (4th edition). University of Chicago Press.

Lanoszka, A. (2016). Russian hybrid warfare and extended deterrence in eastern Europe. *International Affairs*, 92(1), 175–195. https://doi.org/10.1111/1468-2346.12509.

Laruelle, M., & Limonier, K. (2021). Beyond "hybrid warfare": A digital exploration of Russia's entrepreneurs of influence. *Post-Soviet Affairs*, 37(4), 318–335. https://doi.org/10.1080/1060586X.2021.1936409.

Lawson, E. (2021). We need to talk about hybrid. *The RUSI Journal*, 166(3), 58–66. https://doi.org/10.1080/03071847.2021.1950330.

Leimbach, W. B., Jr., & Levine, S. D. (2021). Winning the gray zone: The importance of intermediate force capabilities in implementing the national defense strategy. *Comparative Strategy*, 40(3), 223–234. https://doi.org/10.1080/01495933.2021.1912490.

Libiseller, C. (2023). 'Hybrid warfare' as an academic fashion. *Journal of Strategic Studies*, 46(4), 858–880. https://doi.org/10.1080/01402390.2023.2177987.

Lim, K. S. (2018). *China's Arctic policy and the Polar Silk Road vision* (SSRN Scholarly Paper 3603710). https://papers.ssrn.com/abstract=3603710.

Lupion, M. (2018). The gray war of our time: Information warfare and the Kremlin's weaponization of Russian-language digital news. *The Journal of Slavic Military Studies*, 31(3), 329–353. https://doi.org/10.1080/13518046.2018.1487208.

Mastriano, D. (2017). Putin—the masked nemesis of the strategy of ambiguity. *Defense & Security Analysis*, 33(1), 68–76. https://doi.org/10.1080/14751798.2016.1272175.

Matlary, J. H., & Heier, T. (2016). *Ukraine and beyond: Russia's strategic security challenge to Europe*. Palgrave Macmillan UK. https://doi.org/10.1007/978-3-319-32530-9.

McGlinchey, S. (2022). *International relations*. https://www.e-ir.info/publication/beginners-textbook-international-relations/.

Mittelmark, C. (2021). Playing chess with the Dragon: Chinese-U.S. competition in the era of irregular warfare. *Small Wars & Insurgencies*, 32(2), 205–228. https://doi.org/10.1080/09592318.2021.1870423.

Monsees, L. (2021). Information disorder, fake news and the future of democracy. *Globalizations*, 0(0), 1–16. https://doi.org/10.1080/14747731.2021.1927470.

Mueller, R. S. (2019). *Report on the investigation into Russian interference in the 2016 presidential election*. U.S. Department of Justice. https://www.justice.gov/archives/sco/file/1373816/download.

Mumford, A. (2020). Understanding hybrid warfare. *Cambridge Review of International Affairs*, 33(6), 824–827. https://doi.org/10.1080/09557571.2020.1837737.

Muradov, I. (2022). The Russian hybrid warfare: The cases of Ukraine and Georgia. *Defence Studies*, 22(2), 168–191. https://doi.org/10.1080/14702436.2022.2030714.

NATO. (n.d.). *Countering hybrid threats*. NATO. Retrieved December 14, 2023, from https://www.nato.int/cps/en/natohq/topics_156338.htm.

Naydenov, M. (2018). The subversion of the Bulgarian defence system—the Russian way. *Defense & Security Analysis, 34*(1), 93–112. https://doi.org/10.1080/14751798.2018.1421408.

Ong, W. (2018). The rise of hybrid actors in the Asia-Pacific. *The Pacific Review, 31*(6), 740–761. https://doi.org/10.1080/09512748.2018.1513549.

Oren, E. (2016). A dilemma of principles: The challenges of hybrid warfare from a NATO perspective. *Special Operations Journal, 2*(1), 58–69. https://doi.org/10.1080/23296151.2016.1174522.

Østensen, Å. G., & Bukkvoll, T. (2022). Private military companies—Russian great power politics on the cheap? *Small Wars & Insurgencies, 33*(1–2), 130–151. https://doi.org/10.1080/09592318.2021.1984709.

Page, M. J., McKenzie, J. E., Bossuyt, P. M., Boutron, I., Hoffmann, T. C., Mulrow, C. D., Shamseer, L., Tetzlaff, J. M., Akl, E. A., Brennan, S. E., Chou, R., Glanville, J., Grimshaw, J. M., Hróbjartsson, A., Lalu, M. M., Li, T., Loder, E. W., Mayo-Wilson, E., McDonald, S., . . . Moher, D. (2021). The PRISMA 2020 statement: An updated guideline for reporting systematic reviews. *Systematic Reviews, 10*(1), 89. https://doi.org/10.1186/s13643-021-01626-4.

Pasitselska, O. (2017). Ukrainian crisis through the lens of Russian media: Construction of ideological discourse. *Discourse & Communication, 11*(6), 591–609. https://doi.org/10.1177/1750481317714127.

Patalano, A. (2018). When strategy is 'hybrid' and not 'grey': Reviewing Chinese military and constabulary coercion at sea. *The Pacific Review, 31*(6), 811–839. https://doi.org/10.1080/09512748.2018.1513546.

Ploumis, M. (2022). Comprehending and countering hybrid warfare strategies by utilizing the principles of Sun Tzu. *Journal of Balkan and Near Eastern Studies, 24*(2), 344–364. https://doi.org/10.1080/19448953.2021.2006005.

Polese, A., Kevlihan, R., & Ó Beacháin, D. (2016). Introduction: Hybrid warfare in post-Soviet spaces, is there a logic behind? *Small Wars & Insurgencies, 27*(3), 361–366. https://doi.org/10.1080/09592318.2016.1151660.

Poulsen, R. W. (2020, July 12). Forget Greenland, the Faroe Islands are the new strategic gateway to the Arctic. *Foreign Policy*. https://foreignpolicy.com/2020/12/07/forget-greenland-faroe-islands-new-strategic-gateway-to-the-arctic/.

Prykhodko, I., Matsehora, J., Lipatov, I., Tovma, I., & Kostikova, I. (2019). Servicemen's motivation in the National Guard of Ukraine: Transformation after the 'revolution of dignity'. *The Journal of Slavic Military Studies, 32*(3), 347–366. https://doi.org/10.1080/13518046.2019.1645930.

Schroefl, J., & Kaufman, S. J. (2014). Hybrid actors, tactical variety: Rethinking asymmetric and hybrid war. *Studies in Conflict & Terrorism, 37*(10), 862–880. https://doi.org/10.1080/1057610X.2014.941435.

Seely, R. (2017). Defining contemporary Russian warfare. *The RUSI Journal, 162*(1), 50–59. https://doi.org/10.1080/03071847.2017.1301634.

Slavova, E. (2022). Globalising genderphobia and the case of Bulgaria. *European Journal of English Studies, 26*(2), 176–196. https://doi.org/10.1080/13825577.2022.2091281.

Sopilko, I., Svintsytskyi, A., Krasovska, Y., Padalka, A., & Lyseiuk, A. (2022). Information wars as a threat to the information security of Ukraine. *Conflict Resolution Quarterly, 39*(3), 333–347. https://doi.org/10.1002/crq.21331.

Stenslie, S., Haugom, L., & Vaage, B. H. (2019). *Etterretningsanalyse i den digitale tid*. Fagbokforlaget.

Stoddard, E. (2020). Maoist hybridity? A comparative analysis of the links between insurgent strategic practice and tactical hybridity in contemporary non-state armed groups. *Studies in Conflict & Terrorism, 0*(0), 1–25. https://doi.org/10.1080/105 7610X.2020.1792724.

Suchkov, M. A. (2021). Whose hybrid warfare? How 'the hybrid warfare' concept shapes Russian discourse, military, and political practice. *Small Wars & Insurgencies, 32*(3), 415–440. https://doi.org/10.1080/09592318.2021.1887434.

Supa, M., Šťastná, L., & Jirák, J. (2020). Media education policy developments in times of "fake news". In *The handbook of media education research* (pp. 373–379). https://doi.org/10.1002/9781119166900.ch35.

Takahashi, S. (2018). Development of gray-zone deterrence: Concept building and lessons from Japan's experience. *The Pacific Review, 31*(6), 787–810. https://doi.org/10.1080/09512748.2018.1513551.

Thomas, T. (2016). The evolution of Russian military thought: Integrating hybrid, new-generation, and new-type thinking. *The Journal of Slavic Military Studies, 29*(4), 554–575. https://doi.org/10.1080/13518046.2016.1232541.

Thompson, M. C. A. (2020). SOF utilization in contemporary competitive spaces. *Special Operations Journal, 6*(2), 95–107. https://doi.org/10.1080/23296151.20 20.1813367.

Veljovski, G., Taneski, N., & Dojchinovski, M. (2017). The danger of "hybrid warfare" from a sophisticated adversary: The Russian "hybridity" in the Ukrainian conflict. *Defense & Security Analysis, 33*(4), 292–307. https://doi.org/10.1080/14 751798.2017.1377883.

Voo, J., Hemani, I., & Cassidy, D. (2022). *National cyber power index 2022.* Cyber Power.

Watanabe, K. (2017). Measuring news bias: Russia's official news agency ITAR-TASS' coverage of the Ukraine crisis. *European Journal of Communication, 32*(3), 224–241. https://doi.org/10.1177/0267323117695735.

Wigell, M. (2019). Hybrid interference as a wedge strategy: A theory of external interference in liberal democracy. *International Affairs, 95*(2), 255–275. https://doi.org/10.1093/ia/iiz018.

Wirtz, J. J. (2017). Life in the "gray zone": Observations for contemporary strategists. *Defense & Security Analysis, 33*(2), 106–114. https://doi.org/10.1080/1475 1798.2017.1310702.

Wither, J. K. (2020). *Defining hybrid warfare* [Viewpoint]. https://www.marshall-center.org/sites/default/files/files/2020-05/pC_V10N1_en_Wither.pdf.

Yan, G. (2020). The impact of artificial intelligence on hybrid warfare. *Small Wars & Insurgencies, 31*(4), 898–917. https://doi.org/10.1080/09592318.2019.1682908.

Zhou, Y. (2019). A double-edged sword: Russia's hybrid warfare in Syria. *Asian Journal of Middle Eastern and Islamic Studies, 13*(2), 246–261. https://doi.org/10.108 0/25765949.2019.1605570.

3

HYBRID THREATS AS A THREAT TO DEMOCRACY

Soft and smart tactics and three dimensions of democracy

Tanja Ellingsen

Introduction

Fukuyama's (1989) optimistic scenario of "the End of History" and the triumph of liberal democracy has been put on hold as democracy seems to be in retreat across much of the world (Papada et al., 2023). Although there are various explanations to the democratic decline (Levitsky & Ziblatt, 2018), involving both internal and external factors, it is fair to say that hybrid threat is a major obstacle to liberal democracies (Wigell, 2021). Free media, pluralism, and economic openness—some of the core characteristics of Western democracies—have become opportunities for foreign and hostile non-state actors to undermine political trust, social cohesion, and with that potentially democracy itself.

NATO Secretary General Jens Stoltenberg stated in an address in 2015, "Of course, hybrid warfare is not new. It is as old as the Trojan Horse. What is different is its scale, its speed and its intensity. And that it is right at our borders" (Stoltenberg, 2015). Also, the European Union (EU) has put hybrid threats as a serious challenge to democracies that needs to be countered, and in 2016, NATO and the EU formed a Joint Framework for Countering Hybrid Threats. Increased global interdependencies and geopolitical turmoil have magnified the need for more knowledge and increased understanding of these threats.

The purpose of this chapter is to give an overview of how hybrid threats to democracy can be described. By separating between electoral, liberal and deliberative aspects of democracies, and how hybrid threats target these, the main contribution of the chapter is to give a more comprehensive and systematic understanding of how and why hybrid threats constitute such

DOI: 10.4324/9781032617916-4

a detrimental threat to the very core of democracies. As the potential list of hybrid threat instruments is exhaustive, I have here limited the analysis to soft (and smart) power tactics targeting democracies in Europa and the United States in the last decade or so. This includes factors such as disinformation campaigns, election interference, conspiracy theories, cyberattacks, fake news, and exacerbating ethnic and religious differences, but excludes economic (i.e. sanctions, corruption) and military factors (i.e. physical attacks or interference). The chapter is organized as follows. First, a theoretical framework for democratic features is described. The framework is then used to organize a three-fold description of hybrid threats to democracy. Finally, conclusions are deduced by summing up threats and proposing plausible remedies for further research.

Theory: democratic features and potential hybrid threat tactics

Democracy

The term "democracy" comes from the combination of two Greek words: demos (people) and kratos (rule). Democracy is a form of government in which the people rule. However, this definition raises several questions like: who are "the people"? and what is meant by "rule"? This is probably the reason why one finds that there are almost as many definitions of democracy as there are articles on democratization (Knutsen et al., 2023). Nevertheless, there are certain features of democracy around which there are significant consensus and are thus oftentimes used as reference points.

Within the democratic theory literature, it is common to distinguish between a "narrow" (thin) and a "comprehensive" (thick) conception of democracy—or what Diamond (1996) has referred to as a distinction between electoral and liberal definitions of democracy. Adherents of the narrow notion define democracy according to factors like voting rights and party competition in elections (Schumpeter, 1942). The more comprehensive conception of democracy is found in Dahl (1971). He emphasizes the responsiveness of the government to the preferences of its citizens, considered as political equals, as a key characteristic of democracy. This in turn relies on various institutional guarantees; most importantly, freedom of expression, alternative sources of information, free and fair elections, and institutions that are dependent upon votes and the expression of preferences. Thus, in Dahl's (ibid.) view, a central quality of democracy is enabling minorities to organize and lobby as well as the presence of elected representatives responding to them. Thus, Dahl adds a liberal aspect to the definition of democracy.

Another common distinction is found between what is oftentimes referred to as minimalist versus maximalist definitions of democracy, where the minimalist definition is parallel to the narrow definition of democracy

(Schumpeter, 1942) and liberal democracy, and the focus on political rights and civil liberties is typically placed in the middle, while the maximalist definitions of democracy (Habermas, 1991; Pateman, 1971) consider civic participation and political deliberation to be essential (Beetham, 1999). Moreover, for participation to be real, resources have to be allocated equally (Knutsen, 2021). V-Dem's classification of central dimensions of democracy (Coppedge et al., 2023) thus distinguishes between all five dimensions of democracy, namely electoral democracy, liberal democracy, participatory democracy, deliberative democracy, and egalitarian democracy. Although I agree with all these dimensions being aspects of a democracy, some of these dimensions are partly in contrast to one another, so no country has a full score in all five of them. In particular, the participatory (channels for direct democracy and civil society) and egalitarian aspects (all social groups equal power and resources) vary across democracies (Knutsen, 2021). I therefore limit my focus to three dimensions: electoral democracy, liberal democracy, and deliberative democracy.

Electoral democracy deals with the existence of elections between at least two parties or candidates for all important political positions, at different levels of governance, as well as the securing of participation rights for the entire adult population in these elections. Furthermore, the elections should open up real competition between different parties and candidates, in the sense that different participants, also in practice, have an opportunity to win the positions that are up for election. The election should also be free and fair.

The central cornerstones of a **liberal democracy** are, as already mentioned, the protection of various individual rights and freedoms, as well as political institutions that limit the power of the executive. The rights not only are limited to political rights such as freedom of speech and assembly but also include minority rights such as freedom of religion and freedom of movement, as well as a free press.

Finally, **deliberative democracy** refers to reasoned discussion, which promotes social cohesion in political life. Thus, the use of logic and reason as opposed to power-struggle is central. This means, among other things, that both traditional media (such as newspapers and TV) and social media facilitate a free, critical, factual, and constructive discussion, where citizens have access to varied information. Group decisions are generally made after deliberation through a vote or consensus of those involved (Knutsen, 2021).

Hybrid threats—and how they target the democratic dimensions

As Chapter 2 of Bjørge and Høiby has shown, the concept of "hybrid threats" has become a buzzword of our times, in particular after the Russian invasion of Crimea in 2014. Although definitions vary, most scholars agree that the phenomena are exacerbated as a result of globalizations, technological innovations, and digitalization (Giannopoulos et al., 2020). Moreover, they all describe hybrid threats as a combined use of military and non-military means with malicious intents to weaken and undermine societies actors (Fiott & Parkes, 2019; Szymanski, 2020, p. 2; Heap, 2019, p. 18). For my purpose, I will rely on the European Centre of Excellence for Countering Hybrid Threats in Helsinki, who defines hybrid threats as follows (Hybrid CoE, 2019, p. 10):

> Coordinated and synchronized action that deliberately targets democratic states' and institutions' systemic vulnerabilities through a wide range of means; Activities that exploit the thresholds of detection and attribution, as well as the different interfaces (war-peace, internal-external security, local-state, and national-international); Activities aimed at influencing different forms of decision-making at the local (regional), state, or institutional level, and designed to further and/or fulfil the agent's strategic goals while undermining and/or hurting the target.

These actions can include disinformation campaigns, election interference, conspiracy-theories, manipulating ethnic, religious, cultural, or other social cleavages including diasporas, cyberattacks, inducing political or economic corruption, infiltrating agents of influence, media control or media interference, as well as economic sanctions, clandestine and paramilitary operations, airspace violations, military exercises, weapons proliferation, attacks on critical infrastructure, etc. This is also further elaborated upon in Chapter 14 by Akrap and Kamenetskyi.

Although many of the tactics used are considered typical soft power and/or foreign power instruments (Nye, 1990), its malicious intent separates hybrid threats from typical soft power politics. Some scholars have thus referred to hybrid threats as a combination of hard power, soft power, and smart power (Schmid, 2022). This chapter focuses mainly on soft (or smart)-power instruments targeting the political, societal, cultural, and information domain (i.e. disinformation, electoral interference, conspiracy-theories).

Methods

To investigate how hybrid threats are a threat to democracy, I rely on a number of published reports and known examples of hybrid threat tactics that have been used throughout the last ten-year period. I then show how each of them is a threat to the three dimensions associated with democracy—electoral, liberal, and deliberative democracy—and their main characteristics. In this sense, my methodological approach is best described as a literature review anchored in a theoretical framework (Hart, 1998). Often, a theoretical framework can be used as a guide for logically developing and understanding the different, yet interconnected, parts of the literature review. The theoretical framework and literature can develop synchronically and then be used to support the data, interpret the findings, and underlie the recommendations (Grant & Osanloo, 2014). In this chapter, the theory is used as a way to organize the empirical cases between tactics (independent variable), what dimensions of democracy they stir/target (intermediate variable), and their anticipated effects/implications (dependent variable).

Describing hybrid threats to democracy

Most scholars would agree that hybrid threats pose a challenge to democracies; however, systematic knowledge about how and what parts of democracy are targeted is less common. In the following section of this chapter, I separate between hybrid threat tactics targeting electoral democracy, liberal democracy, and finally deliberative democracy.

Tactics targeting electoral democracy

Tactics within the realm of hybrid threats that undermine electoral democracy can all be labeled as "election interference". This includes interfering with the actual result of the election, either directly or indirectly through the aspect of competition. Both instances undermine free and fair elections.

An early example of this took place in Estonia in 2007, where the websites of the Estonian Prime Minister, President and Parliament were repeatedly attacked over three weeks. As a result, the political system of the country had limited function ability, thereby depriving Estonia of its right to exercise its sovereign functions (Haataja, 2017).

Another example is Ukraine in the final days before the 2014 election, when the central election system was infiltrated, and the vote-counting system was left out of function. Malicious software was installed, which ensured that the election results would make the ultra-nationalist candidate Dmytro Yarosh the winner. Finally, the website of the Central Election Commission was also shut down.

Then in 2015, the German Bundestag became the target of a massive cyberattack, data was again stolen, including from the computers of Chancellor Angela Merkel and numerous other members of parliament. Again Russia, in particular the intelligence service GRU, was quickly suspected as the perpetrator (Fischer, 2019). An additional element to the German election was that Russia used its own state-controlled media to intensify conflicts between the German right-wing and left-wing radical groups. At the same time, right-wing radical groups within Germany were also supported by international right-wing radical networks (Martin et al., 2019).

Despite these early examples, it is however the US presidential 2016 election that have gained the most attention, where Russian actors associated with the Russian Intelligence Service (GRU) infiltrated the information and e-mail systems of the Democratic National Committee (DNC) and the Clinton campaign. Politically damaging information was then released on the internet and spread propaganda on Twitter, Facebook, YouTube, and Instagram through thousands of fake accounts linked to the Russian troll-farm—the Internet Research Agency (IRA) (Gadde & Roth, 2018). Typically, these bots were purporting as supporting radical political groups within the United States planned or promoted events in support of Trump and against Clinton. It reached millions of social media users between 2013 and 2017, with the goals of harming the campaign of Hillary Clinton, boosting the candidacy of Donald Trump and increasing political and social discord in the United States (US Senate Select Committee on Intelligence, 2020). The extent to which these bots actually did boost the votes for Trump is unclear though—some arguing that it had a small but decisive effect (Gorodnichenko et al., 2018), while other studies (Eady et al., 2023) conclude that the exposure to Russian coordinated influence accounts was heavily concentrated among a small portion of the electorate.

Many analysts and experts also argue that Russia attempted to help influence public opinion in the 2016 British Brexit referendum to swing the Brexit vote in favor of the "Leave" campaign (Mackinnon, 2020; Gorodnichenko et al., 2018). However, these allegations are still unproven, but an independent committee—Ferr's Intelligence and Security Committee—concluded in 2020 that although Russia probably did not have a direct involvement in the voting process, it is certainly possible that Russia indirectly may have been a significant factor. Moscow-based information operations through social media and Russian state-funded broadcasters, like Sputnik and RT, backed up by targeted support to influential voices within UK politics are potential aspects to consider. The report specifically addresses the amount of Russian donations and other economic ties to various political candidates and criticizes the UK government for not having taken steps to investigate the allegations properly. Further, that the influence of Russian business was so deeply embedded in the British financial system that it "cannot be untangled" (Intelligence and Security Committee of Parliament, 2020). This is particularly evident in the city of London, which has resulted in some referring to it as "Londongrad".

Also, in the French election in 2017, there were signs of election interference. Very similar to what happened during the U.S. election in 2016, the headquarters of presidential candidate Emmanuel Macron's campaign in France was hacked and data and e-mails were stolen. Together with a widespread disinformation campaign, consisting of a bogus website resembling the site of Belgian newspaper *Le Soir* reporting that Saudi Arabia was financing his campaign, this hack formed the basis of the Russian influence operation (Kranefeld, 2023). An interesting case is also Germany, where Turkish President Erdogan in 2017 urged German voters of Turkish background not to vote for Angela Merkel (Deutsche Welle, 2017). The electoral interferences failed, though, as both Macron and Merkel won the ballot.

In Sweden the same year—2017—the number of automated Twitter accounts supporting the populist anti-immigration Sweden Democrats had surged as the election got closer, pushing anti-immigrant rhetoric's which allegedly boosted the votes of Sweden Democrats (Fernquist et al., 2018). In search of foreign influence in the Norwegian elections, SINTEF carried out an analysis of the Norwegian municipal council and county council elections in 2019, followed by a new report related to the governmental election in 2021. Both concluded, however, that no clear signs of foreign influence were found (Grøtan et al., 2020; Sivertsen et al., 2021, 2022).

The U.S. presidential election in 2020, on the other hand, was again victim of electoral interference from Russia (National Intelligence Council, 2021). Ironically, some of the actions of President Trump himself were probably more worrisome with regard to interfering the election (Ferrara et al., 2020). Being responsible for dozens of false and misleading claims about the prevalence of election fraud in the United States and the outcome of the 2020 election (Funke, 2021; Kessler et al., 2020), Trump targeted citizens' trust with the electoral system itself. According to Pennycook and Rand (2021), examples included false claims that voting machines fraudulently switched votes to Biden, that large numbers of Trump ballots were destroyed, and that Republican election officials were unduly restricted from observing polling stations. These claims were also parroted by a number of media outlets (e.g., Breitbart, Newsmax, One America News Network), including some of the hosts on Fox News (Darcy, 2020). The claim that the election was "stolen" is a claim that has persisted even long after the 2020 election (Wang & van Prooijen, 2023). In addition, right-wing advocates within U.S. programmed numerous false robocalls to intimidate people from voting with the message "Stay safe, stay home". Given that the election took place during the Covid-19 pandemic, it multiplied its potential effect on voter turnout.

The German election of 2021 was also victim to at least three cyberattacks on parliamentarians at the federal and state levels, as well as on the office of the Federal Election Commissioner and civil society organizations. In addition, the disinformation spread by Russian state media "RT DE" was primarily directed against the Green Party and its candidate Annalena Baerbock. These narratives referred in particular to Covid-19, the legitimacy of postal voting, and a major flood in the summer of 2021. Much of the content would also classify as "misogyny"—including fake nude photos and hate speech directed towards Baerbock. In general, this type of tactic—also referred to as "gendered disinformation"—seems to have become increasingly popular and target in particular young women running for office (Judson et al., 2020).

The 2023 election in Slovakia (September 2023) and Poland (October 2023) was also flooded by disinformation: about NATO/EU, the "election being stolen", what started the war in Ukraine in 2014, anti-immigrant and anti-LGBTQ rhetoric, etc. (CEDMO, 2023). For instance, a London-based non-profit organization Reset claims to have recorded more than 365,000 election-related disinformation messages on Slovak social networks in the first two weeks of September. In many cases, it has been the politicians themselves who are spreading the disinformation (CEDMO, 2023).

Obviously, despite not all countries being subject to electoral interference, the examples are many. Numbers from the Oxford Internet Institute indicate that in 2020 alone, the spread of political propaganda and disinformation was mapped in 81 countries (Bradshaw et al., 2021, pp. 1–5). In the case of Europe and the United States, Russia is behind a majority of them (Martin et al., 2023), and many of the operations have been going on since 2014 (Aleksejeva et al., 2019, p. 3). This clearly affects the right to hold opinions without interference; and saps trust in democratic institutions and distorts electoral processes (Colomina et al., 2021).

As pointed out in Chapter 6 by Mahda and Semenenko and Chapter 8 by Heier, there is a reason to believe that hybrid attacks will continue as Russia does not have the military means to go to direct warfare with more than Ukraine. The tactics used for election interference seem to include everything from hacking/cyberattacks, bribing, pressure/intimidation, false claims, bots, pages and videos, playing on diasporic ethnic or religious identities, nurturing ethnic and religious differences, or issues regarding migration, climate, vaccines, gender, and sexuality. NATO/EU, as well as war in Ukraine and between Israel and Gaza, also seem to polarize.

Tactics targeting liberal democracy

Pressure/intimidation, playing on diasporic identities (ethnic and/or religious minorities who maintain connections with their place of origin) or nurturing ethnic and religious differences, and disinformation campaigns also target the dimension of liberal democracy, such as freedom of speech, minority rights/freedom of religion, pluralism, and a free and independent press.

A prominent example of how freedom of speech is under attack is the issue of Quran burnings, which received special attention after the Danish-Swedish right-wing extremist Rasmus Paludan went on a tour to Sweden to burn the Quran. Over time, it became obvious that the act was not only an expression of freedom but also a strong igniter of social unrest as the burnings led to riots in several Swedish cities. Anger was also directed towards the police who protected the right of Paludan to burn the Quran due to the freedom of expression in Sweden (Boxerman & Kwai, 2023). Since then, the number of Quran burnings has increased in Sweden (and Denmark), and in particular after the Russian military intervention in Ukraine 2022 and the following application of NATO membership for Sweden (and Finland).

Being condemned by Turkish President Erdogan and receiving massive attention through social media have in turn enraged much of the Muslim world, resulting in protests outside the Swedish embassy among others in Iraq and in Tehran, Iran, burning the Swedish flag. Iran also argued that Sweden, with the Quran burning, waged a war against the whole Muslim world. As a result, the security situation within Sweden has deteriorated, and in Belgium on October 16, 2023, two Swedish football fans were killed and a third was injured. Sweden is currently considering changing their laws so that Quran burnings become illegal, like in Finland, a decision Denmark recently also reached. Freedom-of-speech contenders within the Nordic countries warn against this, seeing it as a way of giving in to dictators and decreasing freedom of expression. Freedom of expression itself is thus potentially used by actors as a tactic to weaken it.

Turning to liberal aspects that are specifically related to individual/minority rights—such as freedom of religion or sexuality, as well as non-discrimination against ethnic, religious or other political minorities)—the spread of conspiracy theories is considered a major threat. Ren et al. (2022), for instance, argue that conspiracy theories legitimize false narratives about powerful elites as well as outgroups fueling prejudice towards minority groups. Although conspiracy theories used to be something that only a few people would adhere to, the spread of Internet and social media has brought them to everyone's doorstep. The most obvious and prominent example of this is of course the riots of January 6, 2021, at the U.S. Capitol. Many in the mob on January 6, 2021, believed that there was a "deep state" in control of their country, which had taken over powerful positions and were making decisions. Similar narratives are found within the Reichbürger movement, an anti-constitutional revisionist group in Germany, several facing arrest in 2022 and later in October 2023 for having planned a coup. Theories such as "Eurabia" and "the Great Replacement" are also a threat to liberal democracy, as they insist that entire categories of human beings (oftentimes Muslims, blacks, Jews, but also women and LGBTQ) can or should be excluded from democratic rights and protections. The fact that these types of ideas have also been an inspiration for solo-terrorists and attacks such as Anders Behring Breivik on July 22, 2011, Brenton Tarrant in Churchill, New Zealand, 2019, and Payton Gendron in Buffalo, New York, 2022, to mention a few, a bare witness of its violent and discriminate potential.

A free and independent media is a third key feature of liberal democracy. The spread of disinformation and fake news leads to the information environment becoming toxic, and people struggle to separate facts from fiction and malicious lies, leading to mistrust to media (Kalniete, & Pildegovičs, 2021). Moreover, several media outlets and various social media platforms, such as Tik Tok or Twitter (X), are owned and controlled by authoritarian governments (China) or individuals (i.e. Elon Musk). What kind of news and ideas are spread through these platforms is hard to control. Elon Musk has, for instance, been accused of tweeting conspiracy theories on Twitter (X) (Hickey et al., 2023; *Washington Post*, November 28, 2023). The fact that there are 96.5 million TikTok users in Europe and 80 million in the United States, of which around 71% are between the ages of 18 and 34 or even younger, gives food for thought with regard to potential effects (Pew Research Center, 2024), especially as the younger generation seems to be more likely to believe in conspiracy theories than elders (Weimann & Masri, 2023).

Tactics targeting deliberative democracy

A hybrid tactic that seems to be commonly exploited is to exacerbate existing ethnic, religious, political, or economic fault-lines, thereby undermining societal cohesion. Once social cohesion is undermined, the stage is set for further hybrid activities to deteriorate the lack of unity and thus democracy even more (Kalniete & Pildegovičs, 2021). According to Wigell (2019, p. 270), this tactic is aimed at fomenting polarization and radicalization to the point that the principles of democratic societies are stretched to their extremes. According to EEAS (2020), this was particularly evident during the COVID-19 pandemic as a reaction to various governments' crisis response, exemplified by large demonstrations opposing government restrictions throughout most of Europe as well as vaccination skepticism among certain parts of the population. Other examples of exacerbating ethnic, religious, political, or economic fault lines are for instance Georgia 2008. Russia's forces invaded Georgia in August 2008. Before the attack, Russia had falsely accused the Georgian government of committing grave crimes against citizens. Moscow said Russians were living in Georgia's South Ossetia region. Moscow said it was intervening to prevent genocide (Allison, 2009). Other examples of incidents where cultural, political, or economic differences have been exploited to damage social cohesion are Brexit, where Russian media outlets pushed hard to weaken the EU within the British population (Kirkpatrick, 2016). Also, the situations with the Yellow Vests in France in 2018 (Chamorel, 2019) and the Spanish Constitutional Crisis (Catalonian Crisis) in 2017–2018 (Casañ et al., 2022) have been argued to have been influenced by media outlets and various actors and parties exacerbating differences, resulting in further polarization and social unrest.

Moreover, certain media outlets and webpages have become deemed as spreading false propaganda. Fox news (in the United States), Tsar TV (in Russia), and Steigan (in Norway) are some examples of this (Stelter, 2020; Bonde, 2021). Although media awareness is an important countermeasure against disinformation and fake news, in the end, it can also end up in a situation where citizens only adhere to various echo-chambers, thus putting an end to public deliberation. According to Furman and Tunc (2020), an uninhibited exchange of information and ideas leads to reasoned assessment, both of the multi-faceted complexity of the social issue(s) and their potential solutions. Without this, decisions rest upon emotions and partisan issues (ibid., 2020), which threaten the deliberative aspect of democracy.

Anticipated effects (implications) and protective measures

In sum, by distorting electoral processes, put pressure on individual and minority rights such as freedom of thought and minority protection, fostering polarization and conspiracy theories, these different tactics are able to erode political trust—the very core element that well-functioning democracies rests upon (Warren, 2017). Studies by the OECD (2022) show that trust in governments has declined in most OECD countries in the last decade. In 2021, only 49% of people, on average, trusted their governments (OECD, 2022). Further deterioration of political trust can be detrimental as lack of political trust, combined with polarization and "scapegoating, eventually also can become a formula for social unrest and political violence. The turmoil during the US election 2020 and aspirations of social unrest across Europe during the Covid-19 pandemic are examples of this. The war between Russia and Ukraine as well as between Israel and Gaza could potentially also ignite the fire again, especially as social inequality throughout Europe and the United States is increasing (World Bank, 2023). So how can viable democracies defend themselves towards these types of hybrid threats?

Based on the review, three factors seem particularly important. First and foremost, there is a need to strengthen public vigilance on hybrid threats. A critical and educated population that can distinguish fact from fiction is an extremely important protection mechanism. Investing in civic awareness and media literacy through education is thus key. Finland—ranked no. 1 in Media-literacy Index 2023—has a long history of promoting media literacy as a critical tool for a stable democracy and a healthy society, starting in kindergarten.

Second, there is a need to strengthen judicial agility by broadening the mandate of the EU Disinformation Act, which allows for investigations of foreign political funding and close relations between businesses and politics, as well as mapping cultural and social ties to political candidates. Moreover, developing laws and finding methods for deterring and sanctioning this type of behaviour are important. In this regard, demanding greater transparency and accountability within the digital and economic-political domain is key.

Third, ensuring media safety and empowering a quality and independent media are also detrimental. Journalists and the news media play a critical role in ensuring the integrity and operational capacity of democratic institutions and processes. A stronger emphasis on editor-controlled media is thus key, along with increasing the number of fact-finding media and journalism. There is also a need to take more control over the content—and the number of—actual social media platforms available. This can involve everything from prohibiting or warning about certain platforms, debunking disinformation, to monitoring social media.

However, increased monitoring, mass surveillance, stricter laws, and even fact-finding, as suggested earlier, might interfere with individual freedoms and freedoms of speech, which are crucial for democracies. Extending state control over civil society is thus not a viable liberal democratic strategy (Wigell, 2021) and should thus be used with caution. A more democratic method would thus be to enhance social cohesion by limiting social and economic grievances.

Conclusion

The aim of this chapter has been to contribute to our understanding of hybrid threats to democracy. This has been done through a comprehensive review of hybrid threats to countries in Europe and the United States of how they target three core democratic dimensions—electoral, liberal, and deliberative democracy. While the discussion is limited to the use of soft (or smart) hybrid threat tactics within Europe and the United States and should not be generalized, a few observations are worth mentioning. First of all, foreign and hostile actors use a variety of soft power tactics against democracies. While some of the tactics aim at targeting the election results directly or indirectly (electoral democracy), others put pressure on the civil liberties of individuals and minorities (liberal democracy). This, combined with exacerbation and manipulation of ethnic, religious, racial, and/or other differences, restrains the deliberative discussion that democracies are dependent upon and, in the end, also political trust. In a worst-case scenario, this leads to democratic decline and social unrest. Building resilience towards these types of tactics is thus important and demands measures both within the educational, judicial, and media-oriented sectors of society but should be carefully considered. Raising this discussion not only within but also across these sectors is a first step in the right direction. Preventing a hostile actor from exploiting European democratic pluralism involves taking a whole-of-society approach to security. Creating a forum where media companies, first responders, and teachers can meet and discuss these issues and develop educational programs and public campaigns is thus important. Ensuring and broadening social cohesion through integration and welfare, in particular towards diasporas and ethnic and religious minorities, are even more important. At the moment, however, there is only a limited amount of knowledge and research on which individuals, groups, local communities, and nations are particularly vulnerable and/or resilient to the various tactics. Comparative and large N-studies across countries, groups, and individuals and counter-measures should thus be a particular priority for future studies of hybrid threats.

References

Aleksejeva, N., Andriukaitis, L., Bandeira, L., & Barojan, D. (2019, August). *Operation "Secondary Infektion": A suspected Russian intelligence operation targeting Europe and the United States*. DFRLab.

Allison, R. (2009). The Russian case for military intervention in Georgia: International law, norms and political calculation. *European Security, 18*(2), 173–200. https://doi.org/10.1080/09662830903468734.

Beetham, D. (1999). *Democracy and human rights* (Vol. 249). Polity Press.

Bonde, A. (2021). Grønn politikk og falske nyheter. *Stat & Styring, 31*(1), 10–13. https://doi.org/10.18261/ISSN0809-750X-2021-01-05.

Boxerman, A., & Kwai, I. (2023). What's happening with the Quran burnings in Sweden. *The New York Times* (Digital Edition), NA-NA.

Bradshaw, S., Campbell-Smith, U., Henle, A., Perini, A., Shalev, S., Bailey, H., & Howard, P. N. (2021). *Country case studies industrialized disinformation: 2020 Global inventory of organized social media manipulation*. Oxford Internet Institute.

Casañ, R. R., García-Vidal, E., Grimaldi, D., Carrasco-Farré, C., Vaquer-Estalrich, F., & Vila-Francés, J. (2022). Online polarization and cross-fertilization in multi-cleavage societies: The case of Spain. *Social Network Analysis and Mining, 12*(1), 79. https://doi.org/10.1007/s13278-022-00909-5.

CEDMO. (2023). Central European digital media observatory. *CEDMO Fact-Checking Summary*, Q.83: 3.

Chamorel, P. (2019). Macron versus the yellow vests. *Journal of Democracy, 30*, 48.

Colomina, C., Sánchez Margalef, H., & Youngs, R. (2021, April). *The impact of disinformation on democratic processes and human rights in the world*. European Parliament coordinator: Policy Department for External Relations Directorate General for External Policies of the Union. PE 653.635. The impact of disinformation on democratic processes and human rights in the world (europa.eu).

Coppedge, Michael, Gerring, J., Knutsen, C. H., Lindberg, S. I., Teorell, J., Altman, D., Bernhard, M., Cornell, A., Fish, M. S., Gastaldi, L., Gjerløw, H., Glynn, A., Grahn, S., Hicken, A., Kinzelbach, K., Marquardt, K. L., McMann, K., Mechkova, V., Neundorf, A., Paxton, P., Pemstein, D., Rydén, O., von Römer, J., Seim, B., Sigman, R., Skaaning, S.-E., Staton, J., Sundström, A., Tzelgov, E., Uberti, L., Wang, Y.-T., Wig, T., & Ziblatt, D. (2023). *V-Dem codebook v13*. Varieties of Democracy (V-Dem) Project.

Dahl, R. (1971). *Polyarchy: Participation and opposition*. Yale University Press.

Darcy, O. (2020, November 5). Fox News hosts sow distrust in legitimacy of election. *CNN Business*. https://www.cnn.com/2020/11/05/media/fox-news-prime-time-election/index.html.

Deutsche Welle. (2017, August 18). *Erdogan tells Turks in Germany to punish Merkel*. Erdogan tells Turks in Germany to punish Merkel—DW. https://www.dw.com/en/erdogan-tells-german-turks-not-to-vote-for-angela-merkel/a-40149680.

Diamond, L. (1996). Is the third wave over? *Journal of Democracy, 7*, 20–37.

Eady, G., Paskhalis, T., Zilinsky, J., Bonneau, R., Nagler, J., & Tucker, J. A. (2023). Exposure to the Russian internet research agency foreign influence campaign on Twitter in the 2016 US election and its relationship to attitudes and voting behavior. *Nature Communications, 14*(62). https://doi.org/10.1038/s41467-022-35576-9.

EEAS. (2020). *EEAS SPECIAL REPORT UPDATE: Short assessment of narratives and disinformation around the COVID-19/coronavirus pandemic*. https://euvsdisinfo.eu/uploads/2020/05/EEAS-Special-Report-May-1.pdf.

Fernquist, J., Kaati, L., & Schroeder, R. (2018). Political bots and the Swedish general election. In *2018 IEEE international conference on intelligence and security informatics (ISI)* (pp. 124–129). IEEE.

Ferrara, E., Chang, H., Chen, E., Muric, G., & Patel, J. (2020, November). *First Monday*, 25(11–12). https://firstmonday.org/ojs/index.php/fm/article/download/11431/9993, https://dx.doi.org/10.5210/fm.v25i11.11431.

Fiott, D., & Parkes, R. (2019). *Protecting Europe: The EU's response to hybrid threats*. EU Institute for Security Studies.

Fischer, M. (2019). The concept of deterrence and its applicability in the cyber domain. *Connections, 18*(1–2), 69–92. https://doi.org/10.11610/Connections.18.1-2.05.

Fukuyama, F. (1989). The end of history? *The National Interest*, (16), 3–18.

Funke, D. (2021). Global responses to misinformation and populism. In *The Routledge companion to media disinformation and populism* (pp. 449–458). Routledge.

Furman, I., & Tunc, A. (2020). The end of the Habermassian ideal? Political communication on Twitter during the 2017 Turkish constitutional referendum. *Policy & Internet, 12*(3), 311–331. https://doi.org/10.1002/poi3.218.

Gadde, V., & Roth, Y. (2018, October 17). *Enabling further research of information operations on Twitter*. Retrieved June 28, 2023, from https://blog.twitter.com/en_us/topics/company/2018/enabling-further-research-of-information-operations-on-twitter.html.

Giannopoulos, G., Smith, H., & Theocharidou, M., (2020). *The landscape of hybrid threats: A conceptual model, European Commission*. ISPRA. PUBSY No. 123305.

Gorodnichenko, Y., Pham, T., & Talavera, O. (2018). *Social media, sentiment and public opinions: Evidence from #Brexit and #USElection, NBER Working Paper 24631*. National Bureau of Economic Research. http://www.nber.org/papers/w24631.

Grant, C., & Osanloo, S. (2014). Understanding, selecting and integrating a theoretical framework in dissertation research: Creating the blueprint of your "house". *Administrative Issues Journal: Connecting Education, Practice and Research*, 12–26. https://doi.org/10.5929/2014.4.2.9.

Grøtan, T. O., Fiskvik, J., Halland Haro, P., Auran, P. G., Mathisen, B. M., Hågen Karlsen, G., Magin, M., & Bae Brandtzæg, P. (2020). På leting etter utenlandsk informasjonspåvirkning En analyse av det norske kommunestyre-og fylkestingsvalget 2019. *SINTEF report* 2019:01292. SINTEF.

Haataja, S. (2017). The 2007 cyber attacks against Estonia and international law on the use of force: An informational approach, *Law, Innovation and Technology*, 9(2), 159–189. https://doi.org/10.1080/17579961.2017.1377914.

Habermas, J. (1991). *The structural transformation of the public sphere: An inquiry into a category of bourgeois society*. MIT Press.

Hart, C. (1998). *Doing a literature review. Releasing the social science research imagination*. Sage Publications.

Heap, B. (2019). *Hybrid threats. A strategic communications perspective*. NATO Strategic Communications Centre of Excellence.

Hickey, D., Schmitz, M., Fessler, D., Smaldino, P. E., Muric, G., & Burghardt, K. (2023). Auditing Elon Musk's impact on hate speech and bots. *Proceedings of the International AAAI Conference on Web and Social Media, 17*(1), 1133–1137. https://doi.org/10.1609/icwsm.v17i1.22222.

Hybrid COE. (2019). *Countering disinformation: News media and legal resilience*. Hybrid CoE Papers.

Intelligence and Security Committee of Parliament. (2020). *Russia (House of Commons Paper) HC 632*. Dandy Booksellers.

Judson, A., Atay, A., Krasodomski-Jones, A., Lasko, R., & Smith, S. J. (2020, October). *Engendering hate: The contours of state-aligned gendered disinformation online. Demos*. National Democratic Institute.

Kalniete, S., & Pildegovičs, T. (2021). Strengthening the EU's resilience to hybrid threats. *European View*, 20(1), 23–33.

Kessler, G., Rizzo, S., & Kelly, M. (2020). *Donald Trump and his assault on truth: The President's falsehoods, misleading claims and flat-out lies*. Simon and Schuster.

Kirkpatrick, I. (2016). Hybrid managers and professional leadership. In Dent, M. and Bourgeault, I. L. and D. Jean-Louis and Kuhlmann, E. (eds.) *The Routledge companion to the professions and professionalism*. Routledge companions in business, management and accounting. Basindstoke: Taylor & Francis Ltd. (pp. 175–187). ISBN 9781317699484.

Knutsen, C. H. (2021). *Demokrati og diktatur*. Fagbokforlaget.

Knutsen, C. H., Dahlum, S., Allern, E. H., Hagfors, S. B., Klausen, J. E., Søyland, M., & Wig., T. (2023). *Tilstandsrapport for det norske demokratiet (Status report for the Norwegian democracy)*. UiO.

Kranefeld, T. (2023). *The digitalization of disinformation campaigns*. Lit Verlag.

Levitsky, S., & Ziblatt, D. (2018). *How democracies die*. Crown.

Mackinnon, A. (2020, July 21). 4 Key takeaways from the British report on Russian interference. *Foreign Policy*.

Martin, D. A., Shapiro, J. N., & Ilhardt, J. G. (2023). Introducing the online political influence efforts dataset. *Journal of Peace Research*, 60(5), 868–876. https://doi.org/10.1177/00223433221092815.

Martin, D. A., Shapiro, J. N., & Nedashkovskaya, M. (2019). Recent trends in online foreign influence efforts. *Journal of Information Warfare*, 18(3), 15–48.

National Intelligence Council. (2021). https://www.dni.gov/files/ODNI/documents/assessments/ICA-declass-16MAR21.pdf.

Nye, J. S. (1990). Soft power. *Foreign Policy*, 80, 153–171. https://doi.org/10.2307/1148580.

OECD. (2022). *Building trust to reinforce democracy: Main findings from the 2021 OECD survey on drivers of trust in public institutions, building trust in public institutions*. OECD Publishing. https://doi.org/10.1787/b407f99c-en.

Papada, E., Altman, D., Angiolillo, F., Gastaldi, L., Köhler, T., Lundstedt, M., Natsika, N., Nord, M., Sato, Y., Wiebrecht, F., & Lindberg, S. I. (2023). *Defiance in the face of autocratization. Democracy report 2023*. University of Gothenburg: Varieties of Democracy Institute (V-Dem Institute).

Pateman, C. (1971). *Participation and recent theories of democracy* (Doctoral dissertation, University of Oxford).

Pennycook, G., & Rand, D. G. (2021). Research note: Examining false beliefs about voter fraud in the wake of the 2020 presidential election. *Harvard Kennedy School (HKS) Misinformation Review*. https://doi.org/10.37016/mr-2020-51.

Pew Research Center. (2024). *Americans social media use*. Report January 31st, 2024. How Americans Use Social Media | Pew Research Center. https://www.pewresearch.org/internet/2024/01/31/americans-social-media-use/.

Ren, Z. B., Carton, A., Dimant, E., & Schweitzer, M. E. (2022). *Authoritarian leaders share conspiracy theories to attack opponents, promote in-group unity, shift blame, and undermine democratic institutions*. CESifo Working Paper No. 9951. Available at SSRN: https://ssrn.com/abstract=4231779 or http://dx.doi.org/10.2139/ssrn.4231779.

Schmid, J. (2022). Conceptualizing hybrid threats. *HDR*, 1–2, 24–35.

Schumpeter, J. (1942). *Capitalism, socialism and democracy*. Harper and Row.

Sivertsen, E., Bjørgul, L., Lundberg, H., Endestad, I., Bornakke, T., Bæk Kristensen, J., Meldgaard Christensen, N., & Albrechtsen, T. (2022). *Uønsket utenlandsk påvirkning?—kartlegging og analyse av stortingsvalget 2021*. FFI Rapport 22/02746. Forsvarets Forskningsinstitutt.

Sivertsen, E., Hellum, N., Bergh, A, & Bjørnstad, A. L. (2021). *Hvordan gjøre samfunnet mer robust mot uønsket påvirkning i sosiale medier? FFI Rapport 21/01237.* Forsvarets forskningsinstitutt.

Stelter, B. (2020). *Hoax: Donald Trump, Fox News, and the dangerous distortion of truth.* Simon and Schuster.

Stoltenberg, J. (2015, March 25). *Key note speech on NATO transformation seminar, Washington, D.C. NATO—Opinion: Keynote speech by NATO Secretary General Jens Stoltenberg at the opening of the NATO transformation seminar.* https://www.nato.int/cps/en/natohq/opinions_118435.htm.

Szymanski, P. (2020). *Towards greater resilience: NATO and the EU on hybrid threats. OSW commentary 328.* Centre for Eastern Studies.

US Senate Select Committee of Intelligence. (2020). *Russian active measures campaigns and interference in the 2016 U.S. Election. Volume 1: Russian efforts against election infrastructure with additional views.* Report_Volume1.pdf (senate.gov). https://www.intelligence.senate.gov/sites/default/files/documents/report_volume5.pdf.

Wang, H., & van Prooijen, J. W. (2023). Stolen elections: How conspiracy beliefs during the 2020 American presidential elections changed over time. *Applied Cognitive Psychology, 37*(2), 277–289. https://doi.org/10.1002/acp.3996.

Warren, M. E. (2017). What kinds of trust does a democracy need? Trust from the perspective of democratic theory". In Zmerli, S. and TW.G. van der Meer (eds.) *Handbook on Political Trust.* Cheltenham, UK: Edward Elgar Publishing. Retrieved Jul 13, 2024, from https://doi.org/10.4337/9781782545118.00013.

Washington Post. (2023, November 28). *Elon Musk boots Pizzagate conspiracy theory that led to D.C. gunfire.* https://www.washingtonpost.com/technology/2023/11/28/pizzagate-musk-twitter-x-controversy/.

Weimann, G., & Masri, N. (2023). Research note: Spreading hate on TikTok. *Studies in Conflict & Terrorism, 46*(5), 752–765. https://doi.org/10.1080/10576 10X.2020.1780027.

Wigell, M. (2019). Hybrid interference as a wedge strategy: A theory of external interference in liberal democracy. *International Affairs, 95*(2), 255–275. https://doi.org/10.1093/ia/iiz018.

Wigell, M. (2021). Democratic deterrence: How to dissuade hybrid interference. *The Washington Quarterly, 44*(1), 49–67. https://doi.org/10.1080/0163660X.2021.1893027.

World Bank. (2023). *2023 in nine charts: A growing inequality.* World Bank Group. 2023 in Nine Charts: A Growing Inequality (worldbank.org). https://www.worldbank.org/en/news/feature/2023/12/18/2023-in-nine-charts-a-growing-inequality.

4

IDENTIFYING HYBRID THREATS FROM A NATIONAL SECURITY PERSPECTIVE

The case of Chinese hybrid threats in Australia

Patrick Cullen

Introduction

Hybrid threats against other nations are part of a larger design to gain influence and achieve favourable economic, military, and political objectives. An example is China's use of hybrid threats as a part of its international economic strategy within the Belt and Road initiative (Hybrid COE Expert Pool, 2020). In Australia, evidence indicates that CCP hybrid threat operations have been designed to achieve a series of overlapping objectives. First, Beijing has sought to control the Chinese-Australian community in such a way that it aligns with and supports CCP positions and marginalizes opinions hostile to CCP goals. Indeed, the targeting of the ethnic Chinese diaspora living in Australia by Beijing is being replicated worldwide, as Beijing seeks to become the sole arbiter of what it means to be "Chinese", collapsing the categories CCP, China, and Chinese into a single organic whole (Aukia, 2023). The mere existence of an Australian Chinese diaspora whose actions (or lack of action) fail to meet Beijing's criteria of a loyal member of the Chinese Party State is viewed as a serious ideational threat to the CCP simply by virtue of demonstrating a Chinese alternative to active membership and loyalty to the CCP. To this end, Beijing has directly used its own operatives and local proxies in Australia to befriend, influence, monitor, infiltrate, and co-opt wide swathes of Australian Chinese society that do not belong to or fail to demonstrate fealty the Chinese Communist Party State (see "Operation Fox Hunt" discussed later). Failing this, Beijing has used various mechanisms, including threats against family members living in the China, to coerce Chinese-Australian and other overseas ethic Chinese critics of the CCP into compliance while labelling them as unpatriotic (Human Rights Watch, 2021; ASIO, 2019; Whitehouse, 2023; Hamilton, 2018a, July 26; Mahnken et al., 2018; Resnick, 2020).

DOI: 10.4324/9781032617916-5

Another objective of Beijing's Australian hybrid influence campaigns is more recent, ambitious, and long-term. Its goal has been to co-opt and cultivate Australian elites and control the mainstream domestic Australian public discourse on China (Zhang et al., 2023). The purpose is to gradually realign Australia towards Beijing and absorb it into its sphere of influence, while weakening Australia's alliance with the United States. One piece of evidence of this alleged formal strategic objective can be traced to a 2004 decision by the Central Committee of the Party to include Australia in its "overall periphery"—in other words, to treat Australia like it has a land border with China and that it therefore needs to be controlled (Hamilton, 2018b, November). These above political goals aimed at shaping the ideational space in Australia—for example Australian mass and elite opinion, media, educational space, and public discourse—are connected to a much wider CCP effort of human and cyber intelligence collection and penetration of Australian society and government that will be examined later.

The remainder of this chapter is organized as follows. The following section offers a brief discussion of the chapter's theoretical framework and how it is applied. Next, Australian efforts to identify and respond to Chinese hybrid threats are discussed. The next section provides examples of how Chinese strategic culture and CCP political philosophy lend themselves to the use of hybrid threats. Next, the chapter provides an overview of the Chinese security apparatus that conducts hybrid threat operations, followed by examples of how China uses various tools to target Australia across the whole of society.

Theory

The theoretical framework used in this chapter is derived from the author's previous work on the abstract modelling of hybrid threats that has been instrumental in informing counter-hybrid threat efforts at the European Centre of Excellence for Countering Hybrid Threats (Hybrid COE), the European Commission, and various national counter-HT efforts (Cullen & Reichborn-Kjennerud, 2016, 2017; Cullen, 2018; Cullen et al., 2019). Building on this theoretical work, this chapter contextualizes empirical examples of China's hybrid threat campaign as one comprised of the coordinated use of non-military and non-kinetic tools against Australian society at large. For hybrid threats, this "expansion of the battlespace" involves the synchronized use of a wider set of military, political, economic, civil, and informational (MPECI) tools and actors across political, military, economic, social, information, and infrastructure (PMESII) whole-of-society to achieve effects across multiple domains of a target state (Cullen & Reichborn-Kjennerud, 2017).

Crucially, by emphasizing the use of these non-military means and actors, hybrid threats are intentionally designed to operate below and outside of the traditional response and detection thresholds of a state's national security establishment (Cullen, 2018). The intended effect of such persistent low-scale hybrid threats is non-linear and cumulative, creating a net effect of a "death by a thousand cuts". For instance, the cumulative negative effect of university professors self-censoring their research on China for fear of Beijing's financial reprisals against their university has historically fallen below (in terms of traditionally perceived significance and intensity) and outside (in terms of what is monitored) the scope of professional interest by national intelligence services. This example illustrates another component of hybrid threats—their ability to hide in plain sight while damaging national security because they may operate legally—exploiting and manipulating "gaps" within the culture, norms, and laws of liberal democratic societies. As we shall see, this aspect of hybrid threats compromised early efforts of the Australian intelligence community and government to see and understand the problem of Chinese hybrid threats and to correctly gauge their size and scope, thus hampering their response efforts.

Part I Describing the Chinese threat and Australian perceptions

- Australian recognition of Chinese hybrid threats
- Chinese and Chinese Communist Strategic Culture informs Chinese hybrid threats
- Chinese hybrid threat actors

Australian recognition of Chinese the hybrid threats

In the past five years, Australia has emerged on the frontlines of an emerging debate on how China is attempting to use a raft of mechanisms to covertly manipulate and undermine democratic processes and societies worldwide. Indeed, the former Director General of the Australian Security Intelligence Organization (ASIO), Duncan Lewis, has claimed that this interference threat is greater in the last few years than at any time during the Cold War, due to a greater number (and informal civilian types of) intelligence actors operating in Australia, as well as the new role of cyber-technologies in such hybrid interference campaigns. While not naming and shaming Beijing directly, he described foreign interference as "a foreign power using local Australians to observe and harass its diaspora community here in our country through to the recruitment and co-opting of influential and powerful Australian voices to lobby our decision-makers" (Diamond & Schell, 2019, p. 147). The Australian intelligence agency ASIO argued that although the harm from such hybrid threat activities may not be immediately apparent or overt, the consequences are far-reaching and serious. Such consequences include (ASIO, 2019):

- undermining Australia's national security and sovereignty,
- damaging Australia's international reputation and relationships,

- degrading Australia's diplomatic and trade relations,
- inflicting substantial economic damage, and
- compromising nationally vital assets and critical infrastructure.

Australia's Turnbull Government recognized and addressed this problem head-on as early as 2018 by explaining to the Australian public how this (Chinese) hybrid threat was carefully tailored to exploit gaps in Australian law. These gaps allowed it to legally engage in a series of malign influence actions across the whole of society that "falls short of espionage but is intended to harm Australia's national security or influence Australia's political or governmental processes" (Hamilton, Foreign Affairs, 2018a).

Chinese and Chinese communist strategic culture informs Chinese hybrid threats

Hybrid threats involve, among other things, an emphasis on the utility of non-violent and non-kinetic means, asymmetry, the use of indirect methods, exploitation of the "gray zone" between peace and war, and a long-term targeting of an adversary's society as an extended battlespace. The Chinese philosopher Sun Tzu's classic war treatise *The Art of War*, with its emphasis on "doing the unexpected and pursuing an indirect approach" (Griffith & Sun, 2005, p. 9) and winning without resorting to the use of arms aligns perfectly with the contemporary Chinese use of hybrid threats. So, too, does the CCP's intellectual Leninist and Maoist roots emphasizing the permanence of the political struggle between communism and capitalism. More recent CCP writings on warfare interpreted as relevant for Chinese hybrid threats are the late 1990s text *Unrestricted Warfare*, which heavily emphasized the role of non-military tools (e.g. trade, media, finance, crime, psychology, law, economic aid) to attack enemy societies in ways that collapse the distinction between "battlefield and non-battlefield, warfare and non-warfare, military and non-military" (Wang, 1999, p. 120).

A more authoritative contemporary Chinese military concept related to hybrid threats that has been built into the People's Liberation Army's (PLA's) doctrine is the *Three Warfares* that was made official in the PLA revisions of the Political Work Regulations in 2003. *Three Warfares* is comprised of psychological warfare, public opinion warfare, and legal warfare, and each operates during peacetime to achieve effects in the society of other states. For Dean Cheng, peacetime applications of *Psychological Warfare* "involve influencing and altering an opponent's unconscious, implicit views in order to make that opponent more susceptible to coercion" while increasing sympathy for one's own national objectives and creating negative perceptions of one's adversaries (Cheng, 2013). PLA National Defence University texts describe *Public Opinion Warfare* as "using public opinion as a weapon by propagandizing through various forms of media in order to weaken the adversary's 'will to fight'. . . while ensuring strength of will and unity among civilian and military views on one's own side" (Gershaneck, 2020). *Legal Warfare* is used to describe the technique of manoeuvring to gain legal superiority by using or modifying all forms of law—domestic, foreign, and international—for one's own geostrategic advantage. Rather than viewing law as a method of rational order-making, legal warfare looks for ways to use legal advantage to influence targets by delivering the effects of defeat, deterrence, or defence via legal means. The legal protections of citizens against the state common in democratic societies are viewed as opportunities for hybrid threat exploitation by Beijing.

Chinese hybrid threat actors

Today, almost all the functions that are central to Chinese overseas political warfare operations sit at the core of the Communist Party's operating structure (Mahnken et al., 2018). President Xi Jinping chairs all the key functions of the CCP-controlled Chinese state. He is also at the center of the Chinese political warfare apparatus and has direct strategic control over its use. China analysts also point to a speech President Xi gave in September 2014, using Mao's term "Magic Weapon" to describe United Front Work, as evidence of Xi's interest in expanding Chinese influence operations (Groot, 2017). Perhaps for these reasons, President Xi is partly responsible for the increase in current Chinese hybrid influence operations, acting as an advocate for its use and politically elevating the political status of the organizations responsible for its implementation.

Beijing uses a wide range of party, state, and non-state proxy actors in pursuit of its hybrid operations, and no single institution in China's party-state is wholly responsible for carrying out its hybrid activities. Different actors and organizations responsible for carrying out overseas influence operations—referred to as "United Front Work" that is carried out by the eponymous United Front Work Department as well as other organizations—each have varying levels of affiliation to the PRC. In the words of a U.S. Congressional report, "it is precisely the nature of United Front work to seek influence through connections that are difficult to publicly prove and to gain influence that is interwoven with sensitive issues such as ethnic, political, and national identity" (Bowe, 2018, p. 3).

A non-exhaustive list of the main agencies responsible for foreign influence operations include:

- United Front Work Department
- Central Propaganda Department
- International Liaison Department

- State Council Information Office
- All-China Federation of Overseas Chinese
- Chinese People's Association for Friendship with Foreign Countries.

These organizations and others are bolstered by various state agencies such as the Ministry of Foreign Affairs and the PLA—the latter of which is responsible for conducting the *Three Warfares*. The fact that United Front activities may only take up a portion of the work conducted by various Chinese organizations—as opposed to a Western intelligence office solely dedicated to its intelligence portfolio—is another aspect of the ambiguous nature of this hybrid threat. These state agencies are also bolstered by organizations like the Chinese Students and Scholars Associations that have a much more indirect connection to the Beijing party-state. Various CCP-affiliated organizations that operate overseas may be given various strategic communications, surveillance, and collection tasks and are provided with access to local networks and intelligence in support of their United Front Work (Christodoulou et al., 2019). Due to the multitude of organizations that conduct United Front Work—and the differing levels of formal affiliation to Beijing's party-state, it is difficult to find accurate figures relating to the size of Chinese overseas hybrid influence operations—or their operations in any country specifically. However, it is instructive to note the following:

- The Ministry of State Security alone is funded at level comparable to the PLA (Mahnken et al., 2018).
- In 2014, a retired spy stated the PLA had "at least 200,000 agents abroad" (Brady, 2017).
- In 2005, defecting former Chinese diplomat Chen Yonglin said "China had more than 1000 agents in Australia" (Wallace, 2017).

In sum, the Chinese conduct of hybrid threat operations is not a small effort. Under President Xi, the CCP has put tremendous resources into its interference and influence activities overseas, with one estimate placing the cost of such operations at 10 billion USD per year (Shambaugh, 2015).

Part II How can China's hybrid threats be comprehended?

Examining CCP hybrid threats across the Australian whole-of-society

As discussed earlier, Beijing uses a wide variety of institutionalized non-military tools within the MPECI conceptual toolbox to influence and/or interfere with a wide variety of targets across Australian PMESII whole-of-society. This chapter has also described how these tools have been designed to operate in an ambiguous space that is neither peaceful nor warlike and how as a result have fallen below Australia's traditional detection and political response thresholds. Borrowing from the theoretical observations that hybrid threats use MPECI tools to inflict damage on the PMESII whole-of-society, the following section offers a series of empirical examples of how Beijing is reported to have used a variety of tools—organized below into the ideal types political, legal, media, economic, and educational categories—in pursuit of its objectives across Australian society.

Political

Plausible Chinese cyber "priming" operations targeting Australian political parties

The Australian government is reluctant to engage in the attribution of hostile cyber activities to foreign governments. However, scholars from the Australian Political Science Institute have recently suggested that an unattributed—though likely Chinese—2019 hacking of Australia's political parties months before an election may be tied to preparing the groundwork for future interference and influence operations—what the Hybrid Centre of Excellence for Countering Hybrid Threats refers to as "priming". Speculating on the purpose behind the cyber-attack, Cave and Uren write:

> It is worrying, however, that the Liberal, Labor and National parties were targeted. The more that intelligence-gathering extends beyond government and parliament, the less likely it is that any intelligence gained will provide any insight into official government positions, and the more useful it would be for interference activities
>
> *(Cave & Uren, 2019).*

They speculated that this type of priming activity could provide Beijing with stolen emails and other confidential information that could be used in a timed release to impact Australian elections—mirroring Russian interference operations in the 2016 U.S. election. The authors also speculated that the stolen information could be used to identify politicians and staffers who may be susceptible to influence, to assist United Front Work assets to create future relationships with them, and to find points of leverage that might convince, cajole, or coerce them into a supportive position. The in-depth understanding of Australian political parties and the machinations of parliament—the exact targets of this hack—would be far more helpful in enabling this kind of interference than it would be in illuminating Australia's official decision-making and policy-making processes.

Recent evidence suggests that this type of intelligence collection is indeed occurring and even expanding through the use of Chinese private sector companies using artificial intelligence to engage in mass profile-building of

Australian (and other) national citizens of potential interest to the CCP. One such company, Zhenhua Data, with links to Chinese military and intelligence, was exposed in a data leak showing it had collected data on 35,000 Australians—including prominent and influential figures—with 656 Australians being labelled as "special interest" or "politically exposed" (Probyn & Doran, 2020).

"Astro-turfing" grassroots political movements

Beijing has been accused of "astro-turfing" various Australian grassroots political protests by surreptitiously using its influence to organize individuals to drown out opposition voices and artificially inflate perceptions of ethnic Chinese-Australian support for Beijing's policies. Much of this behavior occurs online, but street protests are also organized. In 2008, the Chinese embassy and CCP-affiliated organizations helped mobilize thousands of Chinese in Canberra via online/letter campaigns to stop Falung Gong, free-Tibet, and other protesters from interfering with the Olympic torch as it passed on its way to the Beijing-hosted Olympics. One group of Australian researchers has recently released a report detailing allegations of a "CCP cyber enabled influence operation" tied to the "Spamoflauge network" using fake personas in an instance of digital astroturfing to push misinformation about the United States to foster anti-American sentiment in the Indo-Pacific. APSI researchers geolocated the operators of this operation to Yancheng in Jiangsu Province, arguing some members were plausibly linked to the Yancheng Public Security Bureau (Zhang et al., 2023).

CCP-linked money tied to Australian politicians

Lax Australian campaign finance regulations that have allowed foreigners to donate to Australian political campaigns have created many opportunities for Beijing to gain influence in Australian politics. In 2015, Australian intelligence reached out to Australia's largest political parties to warn them that two of the country's most generous political donors had strong links to CCP and that accepting these donations could make them vulnerable to undue political influence. One of these donors, Chinese Billionaire Huang Xiangmo, threatened to cancel a pledged a 400,000-dollar donation in an alleged effort to influence the Labor Party's position on the South China Sea in Beijing's favour. Huang was the chair of Australian Council for the Promotion of Peaceful Reunification of China, an organization connected to the United Front Work Department, and whose leadership and activity are closely guided by the Chinese embassy in Canberra (Joske, 2020).

Perhaps the most high-profile example of alleged Huang's influence in Australian politics comes from the 2016 case of the former Australian Labor

Party Senator Sam Dastyari who was standing next to Huang when he was caught on tape reciting CCP talking points at a public event on the South China Sea that ran counter to the foreign policy position of his own nation and party. Later, he was also recorded giving counter-surveillance advice to Huang, telling him his phone was likely tapped (McKenzie et al., 2017).

United Front efforts to place individuals in key political positions

Chinese influence operations aimed at Australian politics have allegedly included attempts by United Front organizations grooming members to participate in Australian politics both by running for office and by becoming political staffers in influential positions. Corroborating evidence for this claim came in late 2017 when Australian newspapers reported that the Australian Security Intelligence Organization had identified at least ten political candidates in state and local elections that had close ties to Chinese intelligence services. This was viewed as "a deliberate strategy by Beijing to wield influence through Australian politics" (Maley & Berkovic, 2017). These concerns are bolstered by the case of Yang Jian, a Chinese-born MP of the New Zealand Parliament until 2020, who spent 15 years working in China's military intelligence sector training Chinese spies before becoming a citizen of New Zealand, and who had hid his PLA affiliations on his permanent residence and employment applications. (Grossman, 2022).

China's threats to influence Australian elections used as diplomatic leverage

In 2017, the Australian opposition leader Bill Shorten received a veiled threat from Beijing that it would use its influence to make sure Mr. Shorten's Labor Party would lose votes from Australia's Chinese community if Labor did not support a controversial extradition treaty that Beijing wanted with Australia. This occurred during a meeting where the former Chinese Minister of Public Security Meng Jianzhu reportedly told a Labor Party delegation that "it would be a shame if Chinese government representatives had to tell the Chinese community in Australia that Labor did not support the relationship between Australia and China" (The Australian, 2017).

Legal

Extra-judicial enforcement of Chinese law in Australia

President Xi Jinping has launched an initiative called "Operation Fox Hunt" to use Chinese security agencies to pursue and prosecute Chinese individuals—many former government and business officials believed to be seeking refuge in Australia—deemed criminal by Beijing. Australian media reports revealed that there may be as many as 100 such individuals and perhaps ten of Beijing's highest priority targets identified as residing in Australia, and that many had returned to China after being located and having come under pressure from Beijing. Although Beijing has given some names to Australian authorities and has asked for their cooperation in tracking them down, Australian intelligence sources have told media that China's Ministry of State Security (MSS) hides the full list of Chinese they are targeting because Beijing does not want Australia tracking contact between Beijing's agents and those they want to convince to return (Riordan, 2018). In other words, there have been incidents where Chinese MSS agents have come to Australia (often on tourist visas) without knowledge of the Australian government to help "persuade" these suspects to return to China. This practice has since been revealed to operate globally, involving the use of informal Chinese "police stations" in various cities around the world. For example, charges have been brought against two Chinese defendants for operating such a "police station" for a provincial branch of the CCP's Ministry of Public Security out of a clandestine location in lower Manhattan to harass and intimidate Chinese political dissidents (Thomas & Barr, 2023).

Media

Australian Chinese language newspapers and radio dominated by pro-Beijing interests

In Australia, it is reported that nearly 95% of Chinese language newspapers are controlled by pro-Beijing entities. The Sydney Morning Herald explains the carrots and sticks that Beijing wields in extending its reach in Australian Chinese language media thusly: "it can come in the form of an admonishing phone call, blocking reporters from a public event, via directives for mainland-linked businesses to pull advertising, or even direct investment from Chinese government bodies." As a result, Beijing controls editorial content coverage of China.

Reports indicate that Beijing also now owns and tightly controls almost all Chinese language radio in Australia, which is being bought up by China Radio International (CRI) or its affiliated companies across most major Australian cities. And when the Australian Broadcasting Corporation began pulling out from broadcasting via shortwave in the Pacific region in 2017, CRI wasted little time taking over those very same frequencies to broadcast its own news. According to Peter Cai at Australia's Lowry Institute, CRI "is even using a Melbourne Chinese community radio station CAMG as a front to set up an extensive international network of Chinese and foreign language propaganda outfits".

Opaque pro-Beijing control of wider media

An investigation conducted by Reuters has identified at least 33 radio stations in 14 countries that are part of a global radio web structured in a way that obscures its majority shareholder: state-run China Radio International, or CRI. This use of corporate subsidiaries has obscured their connection to the CCP party-state and has shielded these radio stations from regulatory oversight. The head of CRI Wang Gengnian has described Beijing's media influencing efforts as the "borrowed boat" strategy of using existing media outlets in foreign nations to promote China's narrative (Qing & Shiffman, 2015).

Economic

Leveraging trade and investment dependencies to coerce partners

The CCP's ability to control access to its economy provides Beijing with opportunities to coerce foreign governments and commercial actors into policies they would not otherwise support. This tool can be used at the macro-economic level to threaten large segments of a nation's economy, or used as a tailored tool to target specific corporations or individuals. On numerous occasions, Chinese officials have threatened the Australian government with Chinese faux "consumer-led" boycotts—boycotts clearly incited by Beijing—targeting Australian goods following Canberra's announcements of policies found objectionable in Beijing.

In an example how highly tailored acts of coercion can work at the micro-level, Beijing has used its economic clout to pressure a French hotel chain to cancel its subscription to an independent (i.e. non-Beijing-controlled) Chinese language newspaper at one of its Sydney-based hotels. This occurred two weeks after the independent newspaper published an investigative piece with content considered negative by Beijing. According to a paper spokesperson, "Sofitel received a call from the Chinese Consulate asking them to remove our newspaper or face financial consequences. Sofitel does a lot of business with China" (Munro & Wen, 2016).

Educational

Chinese influence in Australian universities

Some critics have argued that the Australian university funding model is over-reliant on the generation of private donations and grants (and full-fee paying overseas Chinese university students numbering 152,000), which has created a structural weakness that can be exploited by the PRC. The controversial growth of Confucius Institutes in Western universities is a prominent case in point. Although their primary mission is the promotion of Chinese language and culture, various Australian research institutes, civil society watchdog organizations, and members of Australian university faculty have expressed concerns that Confucius Institutes also serve a secondary function of co-opting Australian universities as vehicles for pro-Beijing propaganda and to give Beijing an advocate within these institutions. Beyond their official status as a body of the Ministry of Education, and their ties to the External Propaganda Leading Group of the CCP Central Committee, these concerns are based on a number of observations: the exclusive use of PRC classroom materials and promotion of Beijing viewpoints; their silence on controversial human rights issues such as Tibet, Tiananmen Square, the Uighurs, or Falun Gong; and an overall undermining of Western standards of politically independent university curricula.

Australian university leadership has publicly downplayed concerns about inappropriate Chinese influence on Australian campuses—including concerns raised by their own faculty members—while also engaging in simultaneous public relations missions to mitigate the Beijing-backed economic retaliation that they anticipate as a result of these public faculty complaints. Australian Intelligence has privately reached out to Australian University administrators to address the problem of CCP influence and collection activities on campus (confidential interview, Melbourne).

Encouraging ethnic Chinese students in Australia to
suppress anti-Beijing views

In 2016, the Chinese Ministry of Education issued a directive to its overseas employees emphasizing an increased engagement with Chinese students studying abroad. It required them to "build a multidimensional contact network linking home and abroad—the motherland, embassies and consulates, overseas student groups, and the broad number of students abroad—so that they fully feel that the motherland cares" (Garnaut, 2017). In an Australian context, this applies to approximately 152,000 Chinese students studying in Australia in 2023. As a result, with direct support from Chinese embassies, Chinese students and Chinese Students and Scholars Associations (CSSAs) have reported on, intimidated, and undermined the academic freedom of other Chinese students as well as Australian university faculty (Christodoulou et al., 2019). While it is obvious that not all Chinese students engage in such activity, and may do so out of patriotism, the fact remains they are encouraged to do so by their government, with both carrots and sticks, and examples are easy to find (Resnick, 2020).

Australia has seen multiple instances of Chinese students recording professors' lectures seen as critical of the PRC, and then uploading them onto the internet, resulting in harassment of these academics on social media. Under these circumstances, the on-campus presence of student organizations like the CSSA that are linked to and funded by Chinese embassies creates an understandable concern that faculty lecturing on topics sensitive to Beijing might fear that their lectures are being monitored, leading to self-censorship. This risk is even greater considering it may involve a faculty member who needs to travel to China for research purposes.

Conclusion

This chapter has shown how China aggressively uses hybrid threats as part of a larger economic and political strategy to increase its power and influence within another nation. The sheer scale and scope of the use of hybrid threats by Beijing against Australia—coupled with Australia's commitment to publicly address and counter this threat—has made it a canary in the coal mine for states dealing with the CCP's global hybrid threat campaigns. Although such threats can and do have effects in the military domain, this chapter emphasizes the wider society-centric (PMESII) targets of Chinese hybrid threats, and the non-military and non-state/CCP Party-State proxy tools (MPECI) used to carry them out. This chapter has also illuminated the tools used by the target country to detect and defy political and economic threats. To date, Australia has made progress in identifying and advertising the increasing role played by Chinese hybrid threats. It has also made gains by updating its legislation to close legal spaces for some opaque hybrid influence campaigns to exist. However, care must be taken to ensure that Australian counter-hybrid threat activities do not become a legal "box ticking" exercise, without enforcement or follow-up, as some critics suggest (Blackwell, 2023). Countering Beijing's hybrid threat campaigns must be as comprehensive as the threat itself.

Finally, although this case study is narrowly focused on Chinese hybrid threats in Australia, the fact that China deploys these same hybrid threat actors, using the same techniques, targeting the same places across the whole-of-society, for the same political objectives, demonstrates that Australia's experience provides a template for other states to detect, deter, and respond to hybrid threats in their own countries.

References

Aukia, Jukka. (2023, April 6). *China's hybrid influence in Taiwan: Non-state actors and policy responses.* Hybrid COE Report 9. https://www.hybridcoe.fi/publications/hybrid-coe-research-report-9-chinas-hybrid-influence-in-taiwan-non-state-actors-and-policy-responses/.

The Australian. (2017, December 4). China's veiled threat to Bill Shorten on extradition treaty.

Australian Security Intelligence Organization. (2019). *Australia's security environment and outlook.* https://www.asio.gov.au/AR2018-03.html. Accessed February 20 2020.

Blackwell, Tom. (2023, March 31). Criticized as toothless, Australia's foreign-influence registry a warning as Canada plans its own. *National Post.*

Bowe, Alexander. (2018, August 24). *China's overseas united front work: Background and implications for the United States.* US-China Economic and Security Review Commission Staff Research Report.

Brady, Anne-Marie. (2017). *Magic weapons: China's political influence activities under Xi Jinping.* Wilson Center.

Cave, Danielle & Uren, Tom. (2019, February 21). *Espionage or interference? The attack on Australia's parliament and political parties.* Australian Strategic Policy Institute.

Cheng, Dean. (2013, July 12). *Winning without fighting: The Chinese psychological warfare challenge* Heritage Foundation.

Christodoulou et al. (2019, November 3). *Chinese Students and Scholars Association's deep links to the embassy revealed.* Australian Broadcast Corporation. https://www.abc.net.au/news/2019-10-13/cssa-influence-australian-universities-documents-revealed/11587454.

Cullen, Patrick. (2018, May 25). *Hybrid Threats as a new wicked problem for early warning.* Hybrid COE.

Cullen, Patrick, Monaghan, Sean, & Wegge, Njord. (2019). *Countering hybrid warfare.* MCDC. https://assets.publishing.service.gov.uk/government/uploads/system/uploads/attachment_data/file/784299/concepts_mcdc_countering_hybrid_warfare.pdf

Cullen, Patrick & Reichborn-Kjennerud, Erik. (2016, January). *What is hybrid warfare?* NUPI.

Cullen, Patrick & Reichborn-Kjennerud, Erik. (2017, January). *Understanding hybrid warfare.* Multinational Capability Development Campaign. https://assets.publishing.service.gov.uk/media/5a8228a540f0b62305b92caa/dar_mcdc_hybrid_warfare.pdf

Diamond, Larry & Schell, Orville (eds.). (2019). *Chinese influence and American interests: Promoting constructive vigilance. Appendix 2. Chinese Influence Activities in Select Countries.* Hoover Institute.

Garnaut, John. (2017, August 30). Our universities are a frontline in China's ideological wars. *Australian Financial Review.*

Gershaneck, Kerry K. (2020). *Political warfare: Strategies for combating China's plan to "Win Without Fighting."* Marine Corp University Press.

Griffith, Samuel & Sun, Tzu (trans.). (2005). *The art of war.* Duncan Baird Publishers.

Groot, Gerry. (2017, December 22). United front work after the 19th party conference 17 China brief volume, issue 17. *Jamestown Organization.*

Grossman, D. (2022, December 21). New Zealand is done speaking softly to China. *Rand Blog.* https://www.rand.org/blog/2022/12/new-zealand-is-done-with-speaking-softly-to-china.html.

Hamilton, Clive. (2018a, July 26). Australia's fight against Chinese political interference. *Foreign Affairs.*

Hamilton, Clive. (2018b, November). China's influence activities: What Canada can learn from Australia. *Commentary.*

Human Rights Watch. (2021, June 30). *They don't understand the fear we have: How China's long reach of repression that undermines academic freedom in Australia's universities.* https://www.hrw.org/report/2021/06/30/they-dont-understand-fear-we have/how-chinas-long-reach-repression-undermines.

Hybrid COE Expert Pool. (2020, July). *Trend report 5: Trends in China's power politics.* Hybrid COE. https://www.hybridcoe.fi/wp-content/uploads/2020/07/20200710_Trend-Report-5-China_Web.pdf.

Joske, Alex. (2020, June 1). The party speaks for you: Foreign interference and the Chinese communist party's united front system. *Australian Strategic Policy Institute.*

Mahnken, Thomas, Babbage, Ross, & Yoshihara, Toshi. (2018, May 30). *Countering comprehensive coercion: Competitive strategies against authoritarian political warfare.* Center for Strategic and Budgetary Assessments (CSBA).

Maley, Paul & Berkovic, Nicola. (2017, December 8). Security agencies flag Chinese Manchurian candidates. *The Weekend Australian.*

McKenzie, Massola, & Baker. (2017, November 30). It isn't our place. *The Sydney Morning Herald.* https://www.smh.com.au/politics/federal/it-isnt-our-place-new-tape-of-probeijing-comments-puts-more-heat-on-dastyari-20171128-gzuiup.html.

Munro, Kelsey & Wen, Philip. (2016, July 8). Chinese language newspapers in Australia: Beijing controls messaging, propaganda in press. *Sydney Morning Herald.*

Probyn, Andrew & Doran, Matthew. (2020, September 13). China's 'hybrid war': Beijing's mass surveillance of Australia and the world for secrets and scandal. *Australian Broadcasting Corporation.*

Qing, Koh Gui & Shiffman, John. (2015). Beijing's covert radio network airs China-friendly news across Washington, and the world. *Reuters.*

Resnick, Gabreille. (2020, January 14). Chinese students say free speech in US chilled by China. *Voice of America.*

Riordan, Primrose. (2018, June 16). Beijing's hidden hitlist targets 100 here. *The Australian.*

Shambaugh, David. (2015, July). China's soft-power push: The search for respect. *Foreign Affairs.*

Thomas & Barr. (2023, April 17). DOJ accuses China of using 'police station' to spy on, harass dissidents inside US. *ABC News.* https://abcnews.go.com/Politics/doj-accuses-china-spying-harassing-dissidents-inside-us/story?id=98635039.

Wallace, Charles. (2017, December 3). The art of influence: How China's spies operate in Australia. *Sydney Morning Herald.*

Wang, Qiao. (1999). *Unrestricted Warfare.* PLA Literature and Arts Publishing House. https://www.c4i.org/unrestricted.pdf.

Whitehouse, David. (2023). Proposed US legislation seeks to end foreign government intimidation of diaspora. *The Diplomat.*

Zhang, A., Hoja, T., & Latimore, J. (2023, April 26). *Gaming public opinion: The CCP's increasingly sophisticated cyber-enabled influence operations.* Australian Strategic Policy Institute.

Recommendations for further reading

Bateman, Jon. (2022, April 25). US-China technological 'De-Coupling': A strategy and policy framework. *Carnegie Endowment.*

Cullen, Patrick & Reichborn, Erik. (2017, January). *Understanding hybrid warfare.* MCDC Countering Hybrid Warfare Project.

Cullen, Patrick & Wegge, Njord. (2023). Adapting early warning in an age of hybrid warfare. In Lars Haugom & Brigt H. Vaage (eds.), *Intelligence analysis in the digital age, stig stenslie.* Routledge.

The National Counterintelligence and Security Center. (2022, July). *Safeguarding our future: Protecting government and business leaders at the U.S. State and local level from people's from People's Republic of China (PRC) influence operations.* https://www.dni.gov/files/NCSC/documents/SafeguardingOurFuture/PRC_Subnational_Influence-06-July-2022.pdf.

PART II
The Ukrainian experience

5

HYBRID THREATS IN CYBERSPACE

What do Russia's cyberspace operations in Ukraine tell us?

Mass Soldal Lund

Introduction

Cyberspace operations have, over the last decade, become associated with hybrid threats. Scholars such as Rob Johnsen and Mikael Weissmann extend the "the usual 'DIME' elements of strategy (diplomatic, informational, military, and economic)" (Johnson, 2021, p. 234) with "cyber" in their understanding of hybrid threats and hybrid warfare. Mark Galeotti is less explicit but often refers to cyberattacks in his descriptions of political warfare and "the new way of war" (Johnson, 2021, pp. 231–234; Weissmann, 2021, pp. 63–67; Galeotti, 2019, 2022).

The Russian full-scale invasion of Ukraine on February 24, 2022, was accompanied by what can best be described as a massive campaign of cyberattacks directed toward Ukrainian society, and possibly one of the largest cases of the use of cyberspace operations in conflict to date. The chapter's purpose is to scrutinize the role of cyberspace operations in conflict—and their relation to hybrid threats—using cyber warfare in the first year of the Russian full-scale invasion of Ukraine as the empirical basis. The problem statement is as follows: What characterize Russian cyberspace operations in the Russo-Ukrainian War, how can the operations be interpreted in a hybrid threat context, and what are the implications for national preparedness and response in cyberspace?

The main argument is that both offensive and defensive actors in cyberspace operate—despite the conventional nature of the Russo-Ukrainian war—in a gray zone between war and peace. This is seen best by the continuity of cyberspace operations in the temporal dimension, the crossing of geographical borders, and the blurring of military/civilian and governmental/

DOI: 10.4324/9781032617916-7

private distinctions. This also seems to support a view of cyberspace as a parallel battlefield only loosely related to—though interacting with—the physical battlefield. This loose coupling seems to imply that conventional war like the one experienced in Ukraine is not a necessary condition for the parallel hostilities in cyberspace.

The chapter starts by elaborating the theoretical basis for this argument, as well as methodological considerations. In the chapter's empirical part, Russia's offensive use of cyberspace operations in Ukraine is first described, before Ukraine's countermeasures are explored. Some comparative glimpses of how Estonia, Lithuania, and Norway have been affected by the cyber warfare are also given. Finally, in the chapter's last part, conclusions are deduced.

Theory

As a result of the ever-increasing digitalization of the last decades, critical infrastructure and essential functions of the society are managed by computerized systems. With the realization that inherent vulnerabilities enable attacks on such systems—remotely through computer networks and with potential catastrophic consequences—theories of cyber warfare emerged. Writers such as Clarke and Knake (2010) posit that in future conflict, cyberspace operations—the exploitation of vulnerabilities in computerized systems for the purpose of degrading or manipulating them—will be decisive. By remotely controlling or disabling the computer systems managing your opponent's critical infrastructure, you will be able to destroy his ability and will to fight.

This view has rightly been criticized. From a theoretical point of view, it has been argued that while cyberspace operations certainly can sabotage systems that have computerized components, it will never be possible to fight a war through cyberspace. Because most systems are not made to kill, it is at best very difficult to make them into weapons, even with control over the processes directing them. All damage caused by degrading or manipulating the computer components of a system will be either indirect or on the system itself (Rid, 2013). Furthermore, the empirical evidence shows that hostilities in cyberspace do not take the form of decisive battles but rather attempts at disruptions and subversion (Valeriano & Maness, 2015; Valeriano et al., 2019; Lonergan et al., 2022).

However, both camps of the "cyber war"-debate may perhaps be said to operate with too narrow views of war and warfare. Modern conflicts are not necessarily resolved by decisive battles or strategic attacks on civilian infrastructure, and wars are not fought merely with bullets, bombs, and grenades (Galeotti, 2022). War is conducted by many means, of which not all are violent. With a more pragmatic approach, the dispute may be solved by viewing cyberspace operations used as means in the context of an armed conflict. An example of this is Richards (2014), which—within the context of conflict—characterizes cyber warfare as information operations, tactically enabling operations, and destructive cyber-attacks that are in themselves acts of war.

Known examples of tactically enabling cyberspace operations are few, and examples of cyberspace operations that are independent acts of war are even fewer. The reason for this, Smeets (2022, pp. 7–9) argues, is that to achieve operational goals, cyberspace operation must create effects that are suitable, precise, and timely. This requires that cyber capabilities are developed specifically for their intended targets and effects—but, at the same time, they may be rendered useless by changes or updates to the targets. Thus, succeeding with cyberspace operations is difficult, and maintaining a cyber capability is a costly and intelligence intensive process of continual development.

The idea of tactically enabling cyberspace operations is related to principles of joint operations and combined arms. As such, they can be seen as a "Western" concept. In contrast to this, information operations may be seen as the frame for the "Russian way" of thinking about cyberspace operations (Galeotti, 2019, p. 35). Blank (2017, pp. 81–85), and Lilly and Cheravitch (2020, pp. 133–139) show how information warfare has become an integral part of the Russian government's view of conflict. Information warfare is a continuous battle for domination in the information sphere though technical and psychological means. Cyberspace operations are the technical means to dominate the information sphere by manipulating information systems, while influence operations use psychological means to manipulate public opinion.

How does cyber warfare relate to hybrid threats? On the one hand, it is possible to adopt the view that, in modern warfare, states will utilize any means available that serves to force their will upon their opponents. Offensive cyberspace operations are then one such means, comparable to other "unconventional" means as, for example, deception, sabotage, or stirring of public unrest. Such a view can be developed into a notion of hybrid or unconventional warfare in line with Hoffman's (2010) definition, where cyberspace operations find their place as an unconventional element of warfare. Within such a frame, the rather trivial extrapolation can be made that in future conflict involving states with sufficient "cyber capabilities"—such as Russia—offensive cyberspace operations targeting the infrastructure of the warring parties should be expected. For reasons like this, Galeotti (2019) has criticized the concept of hybrid warfare as encompassing anything and everything, and thus of lacking in explanatory power. On the other hand, hybrid threats signify means below the threshold of war. A way to reconcile this with the use of cyberspace operations in an actual shooting war may be to perceive hostilities in cyberspace as "parallel battles . . . with other, more subtle, and non-violent means" (see introductory chapter of this book). In any case, in order to extract more than trivial observations, it seems necessary to delve into the characteristics of the cyberspace operations in the context of the concrete conflicts.

Methodology

The ongoing conflict in Ukraine and the relatively short time span between the events and the time of writing, combined with the general secrecy surrounding both warfare and cyberspace operations, undeniably present a challenge for this study. However, cyberspace is continually monitored by governmental bodies (such as national cybersecurity centers), security companies, software companies, and NGOs. Many of these organizations regularly publish reports on observed cyberspace operations. The main source for the empirical evidence presented in this chapter is the systematic reading of such report detailing offensive cyberspace operations and defensive countermeasures in the context of war in Ukraine (CISA, 2022; Economic Security Council of Ukraine, 2023; Google, 2023; Information System Authority, 2023; Microsoft 2022a, 2022b, 2023; Ministry of National Defence of the Republic of Lithuania, 2023; NSM, 2022, 2023; RCDC, 2022a, 2022b, 2022c, 2023a, 2023b, 2023c; Recorded Future, 2023; SSSCIP, 2023; Thales, 2023).

While the reports included in this study provide much data, using them as the main source in the study is not entirely unproblematic. Several of the organizations behind the reports make independent investigations, but still, several of the reports are partly dependent on open-source intelligence, in particular on media reports. Furthermore, whether the reports come from governmental bodies, private companies, or NGOs, they are not devoid of agenda. A limited number of academic studies and research reports are available (see, e.g., Bateman, 2022; Beecroft, 2022; Mueller et al., 2023; Wilde, 2022). With some exceptions, these tend to be based on the same kind of reports and news stories, and thus provide little independent evidence. To supplement the study, an interview with a Ukrainian subject matter expert (SME) has been made. In addition, minutes from meetings of the Ukrainian National Cybersecurity Cluster have been a useful source (National Cybersecurity Cluster, 2021a, 2021b, 2021c, 2022a, 2022b, 2023).

The reports included in the study cover the period January 2022–March 2023, thus covering roughly the first year of the invasion. Furthermore, the data and analysis are limited to offences perpetrated by actors related to Russia or its ally Belarus, and the defense made by Ukraine and supporting states. However, providing a coherent picture of a year of offensive actions in cyberspace is not easy. The number of offensive cyberspace operations experienced in Ukraine over the course of the invasion is too large to make timelines of operations practical and useful, and details of the events are often unknown or suppressed. Aggregating the data into statistics suffers under the notorious difficulty of counting cyberspace operations (see, e.g., Mueller et al., 2023, n. 52). Several of the reports included in the study provide timelines, aggregated statistics, or both, but the challenge is apparent in that different choices are made as to which operations to include in the timelines, and that different ways of counting the operations are applied. Furthermore, there are different ways in which we may categorize the operations—by target or victim, by perpetrator, by method or tactics, by intentions, or by effects. In our presentation and analysis of the data, our approach has been to identify and describe broad categories of operations and trends, without attempting any rigorous analytical framework of cyberspace operations and the actors behind them.

Data and analysis

From the data, a picture can be formed of Russia's hostilities in cyberspace against Ukraine and supporting countries, and of Ukraine's defensive response to these hostilities. In the following, these two aspects of the conflict are presented and analyzed in two main sections.

Russia's offence in cyberspace

In the study, three broad categories of offences in cyberspace emerge: destructive cyberattacks, cyber espionage, and hacktivist attacks. Destructive cyberattacks are cyberspace operations where malware is used to cause damage to ICT or computerized systems. Cyber espionage is operations where malware is used to gather intelligence from such systems. Hacktivists attacks or hacktivism—described in more detail below—are primitive cyberattacks carried out by more or less organized private citizens for political, patriotic, or nationalist reasons.

In most cases observed in Ukraine, destructive cyberattacks have utilized a type of malware referred to as *wipers*, because they wipe data and software from the affected computers, making them unusable. In cyber espionage operations, usually malware called *rats* (short for Remote Access Trojans or Remote Access Tools) are used for intelligence gathering, such as exfiltration of electronic documents. However, because destructive operations and intelligence gathering operations in many cases share the means of obtaining the initial access to computer systems, and because destructive operations also may utilize rats, the initial stages of destructive operations and intelligence gathering operations cannot always be distinguished. Thus, when an offensive cyberspace operation is detected and averted in its early stages, the goals of the operation—whether it is destruction or espionage—cannot always be decided. Operations may even have several or shifting goals, so that what starts out as an espionage operation may change objectives to become a destructive attack.

During the initial phases of the Russian invasion, Ukraine saw an unprecedented wave of destructive cyberattacks. According to Microsoft, "40 discrete destructive attacks" (Microsoft, 2022b, p. 3) used "at least nine new wiper families and two types of ransomwares against more than 100 Ukrainian organizations" (Microsoft, 2023, p. 5) to "permanently destroyed files in hundreds of systems across dozens of organizations in Ukraine" (Microsoft, 2022b), with most of these attacks occurring within the first couple of weeks. The details are sparse, but other reports have similar assessments. It is assumed that GRU affiliated groups are behind most of these attacks (e.g., Microsoft (2022b, 2023, p. 5); see also section on actors below). Of particular interest is an attack on the communication services of the American satellite communication company ViaSat in the early hours on the day of the invasion. In the attack, wiper malware (AcidRain) destroyed ViaSat modems, "disrupting broadband service to tens of thousands of users in Ukraine and throughout Europe" (Microsoft, 2022b, p. 7). In addition to this being characterized as a sophisticated attack which must have required careful planning and preparations, attention is given to the attack because the services of ViaSat was used by the Ukrainian defense forces, and because of spill-over to other countries, in particular Germany, France, Hungary, Greece, Italy, Hungary, and Poland. It should also be observed that the first of the wiper attacks (WhisperGate) occurred on January 13, in other words before the invasion.

There have been observed destructive cyber-attacks also after the initial wave, but nothing that can compare in numbers. Some of the security and software companies (Microsoft, 2023; Google, 2023) assert that they observed a second wave of destructive attacks in the autumn of 2022, but this seems to be funded on rather few occurrences of such attacks. An assumption made is that the initial wave was based on access to systems obtained before the invasion, and when these were "used up," the Russian actors had to gain new accesses in new phishing campaigns (see section on access below). The Ukrainian government, on the other hand, assert that there was a shift from destructive attacks to intelligence gathering operations, with the main trend in the second half of 2022 being "sophisticated spear phishing campaigns with the objective of data exfiltration and cyber espionage" (SSSCIP, 2023, p. 9). The reports agree, however, that there was a shift in the sectors targeted, from governmental entities, media, and telecommunications in the initial phase of the invasion, to critical infrastructure, especially energy, logistics, and transportation in the later phases of the war. Of notice are attacks with a ransomware (Prestige) against the transportation sectors in Poland and Ukraine in October 2022. The attacks are attributed to GRU affiliated actors and are seen in relation to transport of Western supplies to Ukraine (SSSCIP, 2023, pp. 27–28; Microsoft, 2023, p. 13).

Some of the reports try to establish correlations between actions seen in cyberspace and on the ground, as this will be evidence for close coordination of cyberattacks with the other fronts of the war (e.g. Economic Security Council of Ukraine, 2023; Microsoft, 2022a, pp. 7–8). At a strategic level, the correlation is obvious: the initial surge of destructive cyberattacks coincides with the initial phase of the invasion, and there seems to be a shift in goals and targets of the cyberattacks as the Russian strategic goals changes. However, on any tactical or operational level, the evidence seems to be anecdotal and spurious. A notable exception is the ViaSat attack on the night on the invasion, which targeted a communication platform used by the Ukrainian defense forces.

The only cyberspace operations targeting military systems reported are unsuccessful attacks targeting two Internet exposed command and control systems. Related to the military sector, there are also reports of both destructive attacks and intelligence gathering operations directed at administrative systems in the Ministry of Defense and military commissariats, and operations targeting logistics companies. Furthermore, there are reports of phishing campaigns targeting military and security personnel, for example "Gamaredon . . . constantly going after the Security Service of Ukraine (SBU) personnel, to compromise Signal messenger accounts" (SSSCIP, 2023, p. 18).

An element of surprise is often assumed to be an essential part of a successful cyberattack. However, both reports and our informant emphasize the predictability of the Russian cyberspace operations. Closer inspection of the operations reveals that most of them are similar to earlier attacks, using the same techniques and in some cases variants of the same malware. Again, the ViaSat attack is a notable exception. Cyberspace operations after the invasion in February 2022 is therefore best seen as a continuation of cyberspace operations during the years leading up to the invasion, with cyberattacks in 2022 copying cyberattack going as far back as 2015.

Access

Five methods for gaining initial access and delivering malware to computer systems—shared by destructive and espionage operations—are observed in Ukraine: Exploitation of technical security vulnerabilities, spreading of malware through email (usually in email attachments; often called *phishing*), credential harvesting through email (tricking people to give up their usernames and passwords; another form of *phishing*), spreading of malware through pirated software (free downloads of licensed software), and through supply chains (attacking ICT providers and planting malware in their products or services to target their customers). Google (2023, p. 10) observed a "steady drumbeat" of phishing campaigns targeting Ukrainians, presumed to be the main method for the actors to gain access to target systems, with a significant increase before the invasion, and several peaks during the first year after the invasion. Our informant points to phishing campaigns following kinetic attacks on Ukrainian infrastructure, exploiting (by pretending to be related to) the relief efforts, and emphasizing an opportunistic aspect of the campaigns.

Actors

The reports points to Russian so-called "Advanced Persistent Threats (APTs)" or "nation-state actors" as the adversaries behind the attacks. These actors are usually assumed to be units within the Russian secret services FSB, GRU, and SVR; however, they are given various identifiers or monikers by security and software companies. The most prominent actors observed in Ukraine are Gamaredon (FSB), Turla (FSB), Dragon-Fly (FSB), Sandworm (GRU), APT28/Fancy Bear (GRU), and APT29/Cozy Bear (SVR). In addition, a group assumed to be associated with the Belarusian government, often identified as GhostWriter, has been observed. It is assumed that these groups have different specialties: DragonFly and Sandworm specializing in destructive attacks, Turla and APT29/Cozy Bear specializing in sophisticated cyber espionage, Gamaredon specializing in mass phishing campaigns, and so forth. Most of the actors' ties to Russian secret services are confirmed by Western intelligence agencies (e.g., CISA, 2022). However, the attribution of concrete cyberattacks to actors seems in most cases to be technical, that is, by matching technical indicators (such as characteristics of malware or use of hacking techniques) to what is previously known about the actors.

Hacktivism

Hostilities in cyberspace have also taken the form of persistent *Distributed Denial of Service (DDoS) attacks* (and to a lesser degree cyberattacks referred to as *website defacement* and *hack and leak*), targeting Ukraine and supporting countries. Despite the unprecedented number of destructive cyberattacks of Russian governmental agencies, by far the largest volume of cyberattacks observed during the war are these kinds of simpler attack. A few of these, especially some of the earlier ones, are attributed to nation-state actors. GhostWriter is assumed to be behind the defacement of 70 Ukrainian websites on February 14, while DDoS attack on February 15–16 and 23 are assumed to be of Russian "nation-state" origin. However, the overwhelming majority of the attacks were carried out by "hacktivists," that is, groups of activists with political, patriotic, or nationalist motives, but with seemingly only loose relations to the government.

Reports list up to 70 pro-Russian hacktivist groups (and also a large number of pro-Ukrainian groups targeting Russia and Belarus). In contrast to the so-called nation-state actors, attribution seems mostly to be based on the hacktivists' declarations of responsibility for (or bragging about) the attacks on Telegram and other social media. One of the most active and visible groups has been Killnet. The group came to prominence in March 2022 and communicates through Telegram channels. The reports indicate that over the first year of the invasion, Killnet started functioning as a hub for the consolidation of hacktivist groups, with other groups starting to act as subsidiaries of Killnet. In the same period, an increased sophistication of attacks attributed to Killnet was observed, which was connected to observations that Killnet seems to have established contact with Russian secret services. Killnet conducted DDoS attacks in Ukraine but is notoriously known for its "declaration of cyber war" and launch of DDoS attacks on countries supporting Ukraine (RCDC, 2022b, p. 7).

Lithuania and Estonia were among the most targeted countries. Lithuania reported "a huge wave of DDoS attacks against the public and private sectors . . . aimed at 130 public accessible websites" in June 2022, but that "[t]he attacks did not damage the companies' information systems" (Ministry of National Defence of the Republic of Lithuania, 2023, p. 9). On the other hand, Lithuania experienced fewer incidents by other types of cyberattacks than previous years, assumed to be because "most such attacks targeted Ukrainian state institutions" (p. 11). Estonia reported "an unprecedented number of denial-of-service attacks" in 2022, four times more than in 2021, with "volumes . . . a hundred times higher than [the infamous DDoS attacks] in 2007" (Information System Authority, 2023, p. 8). However, as in Lithuania, the consequences of the attacks were negligible. Norway was less affected, but a wave of DDoS attacks in June and July 2022 made "a large number of Norwegian organizations" [authors translation] (NSM, 2022, p. 16) experience downtime on their webpages and the second half of 2022 saw "six-fold the number of DDoS attacks compared with the three preceding years combined" [authors translation] (NSM, 2023, p. 10). The Norwegian reporting emphasizes that the attacks did not have serious consequences, and that the attention given to the DDoS attack, and thus the effect in the form of unease in the population, was unproportional compared to the consequences of the attacks. In both Estonia and Norway, DDoS attacks have been observed to respond to political decisions, such as the webpages of the Estonian parliament being attacked after the parliament condemned the Russian annexation of eastern Ukraine.

Ukraine's defensive response

Brantly (2022) describes how the Ukrainian cyber defenses in 2014 was in a sorry state. In the aftermaths of the Euromaidan revolution, "Ukraine was in political and bureaucratic disarray" (p. 159), and the governmental organizations responsible for cybersecurity and cyber defense, as other parts of the

Ukrainian government, were plagued by corruption, bureaucratic infighting, fragmentation, underfunding, and Russian infiltration. From 2015, however, systematic efforts to strengthen national cyber capabilities were started. Overall directions for the efforts were laid out in the country's first national cybersecurity strategy, approved by presidential decree in 2016. A revised cybersecurity strategy was approved in 2021 (Brantly, 2022; Shypilova, 2019; Streltsov, 2017; President of Ukraine, 2021).

Over the period from 2014 to 2022, Ukraine has experienced a large number of cyber-attacks attributed to Russian actors, to the degree that the country has been described as "Russia's test lab for cyberwar" (Greenberg, 2017; see also Greenberg, 2019). The national cybersecurity strategy cites "hybrid aggression of the Russian Federation against Ukraine in cyberspace" (President of Ukraine, 2021) as a dimensioning threat, and clearly this has been one of the drivers in the establishment of functioning cyber defense capabilities from 2016 onward. Furthermore, it has enabled Ukrainian cyber defenders to observe and learn Russian tactics.

At the strategic level, cybersecurity is recognized as a part of national security, and the national cybersecurity strategy thus follows the national security strategy. Furthermore, cybersecurity at the strategic level is coordinated by the National Cybersecurity Coordination Center (NCCC), which lies under the National Security and Defense Council (NSDC) led by the Ukrainian president. As the head of the National Security and Defense Council, the constitution of Ukraine gives the president decisive influence over national security. On the other hand, his power is limited by the Ukrainian constitution demanding that all powers and responsibilities given to governmental agencies must be assigned and regulated by legislation. Thus, the 2016 national cyber strategy followed the 2017 Law on cybersecurity, regulating the roles and powers of agencies involved in national cybersecurity and cyber defense (Brantly, 2022; Shypilova, 2019; Streltsov, 2017).

At the operational level, the State Service of Special Communication and Information Protection of Ukraine (SSSCIP) is the coordinator of the Ukrainian national cyber capabilities. The agency reports to and advice the president, the cabinet, and the parliament; coordinates capabilities; and coordinates with governmental agencies, local authorities, and public and private sector. Furthermore, SSSCIP is responsible for detecting, preventing, and responding to cyberattacks through its operation of the State Cyber Protection Centre (SCPC) and the Computer Emergency Response Team of Ukraine (CERT-UA). During the first year of the invasion, SSSCIP had a particular responsibility for Ukraine's critical infrastructure and telecommunication networks. The Security Service of Ukraine (SSU) has a double role as a counter-intelligence agency and a specialized law-enforcement agency. In these two capacities it has the responsibility of combating cyber-espionage and sabotage of critical infrastructure, as well as investigating incidents involving key information infrastructure (Streltsov, 2017, pp. 158–161; SCPC, n.d.).

Ukraine's international support

The establishment, modernization, and professionalization of the cyber defense structures has been helped by substantial support, both financially and in the form of training, from USA, NATO, and EU countries. An example of this is the formation in February 2021 of the National Cybersecurity Cluster as a cooperative effort of NSDC and the NGO U.S. Civilian Research and Development Foundation (CRDF Global), and with support of the U.S. Department of State. The goal of the National Cybersecurity Cluster is to strengthen cybersecurity in Ukraine by activities, such as policy development and awareness training, and by coordination between governmental institutions, international partners, and the private sector. Another example is the activation of EU's Cyber Rapid Response Teams in support of Ukraine on February 22, 2022—two days before the invasion. Support from other nations, in the form of training, analysis and information sharing, continued after the invasion. In Lithuanian reporting, for example, there are more emphasis on support of Ukraine than measures taken in Lithuania in response to the invasion.

The establishment of public-private partnerships within cybersecurity has, for several years, been an ambition but also a challenge for the Ukrainian government. The Ukrainian cybersecurity strategy of 2021 emphasized that "an effective model of public-private partnership remains unresolved" (President of Ukraine, 2021), and the International Foundation for Electoral Systems reported in 2019 that "[r]epresentatives of the SSSCIP acknowledged that private companies are not eager to cooperate with the state authorities predominantly because they do not see the benefits of such cooperation" (Shypilova, 2019, p. 17; see also Streltsov, 2017, p. 161).

However, with the invasion in 2022, this seemed to change. A central element in "NCCC's response to the escalation in cyberspace" was that "private sector quickly became involved" (National Cybersecurity Cluster, 2022b). Furthermore, reports are asserting that substantial support from international private companies strengthened Ukrainian cyber defense capabilities. Large cooperations like Microsoft and Google, as well as smaller cybersecurity companies, have provided licenses to cybersecurity software, cloud services, and detection and analysis of cyberattacks. In March 2022, the Cyber Defense Assistance Collaborative (CDAC) was established on initiative of CRDF Global and independent U.S. cybersecurity experts. CDAC organized support of NCCC with threat intelligence and cybersecurity technologies and services from a number of international cybersecurity companies (CDAC, 2024). By the reports of this study, this international support may have been a decisive factor in the relative success of the defense of the Ukrainian cyberspace. Beecroft (2022) has made similar observations based on interviews with a number of stakeholders.

Conclusion

This chapter has described Russia's cyber offensive during the 2022 full-scale invasion of Ukraine and scrutinized the Ukrainian countermeasures. Based on this, both theoretical and empirical considerations can be made.

Theoretically, cyberspace operations may figure as an integral part of a broader military assault, but only in a supporting capacity. Hostilities in cyberspace may be seen as part of a strategic effort to undermine and disrupt the civil society that, as often postulated, resides in a gray zone between peace and war. As shown in this study, this is true even when cyberspace operations and cyber warfare are conducted in the context of a conventional war. Hostilities in cyberspace are only loosely connected to the military campaign, and not necessarily confined to the hot conflict, neither in time nor in space. Thus, a picture emerges of cyberspace as a parallel battlefield to the war "on the ground." Mueller et al. (2023), though their goal is to counter the idea of decisive cyber-attacks, reach a similar conclusion.

Empirically, as cyberspace operations and cyber warfare blur the distinction between peace and war, there are no fundamental differences in the preparations needed to defend against peacetime cyberspace operations and against cyberspace operations in war. The ability to scale up the cyber defense capacities as conflicts escalate, however, is essential. Without utilizing the capacities of supporting nations and international technology corporations, such a rapid scaling up is not possible. Efficient and effective cyber defense is dependent on threat intelligence, and not the least access to technologies provided by private companies. The implication is twofold. First, national preparedness in cyberspace requires good international relations and mechanisms for coordinating with the private sector. Second, not only the offensive side of cyber warfare but also the cyber defenses contribute to blurring the boundaries and creating a gray zone around the measures taken in cyberspace.

In the analysis of the empirical data, the perhaps most striking observation is the opportunistic and predictable nature of the Russian cyberspace operations. These are contrary to common perceptions about the "advanced and persistent nation-state actors" and unpredictable and surprising cyberattacks. Opportunistic and predictable are traits more often associated with cyber criminals. These two properties are not unrelated, however. The opportunistic nature of the Russian cyber-attacks may at least partly explain their predictability: If the attackers can spot opportunities, then defenders should be able to spot them as well. For the development of defenses against cyber and hybrid threats, this is perhaps where we may find the greatest potential for lessons learned.

References

Documents

Cyber Defense Assistant Collaborative [CDAC]. (2024, January 26). Who we are. *CRDF Global*. Retrieved January 26, 2024, from https://crdfglobal-cdac.org/who-we-are/.

Cybersecurity & Infrastructure Security Agency [CISA]. (2022). *Alert (AA22–110A): Russian state-sponsored and criminal threat to critical infrastructure.* https://www.cisa.gov/uscert/ncas/alerts/aa22-110a.

Economic Security Council of Ukraine. (2023). *Cyber, artillery, propaganda. comprehensive analysis of Russian warfare dimensions.* https://reb.org.ua/storage/163/comprehensive-analysis-of-russian-warfare-dimensio. . . .pdf.

Google. (2023). *Fog of war. How the Ukrainian conflict transformed the cyber threat landscape.* https://services.google.com/fh/files/blogs/google_fog_of_war_research_report.pdf.

Information System Authority. (2023). *Cyber security in Estonia 2023.* https://query.prod.cms.rt.microsoft.com/cms/api/am/binary/RE4Vwwd.

Microsoft. (2022a). *Defending Ukraine: Early lessons from the cyber war.* https://aka.ms/June22SpecialReport.

Microsoft. (2022b). *Special report: Ukraine. An overview of Russia's cyberattack activity in Ukraine.* https://aka.ms/ukrainespecialreport.

Microsoft. (2023). *A year of Russian hybrid warfare in Ukraine. What we have learned about nation state tactics so far and what may be on the horizon.* https://www.microsoft.com/en-us/security/business/security-insider/wp-content/uploads/2023/03/A-year-of-Russian-hybrid-warfare-in-Ukraine_MS-Threat-Intelligence-1.pdf.

Ministry of National Defence of the Republic of Lithuania. (2023). *Key trends and statistics of the national cyber security status of Lithuania 2022.* https://www.nksc.lt/doc/en/2022_key-trends-and-statistics-of-cyber-security.pdf.

Nasjonal sikkerhetsmyndighet [NSM]. (2022). *Nasjonalt digital risikobilde 2022* [In Norwegian]. https://nsm.no/getfile.php/1312007-1667980738/NSM/Filer/Dokumenter/Rapporter/NDIG2022_online.pdf.

Nasjonal sikkerhetsmyndighet [NSM]. (2023). *Nasjonalt digital risikobilde 2023* [In Norwegian]. https://nsm.no/getfile.php/1313382-1697777843/NSM/Filer/Dokumenter/Rapporter/Nasjonalt digitalt risikobilde 2023.pdf.

National Cybersecurity Cluster. (2021a). *Summary report on the national cybersecurity summit.* https://cybersecuritycluster.org.ua/wp-content/uploads/2021/12/1st-cyber-summit-report_eng.pdf.

National Cybersecurity Cluster. (2021b). *Meeting minutes,* February 25, 2021. https://cybersecuritycluster.org.ua/wp-content/uploads/2023/01/cs-cluster-1_feb-25-2021_mm_eng.pdf.

National Cybersecurity Cluster. (2021c). *Meeting minutes,* September 23, 2021. https://cybersecuritycluster.org.ua/wp-content/uploads/2023/01/cs-summit-1_sep-23-2021_mm_eng.pdf.

National Cybersecurity Cluster. (2022a). *Meeting minutes,* June 30, 2022. https://cybersecuritycluster.org.ua/wp-content/uploads/2023/01/cs-cluster-12_jun-30-2022_mm_eng.pdf.

National Cybersecurity Cluster. (2022b). *Meeting minutes*, December 8, 2022. https://cybersecuritycluster.org.ua/wp-content/uploads/2023/01/cs-cluster-16_dec-8-2022_mm_eng.pdf.

National Cybersecurity Cluster. (2023). *Meeting minutes*, October 26, 2023. https://cybersecuritycluster.org.ua/wp-content/uploads/2023/11/23rd-cluster-mm_eng_10.26.23.pdf.

President of Ukraine. (2021). *Decree of the president of Ukraine No 447/2021. On the decision of the national security and defense council of Ukraine of May 14, 2021 "On the cyber security strategy of Ukraine"* [Unofficial translation]. National Cybersecurity Cluster. https://cybersecuritycluster.org.ua/wp-content/uploads/2021/12/cybersecurity-strategy-decree-august-2021_en_unofficial-translation.pdf.

Recorded Future. (2023). *Russia's war against Ukraine disrupts the cybercriminal ecosystem. CTA-RU-2023–0223.* Insikt Group. https://www.recordedfuture.com/russias-war-against-ukraine-disrupts-cybercriminal-ecosystem.

Regional Cyber Defence Centre [RCDC]. (2022a). *1st quarter report.* https://www.nksc.lt/doc/rkgc/CTAC_2022_1st_Quarter_Report.pdf.

Regional Cyber Defence Centre [RCDC]. (2022b). *2nd quarter report, 2022.* https://www.nksc.lt/doc/rkgc/CTAC_2022_2nd_Quarter_Report.pdf.

Regional Cyber Defence Centre [RCDC]. (2022c). *3rd quarter report, 2022.* https://www.nksc.lt/doc/rkgc/CTAC_2022_3rd_Quarter_Report.pdf.

Regional Cyber Defence Centre [RCDC]. (2023a). *4th quarter report, 2022.* https://www.nksc.lt/doc/rkgc/CTAC_2022_4th_Quarter_Report.pdf.

Regional Cyber Defence Centre [RCDC]. (2023b). *1st quarter report, 2023.* https://www.nksc.lt/doc/rkgc/CTAC_2023_1st_Quarter_Report.pdf.

Regional Cyber Defence Centre [RCDC]. (2023c). *Report on the cyber lessons learned during the war in Ukraine.* https://www.nksc.lt/doc/rkgc/report_on_cyber_lessons_learned_during_the_war_in_ukraine.pdf.

State Cyber Protection Centre [SCPC]. (n.d.). *The history of State Cyber Protection Center.* State Sites of Ukraine. Retrieved June 26, 2024, from https://scpc.gov.ua/en/history.

State Service of Special Communications and Information Protection of Ukraine [SSSCIP]. (2023). *Russia's cyber tactics: Lessons learned 2022.* https://cip.gov.ua/services/cm/api/attachment/download?id=53466.

Thales. (2023). *2022–2023: A year of cyber conflict in Ukraine. The extensive analysis by the Thales cyber threat intelligence team.* https://bo-cyberthreat.thalesgroup.com/sites/default/files/2023-03/Brochure-resume-A5-WEB.pdf.

General references

Bateman, J. (2022). *Russia's wartime cyber operations in Ukraine: Military impacts, influences, and implications* (Working paper). Carnegie Endowment for International Peace. https://carnegieendowment.org/2022/12/16/russia-s-wartime-cyber-operations-in-ukraine-military-impacts-influences-and-implications-pub-88657.

Beecroft, N. (2022). *Evaluating the international support to Ukrainian cyber defense.* Carnegie Endowment for International Peace. https://carnegieendowment.org/2022/11/03/evaluating-international-support-to-ukrainian-cyber-defense-pub-88322.

Blank, S. (2017). Cyber war and information war à la Russe. In G. Perkovitch & A. E. Levite (Eds.), *Understanding cyber conflict. 14 analogies* (pp. 81–98). Georgetown University Press.

Brantly, A. (2022). Battling the bear. Ukraine's approach to national cyber and information security. In M. D. Cavelty & A. Wenger (Eds.), *Cyber security politics. Socio-technological transformations and political fragmentation* (pp. 157–171). Routledge.

Clarke, R. A., & Knake, R. K. (2010). *Cyber war. The next threat to national security and what to do about it*. HarperCollins Publishers.

Galeotti, M. (2019). *Russian political war. Moving beyond the hybrid*. Routledge.

Galeotti, M. (2022). *The weaponisation of everything. A field guide to the new way of war*. Yale University Press.

Greenberg, A. (2017, June 20). How an entire nation became Russia's test lab for cyberwar. *Wired*. https://www.wired.com/story/russian-hackers-attack-ukraine/.

Greenberg, A. (2019). *Sandworm. A new area of cyberwar and the hunt for the Kremlin's most dangerous hackers*. Doubleday.

Hoffman, F. G. (2010). 'Hybrid threats': Neither omnipotent nor unbeatable. *Orbis*, *54*(3), 441–455.

Johnson, R. (2021). Military strategy for hybrid confrontation and coercion. In J. H. Matlary & R. Johnsen (Eds.), *Military strategy in the 21st century. The challenges for NATO* (pp. 227–249). Hurst & Company.

Lilly, B., & Cheravitch, J. (2020). The past, present, and future of Russia's cyber strategy and forces. In T. Jančárková, L. Lindström, M. Signoretti, I. Tolga, & G. Visky (Eds.), *2020 12th International conference on cyber conflict* (pp. 129–155). NATO CCDCOE Publications. https://www.ccdcoe.org/uploads/2020/05/CyCon_2020_8_Lilly_Cheravitch.pdf.

Lonergan, E. D., Lonergan, S. W., Valeriano, B., & Jensen, B. (2022, March 7). Putin's invasion of Ukraine didn't rely on cyberwarfare. Here's why. *The Washington Post*. https://www.washingtonpost.com/politics/2022/03/07/putins-invasion-ukraine-didnt-rely-cyber-warfare-heres-why/.

Mueller, G. B., Jensen, B., Valeriano, B., Maness, R. C., & Macias, J. M. (2023). *Cyber operations during the Russo-Ukrainian war: From strange patterns to alternative futures* (Research Report). Center for Strategic and International Studies (CSIS). https://www.jstor.org/stable/resrep52130.

Richards, J. (2014). *Cyber-war. The anatomy of the global security threat*. Palgrave Macmillan.

Rid, T. (2013). *Cyber war will not take place*. Hurst & Company.

Shypilova, Y. (2019). *Ukrainian cybersecurity legal framework: Overview and analysis*. International Foundation for Electoral Systems.

Smeets, M. (2022). *No shortcuts. Why states struggle to develop a military cyber-force*. Hurst.

Streltsov, L. (2017). The system of cybersecurity in Ukraine: Principles, actors, challenges, accomplishments. *European Journal for Security Research*, *2*(2), 147–184.

Valeriano, B., Jensen, B., & Maness, R. C. (2019). *Cyber strategy. The evolving character of power and coercion*. Oxford University Press.

Valeriano, B., & Maness, R. C. (2015). *Cyber war versus cyber realities. Cyber conflict in the international system*. Oxford University Press.

Weissmann, M. (2021). Conceptualizing and countering hybrid threats and hybrid warfare. The role of the military in the grey zone. In M. Weissmann, N. Nilsson, B. Palmertz, & P. Thunholm (Eds.), *Hybrid warfare. Security and asymmetric conflict in international relations* (pp. 61–82). I. B. Tauris.

Wilde, G. (2022). *Cyber operations in Ukraine: Russia's unmet expectations* (Working paper). Carnegie Endowment for International Peace. https://carnegieendowment.org/2022/12/12/cyber-operations-in-ukraine-russia-s-unmet-expectations-pub-88607.

6

THE TRANSFORMATION OF HYBRID THREATS INTO FULL-SCALE WAR

The case of Ukraine, 2014–2022

Yevhen Mahda and Viacheslav Semenenko

Introduction

The level of conflict may influence on an aggressor's use of hybrid threat instruments (see Heier, ch. 8). The unleashing of the large-scale Russian-Ukrainian War caused a new era of geopolitical challenges and served as a push for transformative dynamics as to hybrid threats. In this chapter, the following question is addressed: How have the instruments and targets of hybrid threats evolved over time during Russia's war against Ukraine, especially in response to the increasing intensity of the conflict? What lessons can be drawn for the future? The chapter first describes the methods and targets of hybrid threat used by Russia before the 2022 full-scale invasion, and how the hybrid techniques changed after the full-scale invasion. By analysing the facts and drawing the logical chain between them, the chapter presents a nuanced picture of Russia's operations against Ukrainian sovereignty and independence; first, through an operational phase characterised by hybrid threat techniques; thereafter, by a more conventional *modus operandi*. Through an examination of the Russia's actions, the analysis deduce key lessons on how civilian and military authorities may address hybrid threats in a war-torn European theatre.

More knowledge on how Russia's hybrid operations against Ukraine's sovereignty unfolded is important. This is because such threats represent a contemporary and multifaceted challenge to global security. The blending of conventional military actions, cyber operations, and information warfare defies traditional paradigms of conflict analysis. Understanding the dynamics of hybrid threats in relation to conventional war is crucial for shaping effective counteracting strategies, safeguarding national and international

DOI: 10.4324/9781032617916-8

security, and responding diplomatically to the evolving geopolitical land-scape (Weissman et al., 2021).

How did Russia employ hybrid threats before full-scale invasion?

The Russian Federation's large-scale invasion of Ukraine, which began on February 24, 2022, demonstrated that the Kremlin's arsenal of hybrid threat tools was tried out without the expected effect. During 2014–2022, Russia's Armed Forces failed to gain operational control over Ukraine despite the sei-zure of Crimea, the instigation of the war in Donbas, and the various actions aimed at destabilizing the situation inside Ukraine. Instead, Ukraine's inter-nal stability grew significantly and took numerous important domestic and foreign policy steps to stabilise the situation. The occupation of Crimea and the low-intensity hostilities in the east of Ukraine (however, according to the UN, at least 13,000 people became their victims [1]) paradoxically influenced the strengthening of Ukraine's ability to resist. The Kremlin's choice of hybrid threat tools should be analysed in the following context:

Political speculations on the Minsk Agreements

First, several factors shaped public attitudes. The very fact of the existence of agreements of the Ukrainian political leadership with the Kremlin, despite the presence of mediators, was perceived ambiguously, as well as Russia's desire to hide from responsibility for its own actions behind puppets repre-senting the self-proclaimed Donetsk People Republic (DPR) and Luhansk People Republic (LPR). Both agreements were reached after dramatic events for the Ukrainian Armed Forces—the tragedy near Ilovaisk [2, 3] and fierce battles for Debaltseve [4]. The interest of the European leaders of that time—German Chancellor Angela Merkel and French President Francois Hollande—in minimising hostilities was perceived by Ukrainian society as aiding the Kremlin.

Resource support for pro-Russian politicians, public opinion leaders, and journalists

After Ukraine's independence in 1991, the Russian political management tried to establish effective control over the political development of Ukraine. Rus-sian political technologists—people who organise political processes in the society and organise communication between authorities and people—took an active part in election campaigns in Ukraine, filling them with the Krem-lin narratives. The situation did not change radically after the 2004-*Orange Revolution* challenged the Kremlin's dominance in the post-Soviet space. Viktor Yanukovych's rather quick return to power (first as prime minister in 2006, and then elected president in the democratic presidential elections of 2010) demonstrated the effectiveness of Russian instruments of influence. Actually, only the 2014-*Revolution of Dignity* in extremely dramatic conditions changed the paradigm of Ukrainian political development, but it did not manage to finally throw pro-Russian politicians out of Ukrainian political life. The Ukrainian efforts to break out of Russia's sphere of influence accelerated significantly by Russia's full-scale invasion.

Several pro-Russian parliamentarians and officials were active in trying to influence the Ukrainian political system and were regarded as political collaborators. Some pro-Russian Ukrainian politicians consistently brought disruption to American-Ukrainian relations on the eve of the large-scale invasion of Russia [5]. [Title and forename or former position, so that the reader understands what you are elaborating on] A politician like Viktor Medvedchuk has been accused of being one of the main ideologists behind the invasion of Ukraine by Russian forces [6].

Creating a parallel information reality through TV channels and internet

Use of media played an important role in facilitating hybrid threats by supporting TV channels controlled by pro-Russian forces and strengthening the information influence on the Ukrainian agenda of anonymous telegram channels controlled by Russian special services.

The creation of the conditional "TV holding of Medvedchuk" on the basis of the TV channels "112. Ukraine", ZIK, and NewsOne created opportunities for conveying the ideological position of the Kremlin not only to the citizens of Ukraine but also to representatives of the political establishment of the Ukrainian state. The studio of the TV channel seems to be an almost ideal platform for this because it allowed the use of the principle of freedom of speech and regulated competition between various actors in the political process. The decision of the National Security Council to ban the broadcasting of these TV channels [7] is viewed by some researchers as a catalyst for a large-scale Russian invasion of Ukraine, as it marked the collapse of a significant segment of the Kremlin's hybrid technologies used in relation to Ukraine.

At the same time, it can be argued that the promotion of the Telegram messenger in the Ukrainian information field, the use of which was characteristic mainly of representatives of the post-Soviet space, allowed Russia to solve several important tasks. Among them is the formation of a parallel information reality using anonymous Telegram channels and the replacement of influential television channels with Telegram channels, which are able to convey information to the information consumer faster. This communication activity testifies to the strategic importance of the information field in modern wars, and the transformation of the methods of influence allows us to talk about Russia's ability to adjust the mechanisms of influence to current tasks.

Consistent dehumanisation of Ukrainians, and creation of an image of Ukraine as a failed state

A range of tools were employed to undermine the Ukrainian state and the image of the Ukrainian population. A textbook example was the "crucified boy from Sloviansk" [8], which became a fake that operated in the information space for a long time. The case of the crucified boy is a message which was broadly spread via Russian media. The information source referred to someone named Galina Pushnyak. The message talked about a small boy from Slavyansk who was crucified by Ukrainian soldiers. This fake is the most famous fake of the hybrid war of Russia against Ukraine. It became an effective example of using stereotypes to mobilise human resources.

Rejecting the very possibility of Ukraine functioning as an independent state is inherent in the Kremlin's policy, despite the recognition of the inviolability of Ukrainian borders in the *Treaty of Friendship and Partnership* [9], which was signed on May 31, 1997, by Russia's President Boris Yeltsin and Ukraine's President Leonid Kuchma. In the practical sphere, this caused Russian propaganda to use the phrase "Kyiv regime" from the first half of 2014, to promote the thesis about a *coup d'état* in Kyiv through all possible information channels, which resulted in the overthrow of Viktor Yanukovych's power. Russia made maximum diplomatic and informational efforts to reduce Ukraine's weight in the international arena for obvious reasons: Moscow is aware of Ukraine's potential and its ability to become a counterweight to Russian influence, at least in the post-Soviet space.

Discrediting Ukraine in the eyes of the world community

Among the advantages of the Kremlin is their understanding of the functioning of Western institutions and the perception by the local elites (politicians, businessmen, and leaders of public opinion) of the situation in the surrounding world. Therefore, the Russian information triad, created on the basis of a private-state partnership (*Russia Today—Sputnik Media—*troll factory in *Olhino*), consistently tried to discredit both the independence of Ukraine and the prospects for its development. Critical assessments of the situation in Ukraine invented problems, and sometimes, elements of a parallel information reality were reflected in the information activities of the Russian propaganda machine [10].

The consistent discrediting of the "Azov" regiment in the eyes of the Western community can be called one of the revealing points in this matter. Russian propagandists accused this unit of the *Defence Forces of Ukraine* of "all mortal sins", but reality proves that these insinuations are far-fetched [11]. The use of provocation with the symbol of "Azov" during the referendum in the Netherlands on the ratification of the Association Agreement between Ukraine and the EU is an example [12]. Maybe brightest example of such provocation—the active naming of Azov as "neo-nazis" in USA [13].

Efforts to use energy leverage to ensure the realization of one's own interests in Russian-Ukrainian relations

An important part of the Russian threat picture was the control of energy resources.

The occupation of the Crimean Peninsula was accompanied by the seizure by Russian troops of drilling rigs on the Black Sea shelf, known in the Ukrainian information space as "Boyko Towers" [14]. (In September 2023, the Defence Forces of Ukraine regained control over these facilities [15].) With the capture of a part of the Donetsk and Luhansk regions, Ukraine began to experience a shortage of coking coal, necessary for metallurgical production, which contributed to the filling of the State Budget of Ukraine. The successful Ukraine consideration of the dispute between Naftogaz of Ukraine and Gazprom in the Stockholm arbitration [16] became the latest example of a review by the counterparties of the Russian energy monopoly of the terms of supply of energy resources. The "Kremlin's gas weapon" was discharged, so Putin switched to other methods of influencing the situation in Ukraine.

Obstruction of efficient logistics of goods of Ukrainian origin

Several tools for obstructing transport were employed. The Kerch (Crimea) Bridge, completed in 2018, is perceived as a symbol of Putin's ambitions for the occupied Ukrainian peninsula. It also connected the occupied Crimea with the Russian mainland, simplifying the transhipment of cargo, and limited the deadweight of ships that took grain, ore, and other export products from Ukrainian Azov ports. This was critical for maintaining the normal functioning of the Ukrainian economy [17].

At the beginning of the conflict on the territory of Ukraine, Russia aimed to create a chain of self-proclaimed "people's republics" that could destroy economic contacts and logistics inside Ukraine. In this context of the events of the spring of 2014, not only the self-proclaimed DPR and LPR but also the unsuccessful attempts to seize power by pro-Russian forces in Kharkiv and Odesa are important [18]. These failures forced Russia to implement an expensive and rather risky project to build the Kerch Bridge.

Search and recruitment of agents of Russian special services in the interests of future Russian aggression

An example of the continuous search and recruitment of agents of Russian special services was the continuous work of the representative office of "Rossotrudnichestvo". This is an organization that takes care of building relations with Russian compatriots (read—forms the structures of the "Russian world" in the post-Soviet space) for a long time, had the status of a diplomatic institution in Ukraine, continued to work even after the annexation of Crimea, and carried out subversive actions, essentially measures against Ukrainian statehood. Only in the spring of 2021, did this organization cease its operation in Ukraine, thereby eliminating the legal channel for promoting the Kremlin narratives. The main target group for the recruitment is the people who actively demonstrated interest in and support for the "Russian world" values and ideas.

Using problems in Ukraine's relations with other neighbouring states in one's own interests

The lack of a consistent policy of official Kyiv to build good-neighbourly relations created a specific environment for provocations. Therefore, the echoes of the Volyn tragedy (the ethnical cleanings, organised by Ukrainian rebels in 1942–1943) were reflected, in particular, in the grenade attack on the building of the Consulate General of the Republic of Poland in Lutsk ([19], the crisis in relations with Hungary—in the burning of the building of the Society of the Hungarian Language and Culture (this crime was discovered by the Polish special services) [20], in the legislation of Ukraine there is a concept "Moldovan language", although it itself is artificial [21]. The activities that were targeted at the sensitive for Ukraine's neighbours' topics, Kremlin created tension along the whole Western and South borders of Ukraine. In addition, it complicated the potential process of European integration of Ukraine.

Identification of the weak points of the domestic economy, sabotage of the implementation of defence programmes and measures to modernise the Armed Forces

Long-term economic and political contacts allowed the Kremlin to establish reflexive control over the development of Ukraine in certain sectors of the economy, sabotage the necessary transformations, and inhibit the development of promising industries. Reflexive control is the concept used by Russia, whose main tool is the influence on the emotions of opponents and further manipulation with the aim of controlling their further activities. All this had a negative impact on Ukraine's defence capabilities and its preparation for repelling aggression. However, it should be emphasised that the Kremlin did not manage to fully achieve its goals, which is evidenced by the destruction of the flagship of the Russian Black Sea Fleet, the missile cruiser *Moskva* by anti-ship missiles *Neptune* of Ukrainian design and production [22].

Striving to use Ukraine's status as a country of transitional democracy in one's own interests

For a long time, Russia tried to manipulate the political processes inside Ukraine, to inhibit its socio-political development. The goal of the Kremlin is obvious: Ukraine should remain in the political "grey zone" for as long as possible, and not to participate in the processes of European and Euro-Atlantic integration. The beginning of the large-scale invasion of Russia proved that Ukraine's course for joining NATO and the European Union had no alternative [23].

It should be emphasized that the Russian political leadership in the implementation of the aforementioned aspirations was distinguished not only by the presence of a significant number of resources but also by an irrational belief in the effectiveness of its own tools, a bet on proven personnel, and a known pattern in the use of these measures. The system of decision-making within the Russian system of government involves minimal discussions about the ways of implementing the plan and, as experience shows, a lack of consideration of Ukrainian realities. This significantly reduced the Kremlin's ability to influence the situation in Ukraine.

Apparently, in Russia, [the former Ukrainian president] Petro Poroshenko's defeat in the 2019 presidential elections [24] was considered a tactical success, and they counted on a change in Ukraine's foreign policy course. This is evidenced, in particular, by Vladimir Putin's agreement to the meeting of the Normandy Four in December 2019 in France [25] and the reaching of agreements on the transit of Russian gas through the territory of Ukraine, subject to the payment of compensation determined by the Stockholm Arbitration in favour of Naftogaz of Ukraine [26]. However, the Kremlin's plans turned out to be unrealized due to the lack of proper support within Ukrainian society.

The stay of President Volodymyr Zelenskyi in power did not justify the Kremlin's calculations to find opportunities for dialogue with him; public curtseys to the President of Ukraine did not change the negative perception of the actions of the Russian Federation. Ukraine, despite the occupation of Donetsk and Luhansk regions by Russia in 2014–2021, did not go the way Georgia, for example, went after Mikheil Saakashvili's departure from power [27].

What changed after the 2022 invasion?

Even though the Russian aggression had been active long before, the large-scale invasion has changed several significant aspects of Russian state policy towards Ukraine. In particular, the following aspects can be defined:

Russia moved to the use of armed forces to the maximum extent possible and launched a military propaganda machine

At first glance, this seems completely logical and predictable, given the previous actions of the Kremlin. However, there are several factors that must be taken into account. Although the Russian propaganda machine managed to rebuild itself during the large-scale hostilities (for example, the case of the speaker of the Ministry of Defence of the Russian Federation Igor Konashenkov ceased to be the only speaker reporting on the situation in the combat zone), the effectiveness of the Kremlin's propaganda did not increase. As a reminder, according to the official rhetoric, military operations against Ukraine, which became the largest war in the 21st century, are defined as "special military operations".

The category "war crimes" returned to widespread public use due to the actions of the occupation forces of the Russian Federation on Ukrainian territory

War crimes are the tool of threatening and pressure not only on the citizens of Ukraine but also on the citizens of other countries. In such way, Russia tries to threaten them and deprive from the will to resist.

The criminal actions of the Russian occupiers became one of the factors that shaped approaches to the perception of this confrontation around the world. The executions of the civilian population in Bucha [28] and other suburbs of Kyiv, the tragedy of the siege of Mariupol [29], the actions of the occupiers in the city of Izyum in the Kharkiv region [30] became examples of crimes against humanity. We can also mention the name of the camp for prisoners of war Olenivka in the Donetsk region, where, in the summer of 2022, as a result of an explosion, more than 50 captured Ukrainian defenders of "Azovstal" were killed [31]. The nature and scale of war crimes, which are based on the dehumanization of Ukrainians as part of the state policy of the Russian Federation, prompted the creation of the *International Group for Documenting Russian War Crimes in Ukraine*, and the process of preparing for the creation of an International Tribunal for their assessment continues [32].

Unprecedented decision of International Criminal Court

The abduction of Ukrainian children can be interpreted as a separate trend of the Russian occupation policy as a separate case of the large-scale migration crisis caused by Russia's aggression against Ukraine. This course of action by the Kremlin was evaluated by the *International Criminal Court*, which on March 17, 2023, issued an arrest warrant for the President of the Russian Federation, Vladimir Putin, and the Children's Ombudsman of the Russian Federation, Mariya Lvova-Belova [33]. Although the *International Criminal*

Court has previously issued arrest warrants for heads of state, the arrest warrant for the head of a nuclear state appears to be an unprecedented decision that forced the Kremlin to significantly limit Putin's international activities, actually leading to his partial isolation. The characteristics of this crime testify to the significant attention not only of the international community but also of lawyers to the events of the Russian-Ukrainian war.

The Nazi-labelling of opponents

With the beginning of large-scale hostilities, Russian officials and the media used a terminology with the ideological colour of World War II—"Nazis", "fascists", etc. A find for Russian propaganda was the participation of Yaroslav Hinka, a former soldier of the Waffen SS "Halychyna" division, in the solemn speech of the President of Ukraine Volodymyr Zelensky in the Parliament of Canada [34]. Official Russian media prefer to use the terms "militants" and "mercenaries" when talking about servicemen of the Defence Forces of Ukraine. In this way, they actually form a parallel information reality and try to level down the significant support of the Defence Forces of Ukraine by the population of the country. It can be noted that the Russian media call the drone attacks on the military infrastructure of the Russian Federation "terrorist" [35], without giving similar assessments to the Russian attacks on residential buildings in Ukraine.

Violation of the rules of war

The Russian authorities systematically and consistently ignore the norms of the Geneva Convention regarding the observance of the rights of prisoners of war, in particular, by holding a show trial against the former defenders of "Azostal" [36]. These actions can be considered as a continuation of the policy defined earlier, aimed at discrediting the "Azov" regiment. However, it is worth noting that Russia suffered a powerful image blow, because the defence managers of "Azovstal" were not only exchanged for Viktor Medvedchuk [37] but also returned home in the summer of 2023, where they continued to serve [38].

The desire to discredit the political leadership of Ukraine was replaced by the Kremlin's attempt to drive a wedge between the political and military leadership of Ukraine and violate the integrity of state governance. Russian propaganda tried to spread information about the serious wounding of the commander-in-chief of the Armed Forces of Ukraine, Valery Zaluzhny, constantly pedalling insinuations about the contradiction between Zelensky and Zaluzhny in anonymous Telegram channels. Russian propagandists expect that these messages will become the basis for contradictions within Ukrainian society, but their aspirations seem futile.

Nuclear blackmail

Nuclear blackmail in various forms (capture of the Zaporizhzhia and Chornobyl NPPs by Russian troops, information operation with the participation of the Russian leadership regarding the "dirty nuclear bomb", mining of the Zaporizhzhia NPP) has become one of the key areas of Russian state policy. Russian troops seized the Chornobyl and Zaporizhzhia nuclear power plants in the first weeks after the large-scale invasion. The first is the most famous nuclear facility in the world, decommissioned more than 20 years ago; the second is the most powerful nuclear power plant in Europe. In March 2022, the occupiers were forced to leave the territory of the Chornobyl NPP; however, the Zaporizhzhya NPP became the subject of constant manipulation. Let's pay attention to them.

Vladimir Putin's decree on the transfer of the Zaporizhzhia NPP to Russian ownership became an example of the world's largest interstate theft of an energy infrastructure object, carried out publicly. Despite the visit to the nuclear power plant (NPP) by the International Atomic Energy Agency (IAEA) delegation led by the president of the organization, Raphael Grossi, IAEA observers remained at the Zaporizhzhia NPP, but this did not prevent the occupiers from mining this energy facility [39].

Russia did not limit itself to measures of a military nature. In the fall of 2022, the Kremlin conducted a large-scale informational and psychological operation aimed at discrediting Ukraine. Russian Defence Minister Sergei Shoigu [40], Foreign Minister Sergei Lavrov, and State Duma Speaker Vyacheslav Volodin accused the Ukrainian leadership of preparing to use a nuclear bomb in southern Ukraine. Corresponding hints were also made by Vladimir Putin. It can be assumed that the ideologue of this operation was the deputy head of the Kremlin administration, Sergei Kiriyenko, who for a long time headed the Rosatom corporation [41] and is currently responsible for the implementation of the occupation policy in the territories of Ukraine seized by Russia.

Although the placement of Russian tactical weapons in Belarus can obviously also represent a danger for Ukraine, it seems that the Kremlin has chosen Lithuania and Poland as the focus of psychological pressure in this case. The official confirmation of this step by the Kremlin came only from the mouth of the head of the Main Intelligence Agency of the Ministry of Defence of Ukraine Kyryl Budanov [42], and NATO representatives have not yet voiced such information. Information manipulation around nuclear weapons on the territory of Belarus, a combination of conventional (tactical nuclear weapons) and non-conventional (prospects of placing units of the "Wagner" PMC on the territory of Belarus) testify to the Kremlin's plans to exert psychological pressure on countries that are quite clearly aware of the scale of the Russian threat. This is evidenced by the publications of one of the ideologues of the current Russian government, Sergei Karaganov [43].

Targeting critical infrastructure

The destruction of the Kakhovska hydroelectric power plant can be assessed as an act of ecocide on the part of Russia, which generally corresponds to the strategy of inflicting as much damage on Ukraine as possible. The destruction of the Kakhovska hydroelectric power plant (HPP) in the Kherson region, which occurred on the morning of June 6, 2023, was not only a large-scale man-made disaster. The consequences of this tragedy for local residents, the environment, and the economy can be compared to the consequences of using a tactical nuclear charge without the direct use of weapons of mass destruction. The destruction of the Kakhovska HPP dam aggravated the problem of nuclear safety at the Zaporizhzhia NPP because the reservoir for cooling the reactors was fed with water from the Kakhovska reservoir.

The list of challenges is obviously not exhaustive, as the fighting, which is the most intense in the 21st century, continues. However, it can be argued that during the period after the beginning of the large-scale invasion of Ukraine, Russia combined the previous developments of anti-Ukrainian policy with the desire to react to the new situation, without going beyond the boundaries of its own ideological paradigm.

Classification of hybrid threats

Based on the previous description of Russia's use of hybrid techniques, and the changing character of this employment after the 2022 invasion, the following question is addressed: How can the totality of the hybrid threat environment in Ukraine be categorised? Two years into the full-scale invasion, a complex of challenges for the Ukrainian society has emerged. Their preliminary classification is as follows:

1) **Transitional operations employed in the 2014–2022 period targeting Ukraine's socio-political stability:**

   - Efforts to discredit the political and military leadership of Ukraine.
   - Disinformation influence on Ukrainian society with the use of various information dissemination technologies.
   - Denial of Ukrainian statehood and the very possibility of the existence of the Ukrainian state.
   - Efforts to shake the foundations of Ukrainian statehood in order to prevent Ukraine from winning over Russia.

2) **Operations in support of Russia's large-scale invasion, aiming to underscore the Kremlin's ideological rationale**

   - Creation of the format of a "special military operation" aimed at subduing Ukraine by force.

- Application of nuclear blackmail technologies in various manifestations.
- Formation of the image of "liberators" for servicemen of the Russian Federation.
- Ideological support of the process of forceful annexation of the occupied Ukrainian territories.
- Replacement of the Russia Today—Sputnik duo, which became the object of Western sanctions, by the Russian state news agencies RIA Novosti and TASS.

3) **Operations targeting Ukraine's international legitimacy**

- Assurance that the supply of Western weapons cannot change the situation in the theatre of operations.
- The statement that the crimes of the Russian troops in the occupied territory of Ukraine are in the nature of insinuations prepared by British and American specialists in information and psychological operations.
- Broadcasting messages about the failure of the Ukrainian State to the countries of the Global South.

Conclusion

This chapter has scrutinised the following question: How have the instruments and targets of Russia's hybrid threats towards Ukraine evolved over time, especially in response to the increasing intensity of the conflict, and what lessons can be drawn for the future?

According to the analysis conducted, Russia's large-scale invasion of Ukraine should be seen as a consequence of the failure of other attempts to bring the Ukrainian government under control. The Kremlin grossly overestimated its own strength and suffered painful defeats in the first weeks of the invasion. More than two years into the full-scale invasion, the war has become protracted, and Russia continues to rely on a combination of military action and measures to destabilize Ukrainian society.

The military confrontation in the 21st century has the character of the War of Independence for Ukraine and the anti-colonial war, the result of which should be the defragmentation of Russia. Ukraine should promote to the world agenda the thesis on the realization of the right of the peoples of Russia to self-determination and the dismantling of Russia in its current form. The creation of the Rammstein format gives reason to assume a possible reformatting of the UN system after the end of the Russian-Ukrainian War.

While this study seeks to shed light on the evolving tools and objectives of Russian hybrid aggression against Ukraine, it is important to outline certain limitations that may affect the scope and generalizability of our findings. First, the dynamic nature of hybrid war and the ongoing nature of the conflict can make it challenging to record events in real time. Additionally, access to complete and accurate data, especially on covert operations and cyber activities, may be limited due to the sensitive nature of the topic. Basing a study on historical context and available information may introduce some retrospective bias. Despite these limitations, this study aims to provide valuable insights into hybrid conflict dynamics, contributing to a broader understanding of contemporary security challenges. By filling in research gaps, researchers and policymakers can expand collective understanding and develop more tailored strategies to counter hybrid threats in a global security landscape.

References

1. Office of the United Nations High Commissioner for human rights: Speakers urge peaceful settlement to conflict in Ukraine, underline support for sovereignty, territorial integrity of Crimea, Donbas region. February 20, 2019. Access mode: https://press.un.org/en/2019/ga12122.doc.htm.
2. Minsk agreements: Set of measures to implement the Minsk agreements. Access mode: https://www.osce.org/files/f/documents/5/b/140221.pdf.
3. Inna Semenova. The KamAZ truck with the wounded was the first to be hit. Eight years ago, the treachery of the Russian Federation led to the largest ATO/JFO disaster—the Ilovaisk tragedy. *NEW VOICE*. August 19, 2022. Access mode: https://nv.ua/ukraine/events/ilovayskiy-kotel-7-let-posle-krupneyshey-katastrofe-ukrainskih-voysk-na-donbasse-novosti-ukrainy-50176008.html.
4. Alec Luhn. Ukraine troops withdrawing from key town of Debaltseve. *The Guardian*. February 18, 2015. Access mode: https://web.archive.org/web/20150218143926/http://www.theguardian.com/world/2015/feb/18/ukraine-debaltseve-troops-withdraw-fighting-rebels-russia.
5. BBC News. Derkach's films: Ukraine was made an "instrument of discord" in the USA. *BBC News*. May 27, 2020. Access mode: https://www.bbc.com/ukrainian/press-review-52816947.
6. Matthew Ganapolsky. Putin needs Medvedchuk as a symbol of loyalty to "Russian peace". *Deutsche Welle*. September 23, 2022. Access mode: https://www.dw.com/uk/komentar-medvedcuk-potriben-putinu-ak-simvol-virnosti-russkomu-miru/a-63219139.
7. Dmytro Barkar. How did Zelensky dare to "turn off Medvedchuk's TV channels" and what will happen next? *Radio Svoboda*. February 4, 2021. Access mode: https://www.radiosvoboda.org/a/medvedchuk-kozak-telekanaly-sanktsiyi/31084755.html.
8. Olga Musafirova, Victoria Makarenko. The "boy" was not there, but he lives: Who came up with the bloodiest fake of the war in Donbass. *Новая газета*. July 15, 2015. Access mode: https://novayagazeta.ru/articles/2015/07/15/64898-171-malchika-187-ne-bylo-no-on-zhivet.
9. THE LAW OF UKRAINE: Agreement on friendship, cooperation and partnership between Ukraine and the Russian Federation. *Information of the Verkhovna Rada of Ukraine (VVR)*, 1998, N 20, Article 103. Access mode: http://zakon5.rada.gov.ua/laws/show/643_006.
10. EurAsiaDaily. Hunger and cold await Ukrainians. *EurAsiaDaily*. October 26, 2021. Access mode: https://eadaily.com/ru/news/2021/10/26/ukraincev-zhdet-golod-i-holod-deputat-rady.
11. Media Sapiens. Why Azov is not a "neo-nazi battalion". *Media Sapiens*. June 10, 2022. Access mode: https://ms.detector.media/manipulyatsii/post/29642/2022-06-10-why-azov-is-not-a-neo-nazi-battalion/.

12. Radio Svoboda. Fake video on behalf of "Azov": Denied by the Ministry of Internal Affairs and "Azov" itself. *Radio Svoboda*. January 19, 2016. Access mode: https://www.radiosvoboda.org/a/27497468.html.

13. The Daily Beast. Is America Training Neonazis in Ukraine? *The Daily Beast*. December 8, 2019. Access mode: https://www.thedailybeast.com/is-america-training-neonazis-in-ukraine.

14. Apostrophe. "Boika Towers": What is it and how did they appear in the Black Sea. *Apostrophe*. June 20, 2022. Access mode: https://apostrophe.ua/ua/news/politics/2022-06-20/vyishki-boyko-chto-eto-takoe-i-kak-oni-poyavilis-v-chernom-more/272301.

15. UKRINFORM. Ukraine returned to control "Boyko Towers". *UKRINFORM*. September 11, 2023. Access mode: https://www.ukrinform.ua/rubric-ato/3759790-ukraina-povernula-pid-kontrol-viski-bojka-gur.html.

16. SLOVO_I_DILO. "Naftogaz" vs "Gazprom": History of litigation in Stockholm arbitration. *SLOVO_I_DILO*. Analytical Portal. December 3, 2019. Access mode: https://ru.slovoidilo.ua/2019/12/03/infografika/politika/naftogaz-vs-gazprom-istoriya-sudebnyx-tyazhb-stokgolmskom-arbitrazhe.

17. Pavel Novikov. Trap for ships: what Ukraine is losing because of the Kerch Bridge. *Крым. Реалии*. February 6, 2018. Access mode: https://ru.krymr.com/a/29020748.html.

18. Deutsche Welle. What really happened in Odessa on May 2, 2014. *Deutsche Welle*. April 30, 2015. Access mode: https://www.dw.com/ru/dose-odessa-cto-na-samom-dele-proizoslo-2-maa-2014-goda/a-18418921.

19. Radio Svoboda. The shelling of the Polish consulate in Lutsk with a grenade launcher is a provocation by the "third force"—the Polish reaction. *Radio Svoboda*. March 29, 2017. Access mode: https://www.radiosvoboda.org/a/28398281.html.

20. BBC. The arson of the Hungarian center in Uzhgorod could have been coordinated by a German journalist—mass media. *BBC*. January 14, 2019. Access mode: https://www.bbc.com/ukrainian/news-46867548.

21. RadioSvoboda. In all laws of Moldova, the "Moldovan language" will be replaced by Romanian. *RadioSvoboda*. March 16, 2023. Access mode: https://www.radiosvoboda.org/a/news-moldova-mova-rumunska/32321424.html.

22. RadioSvoboda. "Ukraine hit the jackpot". What happened to the cruiser "Moscow"? *RadioSvoboda*. April 14, 2022. Access mode: https://www.svoboda.org/a/ukraina-sorvala-dzhekpot-chto-sluchilosj-s-kreyserom-moskva/31803344.html.

23. UKRINFORM. Ukraine has a record level of support for joining NATO. *UKRINFORM*. December 27, 2023. Access mode: https://www.ukrinform.ua/rubric-polytics/3584786-v-ukraini-rekordnij-riven-pidtrimki-vstupu-do-nato.html#:~:text=%D0%9F%D1%96%D0%B4%D1%82%D1%80%D0%B8%D0%BC%D0%BA%D0%B0%20%D0%B2%D1%81%D1%82%D1%83%D0%BF%D1%83%20%D0%B4%D0%BE%20%D0%84%D0%A1%20%D0%BE%D0%B4%D0%BD%D0%BE%D1%81%D1%82%D0%B0%D0%B9%D0%BD%D0%B0,%D0%BF%D1%96%D0%B4%D1%82%D1%80%D0%B8%D0%BC%D1%83%D0%B2%D0%B0%D0%BB%D0%B8%20%D0%B2%D1%81%D1%82%D1%83%D0%BF%20%D0%B4%D0%BE%20%D0%90%D0%BB%D1%8C%D1%8F%D0%BD%D1%81%D1%83%2076%25.

24. ОПОРА. Election history: How many records did the 2019 presidential campaign set? *ОПОРА*. May 27, 2021. Access mode: https://www.oporaua.org/vybory/vybory_prezydenta_2019-23138.

25. BBC. Normandy meeting: Zelensky passed Putin's test, experts say. *BBC*. December 10, 2019. Access mode: https://www.bbc.com/ukrainian/features-50728010.

26. InterFАx. Gazprom paid Naftogaz $2.9 billion awarded by the Stockholm arbitration. *InterFАx*. December 27, 2019. Access mode: https://www.interfax.ru/business/689697.

27. Настоящее время. "20% of our territory is occupied. How much more do we need to give up?" Special Representative of Georgia on the conflict with Russia. *Настоящее время*. August 8, 2018. Access mode: https://www.currenttime.tv/a/29417270.html.

28. BBC. Massacres of civilians in Bucha: what is known so far. *BBC*. April 4, 2022. Access mode: https://www.bbc.com/russian/news-60986588.

29. Meduza. Six months ago, the siege of Mariupol ended. Once a busy port, under Russian control the city looks more like a cemetery. *Meduza*. November 30, 2022. Access mode: https://meduza.io/feature/2022/11/30/polgoda-nazad-zakonchilas-osada-mariupolya-kogda-to-eto-byl-ozhivlennyy-port-pod-kontrolem-rossii-gorod-bolshe-pohozh-na-kladbische.

30. BBC. Ukraine: Hundreds of unmarked graves were found near Izyum after the Russian occupation, some of the bodies showed signs of torture. *BBC*. September 16, 2022. Access mode: https://www.bbc.com/russian/news-62891856.

31. Suspilne. 40 days have passed since the terrorist attack in Olenivka. What is known. *Suspilne*. September 6, 2022. Access mode: https://suspilne.media/278855-minulo-40-dniv-pisla-teraktu-v-olenivci-so-vidomo/.

32. ZN.UA. "Legal Ramstein": A conference with the participation of prosecutors general and ministers of justice started in Lviv. *ZN.UA*. March 3, 2023. Access mode: https://zn.ua/war/juridicheskij-ramshtajn-vo-lvove-nachalas-konferentsija-s-uchastiem-henprokurorov-i-ministrov-justitsii-mira.html.

33. BBC. How does the International Criminal Court order threaten Putin? Experts explain. *BBC*. March 17, 2023. Access mode: https://www.bbc.com/russian/news-64990680.

34. RadioSvoboda. The authorities of Ukraine reacted to the scandal with the soldier of the SS division "Halychyna" in Canada: Why were they silent and what are they saying. *RadioSvoboda*. September 29, 2023. Access mode: https://www.radiosvoboda.org/a/vlada-ukrayiny-reaktsiya-skandal-voyak-dyviziyi-ss-halychyna-/32616187.html.

35. EurAsiaDaily. Zakharova: Kyiv will send ATACMS missiles to Crimea and Donbass for terrorist purposes. *EurAsiaDaily*. September 27, 2023. Access mode: https://eadaily.com/ru/news/2023/09/27/zaharova-kiev-napravit-rakety-atacms-na-krym-i-donbass-v-terroristicheskih-celyah.

36. RadioSvoboda. "Emaciated, as if from a concentration camp." "Trial" of "Azovites" in Russia. *RadioSvoboda*. June 17, 2023. Access mode: https://ru.krymr.com/a/ukrainskiye-voyennoplenniye-brigada-azov-rossiya-sud/32462473.html.

37. RadioSvoboda. Intelligence explained who in Russia would benefit from Medvedchuk's exchange. *RadioSvoboda*. September 22, 2022. Access mode: https://www.radiosvoboda.org/a/news-fsb-medvedchuk/32046744.html.

38. NewVoice. Back in action. Azov commander Redis took part in military tactical exercises. *NewVoice*. August 4, 2023. Access mode: https://nv.ua/ukraine/politics/denis-prokopenko-redis-vernulsya-v-stroy-azov-pokazal-foto-novosti-ukrainy-50343926.html.

39. EPravda. The occupiers additionally mined the Zaporozhye nuclear power plant and cooling pond—Budanov. *EPravda*. June 20, 2023. Access mode: https://www.epravda.com.ua/rus/news/2023/06/20/701374/.

40. AA. Shoigu discussed the topic of "dirty bombs" in Ukraine with colleagues from India and China. *AA*. October 26, 2022. Access mode: https://www.aa.com.tr/ru/%D0%BC%D0%B8%D1%80/%D1%88%D0%BE%D0%B9%D0%B3%D1%83-%D0%BE%D0%B1%D1%81%D1%83%

D0%B4%D0%B8%D0%BB-%D1%82%D0%B5%D0%BC%D1%83-
%D0%BE-%D0%B3%D1%80%D1%8F%D0%B7%D0%BD%
D1%8B%D1%85-%D0%B1%D0%BE%D0%BC%D0%B1%D0%B
0%D1%85-%D1%83%D0%BA%D1%80%D0%B0%D0%B8%D0%
BD%D1%8B-%D1%81-%D0%BA%D0%BE%D0%BB%D0%BB%D
0%B5%D0%B3%D0%B0%D0%BC%D0%B8-%D0%B8%D0%B7-
%D0%B8%D0%BD%D0%B4%D0%B8%D0%B8-%D0%B8-
%D0%BA%D0%BD%D1%80/2721223.

41. Meduza. A little scary for the future. Sergei Kiriyenko has been ruling Russian politics for three years—and is quietly becoming an increasingly influential figure. *Meduza*. October 7, 2019. Access mode: https://meduza.io/feature/2019/10/07/nemnogo-strashnovato-za-buduschee.

42. RadioSvoboda. Belarus received nuclear warheads from the Russian Federation, but it will not be possible to use them—Budanov. *RadioSvoboda*. August 31, 2023. Access mode: https://www.radiosvoboda.org/a/news-budanov-yaderna-zbroya-rosiya-bilorus-viyna/32573223.html.

43. Russia in the global politics. Why we can't sober the West with the help of nuclear bomb. *Russia in the in the global politics*. June 21, 2023. Access mode: https://globalaffairs.ru/articles/otrezvit-zapad/.

7

COUNTERING RUSSIA'S HYBRID WAR

The orchestration of Ukraine's national resilience

Valerii Hordiichuk, Nina Andriianova and Andrii Ivashchenko

Introduction

Facing a broad range of hybrid threats, it is paramount for national authorities to increase their ability to sustain territorial sovereignty through national security efforts. The concepts of *national resilience* and *national security* are therefore inextricably linked. But, in contrast to national security, which is more military-oriented, national resilience is a concept that encompasses countering threats in all spheres: political, economic, military-political, informational, ethno-cultural, social, ecological, etc. (Pyrozhkov, 2022).

According to the concept of ensuring the national resilience system put into effect by the Decree of the President of Ukraine dated September 27, 2021, No. 479/2021, *national resilience* is the ability of the state and society to effectively resist threats of any origin and nature, adapt to environment changes, maintain sustainable functioning, and quickly recover to the desired balance after crisis situations (On ensuring the national resilience system, 2021). The *national resilience system* is a complex of purposeful actions, methods, and mechanisms of state authorities' interaction, local self-government bodies, enterprises, institutions, organizations, and civil society institutes, which guarantee the preservation of safety and continuity of functioning of the main spheres of life of society and the state before, during, and after the onset crisis (Peredrii et al., 2023).

In this chapter, the following questions will guide the analysis: What does a national resilience mean for the national security from the Ukrainian point of view, and how is it possible to identify, assess, and prioritize between threats?

DOI: 10.4324/9781032617916-9

There are three reasons why this topic is important. First, it provides new knowledge to academics, defence officials, and policy makers throughout the Western hemisphere as to how Russian hybrid threats have materialized inside a conventional total war paradigm. Second, experiences gained from Ukraine may accelerate new expertise and procedures in European civic communities that prepare for a tenser and unpredictable security environment. Third, scrutinizing Russian hybrid techniques allows Western forces to comprehend more comprehensively the close interaction between kinetic and non-kinetic instruments employed by Russian authorities in times of crisis and war.

The chapter is structured as follows. First, Russian strategies employed towards Ukrainian authorities are described. Thereafter, Ukraine's response to Russia's hybrid war is discussed. Finally, conclusions are deduced. The main argument is as follows: in the context of war and, accordingly, a significant state budget deficit, the issue of prioritizing measures to ensure national security and, consequently, national resilience is becoming acute. To do this, it is necessary to identify critical areas and directions that need to be strengthened.

Literature review

It is worth highlighting a few thorough works that describe national resilience as a social phenomenon in the state in a meaningful way. In the national report "National Resilience of Ukraine: A Strategy for Responding to Challenges and Preventing Hybrid Threats", S. Pyrozhkov (2022) and a team of authors in the current geopolitical environment attempt to substantiate the concept of national resilience as the resilience of a country's civilizational subjectivity.

In the monograph "National Resilience in a Changing Security Environment", O. Reznikova (2022) explores the theoretical and practical aspects of ensuring national resilience in a changing and uncertain security environment. The patterns for national resilience system formation and its interaction with the national security system. The peculiarities of adaptive management and formation of systemic links.

The anthology "Theoretical and applied aspects of the Russian-Ukrainian war: hybrid aggression and national resilience" (Koval, 2023), prepared by the National Defence University of Ukraine, is devoted to the analysis of Russian hybrid aggression and searching for countermeasures to it. Authors emphasize that resistance to the aggressor in all spheres and domains is possible only if there is an effective system of national resilience.

While the previous works, investigate national resilience as a general term based on the experience and significance of this concept for Ukraine, the works of W.-D. Roepke and H. Thankey focus on the study of the concept of resilience based on the experience of NATO (Roepke & Thankey, 2019). This study is devoted to the analysis of NATO's basic requirements for strengthening resilience, identifying risks and weaknesses of modern society where national efforts to strengthen resilience are needed, and ensuring the consistency and sustainability of efforts to strengthen resilience.

The book by G. Rouet and G. C. Pascariu "Resilience and the EU's Eastern Neighborhood Countries From Theoretical Concepts to a Normative Agenda: From Theoretical Concepts to a Normative Agenda" (Rouet & Pascariu, 2019) addresses the issue of resilience as a key concept in EU foreign policy, providing a systematic assessment of specific shocks and risks related to internal vulnerabilities (i.e. structural economic, social, institutional, and political instability) and their long-term and medium-term impact on stability, security, and sustainable development in the European region.

None of these works answers the question of how to adaptively distribute efforts to counter hybrid pressure in the conditions of ongoing armed aggression. In this work, the authors proposed an approach to solving this problem.

Methodology

This study is based on several methods of data collection and analysis:

- collection and processing of statistical and analytical information from open sources;
- cluster analysis method;
- systematic analysis;
- comparative analysis;
- retrospective analysis;
- generalization of personal experience and experience of scientists from the National Defence University of Ukraine;
- method of expert assessments of the Centre for Military and Strategic Studies of the National Defence University of Ukraine specialists, obtained through interviews;
- empirical analysis and synthesis of collected materials;
- methods of abstraction and analogy.

In order to develop ways to increase Ukraine's national resilience, the paper solves interrelated scientific tasks: analysis of the main legislative framework for ensuring Ukraine's national resilience; analysis of the strategy, methods, and tools of the Russian Federation's hybrid aggression during the Russian-Ukrainian war; defining, classifying, and prioritization types of hybrid influence for strengthening national resilience.

Data and analysis

National resilience in Ukraine's defence and security strategic documents

The primary document of the Ukraine defence review is the *National Security Strategy of Ukraine*, enacted by the Decree of the President of Ukraine of September 14, 2020, No. 392/2020 (On the National Security Strategy of Ukraine, 2020); according to this document, resilience is one of the main pillars on which the national security strategy is based, namely:

- deterrence—development of defence and security capabilities to prevent armed aggression against Ukraine;
- resilience—the ability of society and the state to quickly adapt to changes in the security environment and maintain sustainable functioning, in particular by minimizing external and internal vulnerabilities;
- interaction—development of strategic relations with key foreign partners, primarily with the European Union and NATO and their member states, the United States of America, pragmatic cooperation with other states and international organizations based on Ukraine's national interests.

Article 47 of this strategy declares that Ukraine will introduce a national resilience system to ensure a high level of readiness of society and the state to respond to a wide range of threats, which will include:

- risk assessment, timely identification of threats and vulnerabilities;
- effective strategic planning and crisis management, including the implementation of universal protocols for crisis response and recovery, taking into account NATO recommendations;
- effective coordination and clear interaction between the security and defence sector, other state bodies, territorial communities, business, civil society, and the population in preventing and responding to threats and overcoming the consequences of emergencies;
- spreading the necessary knowledge and skills in this area;
- establishing and maintaining reliable channels of communication between state authorities and the population throughout Ukraine.

  Task 1 of the Action Plan for the Implementation of the Concept of the National Resilience System by 2025 (Action Plan, 2023) is to implement a *comprehensive system of identification, assessment, and prioritization of threats and risks by central and local executive authorities.* Further, in this work, an attempt to define an approach to this task will be made.

The range of hybrid threat instruments

In the work "Essential features and functional components of modern unconventional (hybrid) conflict" *theatre of hybrid war* is defined as a multidimensional real (physical) and virtual (informational and mental) space of a state existence (a group of states, a nation, an ethnic group, a human community, etc.), which is purposefully transformed into a time-space for conducting a pre-planned complex of direct and indirect (asymmetric) aggressive actions (Viedienieiev & Semeniuk, 2021).

At the stage of the "hot" phase in a hybrid war, which involves the direct use of armed aggression, there is a need to study the use of kinetic and non-kinetic actions in relation to each other. Therefore, in Table 7.1, the overview of the most significant cases of such hybrid aggression after the start of the 2022 Russian full-scale invasion into Ukraine is presented. The overview rests on the following characteristics: goals, tools (tactics, methods), operational and strategic consequences. The table is based on empirical data from the following sources: Parakhonskyi & Yavorska, 2022; Kozubenko, 2022; Aleksejeva & Carvin, 2023; Kolesnikov, 2023; McDermott & Bartles, 2022; Tatlı, 2023; Frum, 2023; LaBelle, 2023; Ukrainian energy sector evaluation, 2023; Devine & van der Merwe, 2023; Cahill & Palti-Guzman, 2023; Brusylovska, 2023; Halunko et al., 2022; Dolin & Kopylenko, 2022; Hordiichuk et al., 2023; Jayanti, 2022; Syrotenko et al., 2020; War Speeches, 2023; Colibăşanu, 2023; Prokopenko, 2023; Shinkarenko & Bartalev, 2023; Vyshnevskyi et al., 2023; Kupriianova & Kupriianova, 2023; Filatov, 2023; An overview of Russia's cyberattack, 2022; World Energy Outlook, 2022, and other.

The overview is, however, not exhaustive. The Russian Federation also uses other tools of hybrid warfare: controlled migration and deportation of the population; the challenge of the demographic crisis; use of all possible levers of international law and international relations (the war in Ukraine revealed the inefficiency of a number of international organizations and their structural divisions, in particular, the UN Security Council, the International Red Cross and Red Crescent Movement, UNICEF, OSCE, IAEA, etc.); creation and financing of various political entities on the territory of other states, and bribery of officials; impact on the supply chains of a number of raw materials and food (causing fluctuations in prices on world markets and reducing the rate of economic growth), the use of the church and religion in the interest of supporting aggression, etc.

Moreover, internationally legally binding rules are also bent and explicitly violated. Examples are among other things violations of the rules of warfare and the laws of war; the use of prohibited weapons (phosphorus, mines, and other ammunition of prohibited principles of action), prohibited methods of waging war (blocking international maritime routes, covert mass destruction and torture of prisoners and civilians, use of civilians and civilian objects as shields, use of mercenaries, etc.).

A separate case of hybrid aggression for study is "legalized" mercenaries—the targeted use of private military companies.

TABLE 7.1 Significant Russian hybrid campaigns

Hybrid influence campaign	Goals	Tools	Strategic and operational consequences
Campaign of information and psychological influence "special military operation"	Justification and informational support for unprovoked armed aggression against Ukraine.	A system of narratives; distortion of facts; manipulation; rewriting of history; propaganda, etc.	1. Absolute support for armed aggression by the population of the Russian Federation; partial support for armed aggression by the pro-Russian population of Ukraine and the world. 2. Possibility of deploying a wide agent network on the territory of Ukraine from among supporters of Putin's politic; weakening of international support due to the cognitive dissonance created by the common audience; the use of conflicts and imperfections of the norms of international law to "legitimize" armed aggression.
Energy resources as a weapon (genocide)	The political split of the EU; Undermining of international support for Ukraine; Panic among the civilian population of Ukraine and administrative chaos; Economic exhaustion; Reduction of the defence-industrial complex capabilities; Reduction of the Ukrainian Armed Forces capabilities.	Blackmail with energy resources; creating an artificial shortage of natural gas in Europe and increasing its market value; destruction of energy infrastructure of Ukraine.	1. Ukraine's energy, gas and heating infrastructure exceeds $10 billion, 22 out of 36 power plants are damaged, destroyed or unavailable, and around 50% of Ukraine's energy facilities were damaged (as of April 2023). 2. The Eurozone registered an economic recession for a certain period. Influence on the decision of some countries to support Ukraine partially works. Europe is redoubling its efforts to break its dependence on Russian hydrocarbons. 3. Energy continues to be a key factor in Russia's foreign economic and geopolitical influence.

(Continued)

TABLE 7.1 (Continued)

Hybrid influence campaign	Goals	Tools	Strategic and operational consequences
Nuclear intimidation and blackmail	Weakening of military support; Blocking the Euro-Atlantic integration of Ukraine; Forcing Europe to put pressure on Ukraine in exchange for nuclear security; Forcing Ukraine to "peace negotiations" taking into account "territorial realities"; destruction/reduction of the Ukraine's economic; Using of potentially dangerous nuclear facilities as a shield for troops; Seizuring of infrastructure facilities for use in one's own interests.	Political pressure and blackmail; public statements and open intimidation; seizure and keeping under control of nuclear potential facilities; radioactive contamination of territories; conducting military exercises simulating the use of nuclear charges, etc.	1. Excitement among the civilian population of Ukraine and countries bordering Russia. 2. Russia's threats of a nuclear catastrophe allowed to demonstrate its barbarism. This increased the determination of some countries and international organizations to support Ukraine. 3. Captured nuclear infrastructure, in violation of international norms, is successfully used by the occupying forces as a shield, which gives them certain operational and tactical advantages. Russians have not been able to organize the production of electricity from captured nuclear facilities in their interests so far. 4. Given the unpredictability, defeatism and idiotic determination of Putin, and his pocket potentates, it is necessary to consider the fact that the order to use nuclear weapons (tactical or even otherwise) is unlikely, but potentially possible.
Food as a weapon (holodomor). "Grain Deel"+.	Pressure on the world community in order to unblock negotiations with the West to obtain concessions, lift sanctions, and promote one's interests on the world market.	Blackmailing the world community with a food crisis; blocking ports, blocking sea trade routes;	1. Russian food blackmail partially worked. Russia and the UN continue contacts in an attempt to push the connection to Rosselkhozbank's SWIFT. 2. The export of agricultural products by land from Ukraine led to political tension in the transit countries. 3. RF captured the traditional agricultural markets for Ukrainian products.

(Continued)

TABLE 7.1 (Continued)

Hybrid influence campaign	Goals	Tools	Strategic and operational consequences
	Increasing the export of Russian agricultural products and stolen from the occupied territories.	destroying industrial facilities and infrastructure, agricultural land.	4. Using excessive pressure, RF forced Ukraine and its allies to look for ways to unblock the sea and, as a result, lost dominance in the Black Sea. 5. RF dealt a tangible blow to the economy of Ukraine; also it can lead to a global increase in food prices, especially in the poorest countries.
Undermining the Kakhovka Hydroelectric Power Plant (ecocide+).	Disruption of the counteroffensive actions of the Ukrainian troops; Shifting the attention of the authorities and withdrawing forces from organizing and ensuring the conduct of hostilities; Reducing the energy potential of Ukraine; Causing a social and economic crisis in the region.	Physical destruction (mining) of the Kakhovka Hydroelectric Power Plant; Panic among the population, "Scorched earth" tactics.	1. Large-scale flooding of settlements significantly destabilized the situation in the region. Ukraine lost its annual supply of drinking water, large areas turned into a desert, etc. 2. Cut off the Ukrainian troops for a certain time, stopping the counteroffensive in the southern direction. Environmental and potential threats of a nuclear nature due to problems with cooling the reactors of the Zaporizhzhya Nuclear Power Plant.
Cyber terrorism	Support and strengthening of military operations with the use of kinetic weapons; Causing the greatest possible damage to the infrastructure of Ukraine (as a separate task).	Cyber influence, cyber-attacks, destructive impact on software and information, and cyber espionage.	1. In a number of cases, cyber activities caused significant interference in the work of information and communication systems of Ukraine. 2. A powerful cyber campaign provoked the emergence of a phenomenon of quite successful cyber resilience thanks to the consolidation of the Ukrainian and international cyber community. 3. Cyber aggression of the RF did not achieve its goals.

Source: created by the authors

Taking into account the fact that V. Gerasimov, who is considered the "founder" of the Russian "doctrine" of hybrid warfare, was appointed commander of the occupation forces in Ukraine instead of S. Surovikin at a critical moment after the Wagner group's rebellion, and then, after "stabilization", returned to the direct performance of the duties of the Chief of the General Staff of the Russian Armed Forces, it can be fairly assumed that he is the main ideologist of Russia's war strategy in Ukraine. Having studied his works and published statements (Gerasimov, 2013, 2023; Sinenko, 2017), it can be expected that in the future Russia's war strategy in Ukraine will be based on exhaustion—or a war by attrition. This strategy will be characterized by the following features:

1. The use of all possible instruments of hybrid warfare in various combinations.
2. The use of armed aggression as a main type of influence along with hybrid tools.
3. With an advance preparation usage of selected lever of hybrid warfare to achieve a cumulative effect, while the other levers will be used in tandem with the main one to achieve a synergistic effect.

A system of identification, assessment, and prioritization of threats

The national resilience system aims to provide protection against the widest list of threats. This list can be determined by studying the nature and experience of hybrid warfare, hybrid threats, and challenges. However, it is unrealistic to protect effectively against all threats, so there is an issue of clustering and prioritizing challenges in order to rationally allocate available resources and build a balanced protection system—a system of national resilience.

According to the *Action plan for the Implementation of the Concept of the National Resilience System until 2025* (Action Plan, 2023), task No. 1 is to introduce a comprehensive system of identification, assessment, and prioritization of threats and risks by central and local executive authorities.

To fulfill this task, a scientific and methodological apparatus is needed. In Ukraine, the state standard DSTU IEC/ISO 31010:2013 "Risk Management: Methods of general risk assessment" provides systematic methods of general risk assessment. One of the well-known scientific methods for accomplishing this task is cluster analysis—risk mapping.

In order to determine what types of impacts the targeted state or society needs to analyze, a cluster analysis method can be used to identify the main challenges. To do this, it is necessary to analyze the maximum possible set of influence instruments and separate them into groups based on their characteristic features, and build a dendrogram with using the hierarchical method.

In this chapter, for the purpose of simplicity and clarity, several well-known works that study hybrid warfare will be analyzed and summarized in Table 7.2 (Expert assessments on hybrid influence); the table thus proposes an appropriate classification of hybrid influences.

Having studied the works of international researchers, think tanks, and research institutions (Countering hybrid threats, 2023; European External Action Service, 2015; Horbulin, 2017; Koval, 2023; Magda, 2015; Sinenko, 2017; Syrotenko et al., 2020; Terrados, 2019), the following groups of hybrid warfare impacts (also known as operational environments—dimensions) identified for further research: informational-cognitive; cyber; financial-economic; international-political/diplomatic; military; special (environmental, social, religious, etc.).

Let us prioritize them using the simplest method—the method of an expert assessment, based on the analysis in Table 7.1 by simple arrangement.

In Table 7.2, using the method of an expert assessment (involving the Center of Military and Strategic Studies experts, including the authors of this research), we gave scores from 1 to 6, characterizing the intensity of impact (by consequences) in each individual hybrid influence campaign, where "6" is the highest score and the highest impact, and "1" is the lowest one.

Using the data in Table 7.2, let us illustrate the percentage of each hybrid influence in the hybrid warfare (see Figure 7.1).

The figure illustrates the percentage part of each hybrid influence in the hybrid warfare totally.

Thus, it can be concluded that among the analyzed most significant (according to the authors) cases, the most intensive in the hot phase of the Russian hybrid aggression was the informational-cognitive influence, and the hierarchical prioritization of the influences is as follows:

N1. Informational-cognitive;
N2. Military;
N3. International-political/diplomatic;
N4. Cyber and financial-economic;
N5. Special (environmental, social, religious, etc.)

Therefore, during building a national resilience system, in accordance with the defined hierarchy of hybrid influence, the prioritization of resource allocation for ensuring national resilience spheres should be determined accordingly, but it should be taken into account that this is fair while implementing a counter-strategy for a smouldering confrontation, just such a conflict is a war of exhaustion.

TABLE 7.2 Expert assessments on hybrid influence

№	Type of influence	Info-psycho Campaign "Special Military Operation"	Energy resources as a weapon (genocide)	Nuclear intimidation and blackmail	Food as a weapon (holodomor)	Undermining hydroelectric stations (ecocide+)	Cyber terrorism	Total
1.	Informational-cognitive	6	3	6	3	4	5	27
2.	Cyber	5	1	2	2	2	6	18
3.	Financial-economic	1	5	1	5	3	3	18
4.	International-political/diplomatic	4	4	4	6	1	4	23
5.	Military	3	6	5	4	6	2	26
6.	Special (environmental, social, religious, etc.)	2	2	3	1	5	1	14

Source: created by the authors

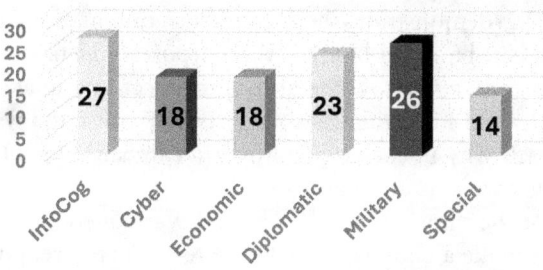

FIGURE 7.1   The Intensity of Russian aggression hybrid influences

The intensity degree of defined hybrid influence types during the Russian aggression totally.

Source: created by the authors

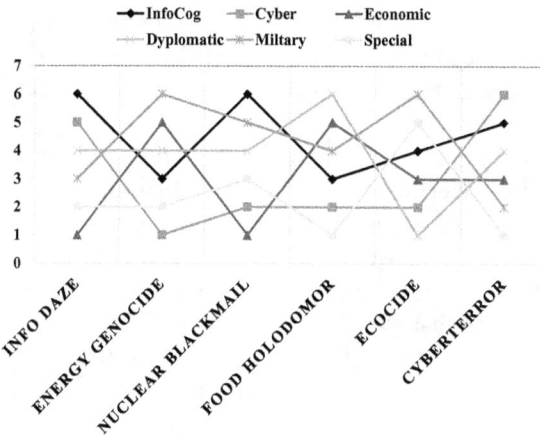

FIGURE 7.2   Impact-type intensity assessment

Graphical representation of the defined influence type intensity in different hybrid campaigns.

Source: created by the authors

Another issue that arises is "firefighting". It is necessary to take separate measures to prevent the cumulative effect of hybrid influence campaigns, such as the one observed during the blowing up of the Kakhovka Hydroelectric Power Plant dam in June 2023. To prevent the consequences of such enemy actions, it is necessary to protect oneself as if this act will definitely be carried out. For example, having information about the possible undermining of the Kakhovka Hydroelectric Power Plant, it is necessary to take all measures to reduce the negative impact: evacuate the population in advance; provide rescue services with appropriate equipment; consider redistributing energy supply from backup sources, etc.

In Figure 7.2, the peaks (the highest score—6) of a certain type of influence in different hybrid campaigns are exactly the "fires" that need to be prepared for.

The Figure 7.2 demonstrates how different combinations of the hybrid influence types can be depending on the purposes and tasks of the hybrid strategy/campaign (we see how the graphs intersect in a quasi-chaotic manner). Different hybrid campaigns has its purposes, tasks, and ways how to be performed. That's why in each type of hybrid campaign each type of hybrid influence has its own role and part as well.

In the case of the "hot" phase of a hybrid war, kinetic actions serve as a detonator to provoke a cumulative hybrid effect. Therefore, physical protection remains the number one task in a war with open armed conflict.

Conclusion

This chapter has scrutinized Russia's hybrid war against Ukraine and the Ukrainian response. Particular emphasis has been given to assessment and priority within the national resilience, as a defensive instrument for state survival.

As it clear from the above analysis, to organize and ensure the conduct of a large-scale hybrid war, the Russian Federation used all executive power bodies and the means by which they manage them. The range of domains and spheres of hybrid threats is extremely broad. In order to contain such attacks and not to violate the critically important foundations of the state functioning, Ukraine needs to create an effective mechanism of resilience. This mechanism should involve all state authorities and relevant means.

The chapter proposes an approach to ensuring the concept of national resilience by fulfilling task No. 1 of the *Action Plan for the Implementation of the Concept of the National Resilience System* by 2025—Introduction of a comprehensive system for identifying, assessing, and prioritizing threats and risks.

The authors propose a prioritization of the spheres of national resilience based on the method of cluster analysis and expert assessments. It should be noted that it is possible to achieve a more accurate prioritization using mathematical analysis, but this requires working with large amounts of data and having the appropriate statistics at your disposal.

Considering the previous one, it is necessary to narrow down the priority of state and society activities in the course of the Ukraine's total defence in the war of exhaustion for the maximum resource concentration in threatening spheres. Based on the conducted analysis and empirical experience, in a framework of further countering to Russian hybrid aggression, the authors propose three main pillars of national resilience of Ukraine:

1. *Ideology* (public recognition of national identity; clearly defined tasks for the liberation war; cognitive motivation; educative and anti-fake activities, etc.).
2. *State administration system and security sector* (personnel management: selection and recruitment of personnel of state authorities, taking into

account the requirements for competencies, including leadership skills; building a clear structure of subordination and interaction; comprehensive support of the Defence Forces of Ukraine, in particular in matters of mobilization, training and supply).

3. *Economy and infrastructure* (protection of critical infrastructure facilities; transition of the economy to wartime operation; broad involvement of the private sector; monitoring, and redistribution of internal reserves, etc.).

Proposed conclusions will suffice for making certain decisions in a time-limited environment. The model of hybrid influences prioritizing is universal and can be extended to a broad range of hybrid threats.

References

Action Plan for the Implementation of the Concept of the National Resilience System by 2025. (2023). Resolution of the cabinet of ministers of Ukraine № 1025-p. November 10, 2023. https://www.kmu.gov.ua/npas/pro-zatverdzhennia-planu-zakhodiv-z-realizatsii-kontseptsii-zabezpechennia-natsionalnoi-systemy-stiikosti-do-2025-roku-i101123-1025.

Aleksejeva, N., & Carvin, A. (Eds.). (2023). Narrative warfare. How the Kremlin and Russian news outlets justified a war of aggression against Ukraine. *Atlantic Council.* https://www.atlanticcouncil.org/in-depth-research-reports/report/narrative-warfare/.

An overview of Russia's cyberattack activity in Ukraine. (2022). *Special report.* Microsoft Publishing, 20. https://query.prod.cms.rt.microsoft.com/cms/api/am/binary/RE4Vwwd

Brusylovska, O. (2023). *Russia's nuclear blackmail as a threat to the global nuclear order. Russia's war on Ukraine.* Switzerland: Springer Nature, 39–52. https://doi.org/10.1007/978-3-031-32221-1_4.

Cahill, B., & Palti-Guzman, L. (2023). The role of Gas in Ukraine's energy future. *Center for Strategic & International Studies.* https://www.csis.org/analysis/role-gas-ukraines-energy-future.

Colibășanu, A. (2023). Are blockade runners challenging Russia's grain gauntlet? https://cepa.org/article/are-blockade-runners-challenging-russias-grain-gauntlet/.

Countering hybrid threats. (2023). NATO. https://www.nato.int/cps/en/natohq/topics_156338.htm.

Devine, K., & van der Merwe, B. (2023). Lights over Ukraine: The energy war. https://news.sky.com/story/the-energy-war-satellite-images-show-the-scale-of-russian-attacks-on-ukraine-infrastructure-12773049.

Dolin, V., & Kopylenko, O. (2022). Global nuclear threats. https://doi.org/10.13140/RG.2.2.25626.31687.

Filatov, S. (2023). Barbarians: Who destroyed the Kakhovka hydroelectric power plant and why. https://doi.org/10.21557/IAF.87320159.

Food-for-thought paper "Countering Hybrid Threats". (2015). *EU.* European External Action Service Working Document EEAS, 731. http://www.statewatch.org/news/2015/may/eeas-csdphybrid-threats-8887-15.pdf.

Frum, D. (2023). Why Putin's secret weapon failed. *The Atlantic.* https://www.theatlantic.com/ideas/archive/2023/06/russia-ukraine-natural-gas-europe/674268/.

Gerasimov, V. V. (2013). The value of science in foresight. *Military-Industrial Courier*, *8*, 1–6 [in Russian].

Gerasimov, V. V. (2023). *Information about the head*. Ministry of Defense of the Russian Federation. https://structure.mil.ru/management/info.htm?id=11113936@SD_Employee [in Russian].

Halunko, V., Buglak, I., & Boiko, V. (2022). Putin's nuclear blackmail. *Advanced Space Law, 9*, 93–107. https://doi.org/10.29202/asl/9/9.

Horbulin, V. P. (2017). World hybrid war: Ukrainian front. *Kyiv: NISS*, 496. http://resource.history.org.ua/item/0013707.

Hordiichuk, V., Snitsarenko, P., Sarychev, Y., & Hrytsiuk, V. (2023). *War in cyberspace as a component of the Russian-Ukrainian hybrid war. Theoretical and applied aspects of the Russian-Ukrainian war: Hybrid aggression and national resilience*. Koval, M. (Ed.). Kharkiv: Technology Center PC, 183–208. https://doi.org/10.15587/978-617-8360-00-9.

Jayanti, S. E.-P. (2022). The complex reality behind Vladimir Putin's nuclear blackmail in Ukraine. *Atlantic Council*. https://www.atlanticcouncil.org/blogs/ukrainealert/the-complex-reality-behind-vladimir-putins-nuclear-blackmail-in-ukraine/.

Kolesnikov, A. (2023). How Putin's "special military operation" became a people's war. *Carnegie Endowment for International Peace*. https://carnegieendowment.org/politika/89486.

Koval, M. (Ed.). (2023). *Theoretical and applied aspects of the Russian-Ukrainian war: Hybrid aggression and national resilience*. Kharkiv: Technology Center PC, 372. https://doi.org/10.15587/978-617-8360-00-9.

Kozubenko, O. (2022). How Russia's "special military operation in Ukraine" is being carried out, or About the Kremlin's failed plans. *ArmyInform*. https://armyinform.com.ua/2022/08/25/yak-vykonuyetsya-speczialna-vijskova-operacziya-rf-na-ukrayini-abo-pro-provaleni-plany-kremlya/ [in Ukrainian].

Kupriianova, L., & Kupriianova, D. (2023). Ecocide as a precursor of a particularly acute and postponed genocide of Ukrainian and European population in 2023. https://doi.org/10.12775/JEHS.2023.37.01.014.

LaBelle, M. C. (2023). Energy as a weapon of war: Lessons from 50 years of energy interdependence. *Global Policy, 14*(3), 531–547. https://onlinelibrary.wiley.com/doi/full/10.1111/1758-5899.13235.

Magda, Y. (2015). *Hybrid war: To survive and win*. Kharkiv: Vivat, 304. https://balka-book.com/files/2017/06_10/12_24/u_files_store_6_68818.pdf

McDermott, R. N., & Bartles, C. K. (2022). Defining the "special military operation". *Russian Studies Series*, 05. https://www.ndc.nato.int/research/research.php?icode=777.

On ensuring the national resilience system. (2021). Decree of the president of Ukraine No. 479/2021. September 27, 2021. https://www.president.gov.ua/documents/4792021-40181.

On the Military Security Strategy of Ukraine. (2021). Decree of the President of Ukraine No. 121/2021. March 25, 2021. https://www.president.gov.ua/documents/1212021-37661.

On the National Security Strategy of Ukraine. (2020). Decree of the President of Ukraine No. 392/2020. September 14, 2020. https://www.president.gov.ua/documents/3922020-35037.

Parakhonskyi, B., & Yavorska, H. (2022). *The war as a result of the fragile peace: The conceptual logic of war*. National Institute for Strategic Studies. https://niss.gov.ua/news/statti/porodzhennya-viyny-z-bezsyllya-myru-smyslova-lohika-viyny

Peredrii, O., Hordiichuk, V., Shchypanskyi, P., & Koretskyi, A. (2023). *Building national resilience system: Ukrainian and international experience. Theoretical and applied aspects of the Russian-Ukrainian war: Hybrid aggression and national resilience*. Koval, M. (Ed.). Kharkiv: Technology Center PC, 100–112. https://doi.org/10.15587/978-617-8360-00-9.

Prokopenko, A. (2023). Is this the end of the road for the Ukraine grain deal? https://carnegieendowment.org/politika/90225.

Pyrozhkov, S. I. (2022). *National resilience of Ukraine: A strategy for responding to challenges and preventing hybrid threats: A national report*. Kyiv: National Academy of Sciences of Ukraine, 551 [in Ukrainian].

Reznikova, O. O. (2022). *National resilience in a changing security environment: Monograph*. Kyiv: NISS, 455 [in Ukrainian].

Roepke, W.-D., & Thankey, H. (2019). Resilience: The first line of defense. *NATO*. https://www.nato.int/docu/review/uk/articles/2019/02/27/stjkst-persha-lnya-oboroni/index.html [in Ukrainian].

Rouet, G., & Pascariu, G. C. (2019). *Resilience and the EU's Eastern Neighbourhood countries: From theoretical concepts to a normative agenda*. Cham, Switzerland: Springer International Publishing, 598. https://doi.org/10.1007/978-3-030-25606-7.

Shinkarenko, S., & Bartalev, S. A. (2023). The consequences of damage to the Kakhovka Reservoir dam on the Dnieper River. https://doi.org/10.21046/2070-7401-2023-20-3-314-322 [in Russian].

Sinenko, S. (2017). Chief of the General Staff Valery Gerasimov on hybrid warfare. *In The Midst of Russia*. https://posredi.ru/nachalnik-generalnogo-shtaba-valerij-gerasimov.html

Syrotenko, A. et al. (2020). *Military aspects of countering "hybrid" aggression: Ukraine's experience: A monograph*. Kyiv: National Defence University of Ukraine. https://nuou.org.ua/assets/monography/mono_gibr_viin.pdf

Tatlı, M. (2023). The energy dimension of hybrid war and the Ukraine crisis. *International Journal of Social Humanities Sciences Research (JSHSR)*. https://doi.org/10.5281/zenodo.8111935.

Terrados, J. J. (2019). Countering hybrid threats. *The Three Swords Magazine*, 35/2019, 43–51. https://www.jwc.nato.int/images/stories/threeswords/Hybrid-War_Dec2019.pdf.

Ukrainian energy sector evaluation and damage assessment—VI. (2023). Cooperation for restoring the Ukrainian energy infrastructure project task force. https://www.energycharter.org/fileadmin/DocumentsMedia/Occasional/2023_01_24_UA_sectoral_evaluation_and_damage_assessment_Version_VI.pdf.

Viedienieiev, D., & Semeniuk, O. (2021). Essential features and functional components of modern unconventional (hybrid) conflict. *Strategic Panorama, 1–2*, 30–47. https://niss-panorama.com/index.php/journal/article/view/112.

Vyshnevskyi, V., Shevchuk, S., Komorin, V., & Oleynik, Y. (2023). The destruction of the Kakhovka dam and its consequences. https://doi.org/10.1080/02508060.2023.2247679.

War Speeches. (2023). Russia swallows the "grain deal" and complains that Patriot is shooting down: "Kinzhals". https://www.pravda.com.ua/eng/columns/2023/05/22/7403278/.

World Energy Outlook. (2022). *Report*. International Energy Agency. https://www.iea.org/reports/world-energy-outlook-2022.

PART III

Response strategies

8

CIVIC COMMUNITIES OR ARMED FORCES AS FIRST LINE OF DEFENCE?

Tormod Heier

Introduction

This chapter analyses how inferior states pursue political objectives without going to war. Staying below the threshold of war is not out of good will but out of strategic necessity. If your political, economic, and military leverage is curtailed, so is your freedom of action. These limitations are imperative when states try to align ends and means into a coherent strategy. Not least if the balance of power is uneven: For inferior states, a systematic mapping and exploitation of opponents' vulnerabilities are both rational and logic.

Historically, asymmetric power balances signify a broader phenomenon: how inferior states always seek to find ways to compensate for own shortcomings (Paret, 1986; Heuser, 2010). Among realists, these strategies are regarded as a precondition for national existence. Pursuing the right strategy is nothing less than question of national survival in a world of anarchy and great power rivalry (Waltz, 1979; Mearsheimer, 2001). This is why threat assessments often tend to be non-linear, complex, and ambiguous: because it is more rational to search for leverage along lines of least expectations in arenas where roles and responsibilities are blurred. Within spheres of ambiguity, inferior parts can exploit operational advantages and roam freely within an opponent's grey zone. This phenomenon grasps the core of a timeless strategy. As pointed out by Sir Lawrence Freedman (2013, p. xii), it is all about ". . . getting more out of a situation than the starting balance of power would suggest. It is the art of creating power".

This timeless wisdom is prevalent in Europe's post-Ukrainian landscape, a continent where a declining power, Russia, is confronted by a mighty U.S.-led coalition of more than 50 likeminded states. The chapter therefore

DOI: 10.4324/9781032617916-11

scrutinizes Russia's art of creating power. Suffering from military inferiority vis-à-vis NATO, the chapter analyses the following question: *What is Russia's most likely course of action towards the Alliance in the years to come?* Is it a strategy where NATO-members should prepare for an existential war against Russia? Or should Russia primarily, although not exclusively, be seen as a peace-time challenge; an aggressive state assertively pursuing its' interests, but without risking a full-scale war with NATO forces?

The puzzle is pertinent. As pointed out by Bjørge and Høiby in Chapter 2, research on hybrid threats have traditionally been militarized. Little emphasis is put on civil perspectives as Europe's most potent threat looms on the Eastern part of the continent. Scrutinizing Russia's most likely lines of operations, and what role civic communities may have, is therefore timely and relevant. Russia's 2022 invasion was also the third time in 14 years that the Kremlin-regime violated a neighbouring states' sovereignty. Since then, debates on how NATO, the United States, and Europe should respond have flourished: Should Europe's military rapidly regain strength to prepare for a possible new wave of Russian missiles, jets, tanks, and artillery conquering more land further West? Or should Europe be more concerned about endemic vulnerabilities thriving within its civic communities; weak spots that Russia exploits as affordable and less risky alternatives to nuclear escalation with U.S. forces? In-depth analysis of both approaches is scrutinized by Mahda and Semenenko in Chapter 6, by Hordiichuk et al. in Chapter 7, and by Akrap and Kamenetskyi in Chapter 14.

The chapter first presents a theoretical framework that builds upon perennial lessons from Sun Tzu and Carl von Clausewitz; the purpose is to explain generically how strategic interaction is shaped by the inferior part's focus on adversary vulnerabilities. Thereafter, Russia's room for manoeuvre vis-à-vis NATO is described and discussed before findings are deduced.

A theory of battle avoidance

In his seminal work, *The Art of War*, the Chinese general and philosopher Sun Tzu codified a timeless logic used by numerous strategists in a European, Russian, American, and Asian context. Being essential reading in the East since ancient times, this theory of war was introduced to Napoleon in 1792, inspiring the Prussian Carl von Clausewitz and the Swiss Antonie-Henri de Jomini in the 1830s. *The Art of War* was also used by Mao Zedong, Fidel Castro, and Vo Ngyen Giap in guerrilla wars waged in China, Cuba, and Vietnam in the 20th century. Even Stalin is said to have been inspired by Sun Tzu, as well as U.S. generals preparing for *Operation Desert Storm* in 1991 (Cawthorne, 2020, p. 6).

In Europe's post-Ukrainian security environment, Sun Tzu's relevance rests on the principle of battle avoidance. Planning and executing wars should emphasize the enemy's critical vulnerabilities. This is particularly so ". . . if the enemy is in superior strength" (Sun Tzu, 2020, p. 19). A synchronised and well-orchestrated planning therefore needs be accomplished in secrecy. If not, opponents may compensate for their vulnerabilities by redirecting attention and resources to mitigate the threat. The divine art of subtle and secrecy therefore demands a *modus operandi* where different courses of action follow different but mutually reinforcing lines of operations. Such an approach will be hard to defend because your freedom of action allows you to operate on arenas "where you are not expected" (ibid., p. 19). You may thus deny the opponent the privilege of exploiting his comparative strengths; his advantages become less useful against your vulnerabilities. The doctrinal imperative, therefore, is "to avoid strength and attack weaknesses, always striking the enemy where they are most vulnerable" (ibid., p. 129).

But what is an opponent's vulnerability? According to Sun Tzu, the enemy's internal cohesion should be the main target. In today's parlour, it means influencing the basic norms, the common values, the trust, and the confidence that nurture solidarity and expectations of reciprocity among likeminded states. Sun Tzu's ideal is as simple as it is logical: "if his forces are united, separate them" (ibid.).

The theory of divide and conquer, codified in China during the fifth century BC, resembles with the Clausewitzian logic of 19th-century Europe. This is because any war, according to Clausewitz, consists of three elements. If these elements are separated rather than united, the opponent will not benefit from the raw emotions and the psychological energy needed to exploit the war's full combat potential. Clausewitz claimed the first element to be violence and passion, and this element clearly belongs to the people. The second element consists of chance and probability. This element belongs to the commander and his forces. The third element consists of political purpose and effect and "is the business of government alone" (Clausewitz, 1976, p. 89). Together, the three elements constitute the power ingrained in real war. But as Peter Paret points out, these three elements are highly subjective:

> They reveal the author of *On War* in his historical posture, a soldier who regards himself as servant of the Prussian state and the protector of a society whose raw emotions must be exploited but also controlled. In his view it was the task of the political leadership to abstract the energies of the society without succumbing to their irrational power: a government channels psychic energy into rational policy, which the army helps carry out.
>
> *(Paret, 1986, p. 201)*

Clausewitz' so-called "remarkable trinity" (Clausewitz, 1976, p. 89) thereby consists of three mutually reinforcing elements: the people, the commander and his forces, and the government. These three elements cannot be separated.

But what if they are, as advocated by Sun Tsu? What if the people are separated from its government and its Armed Forces? In that case, neither the government nor their military commanders will benefit from the raw emotions and the psychic energy—or in contemporary parlour—the peoples' will to defend their values and interests. If legitimacy erodes, confidence evaporates, and public trust dwindles, the stronger part becomes more vulnerable and may even disintegrate from within.

For the inferior part therefore, the social fabric that keeps these three components together as one strong, neatly balanced, and cohesive unit is of particular interest. Clausewitz' theory rests on the rational logic of Sun Tzu: "to fight and conquer in all your battles is not supreme excellence; supreme excellence consists in breaking the enemy's resistance without fighting" (Tzu, 2020, p. 28).

Methodology

Using a deductive research design, the battle-avoidance theory allows us to deduce one empirical expectation: Russia's courses of action will avoid a military confrontation on NATO terms. As the empirics is described and interpreted below, we may expect Russian leverage along lines separating Western citizens from their governments. As the independent variable "course of action" is operationalized, indicators that serve as a guide for data collection go beyond quantitative measures such as a certain number of NATO brigades, jet fighters, or navy vessels. On the contrary, analysing Russia's course of action, qualitative indicators are emphasized. Examples may be the glue that keeps liberal, transparent, and civic communities together: the confidence-based contract between citizens and the state; the relationship between the governed and the governance; the social fabric that provides social order and prevents political chaos. Together, these concepts allow us to exploit a broad array of data material retrieved from a diverse European dataset.

But are the sources reliable? The data material is mainly secondary sources collected from written material inside Western institutions. A certain value-based bias may therefore prevail because many European sources are highly emotional following the Ukraine war. Frustration, anger, and anxiety in the aftermath of Russia's 2022 invasion have intentionally and unintentionally coloured much of the retrieved data material. Governmental reports and policy documents are also biased, politically and ideologically. This leaves the dataset with a typical pro-Western flavour that must be considered. Newspapers and other media articles may, therefore, at least to a certain extent, be a useful corrective to biased estimates and assessments from think tanks and governmental agencies; not least to reveal dilemmas and tension arising from within Western state-driven security and defence processes dealing with Russia.

Russia's room for manoeuvre

Russia's full-scale war in Ukraine changed the European power balance. As Europe's two largest states went to war, the continent witnessed the largest armed conflict since the Second World War; a war that severely has curtailed Russia's room for manoeuvre. Three geopolitical and one military reality can be used to describe Russian inferiority vis-à-vis the West.

American leadership

Russia's full-scale invasion of Ukraine inadvertently brought the United States back to Europe. This time, however, the World's sole military superpower returns with more nuclear and conventional forces than any time since the end of the Cold War. Containing China in South-East Asia is still the most important task. But as the United States accounts for more than 40 per cent of the World's total defence expenditures, Russia's primary rival can also afford to deploy over 100,000 troops to Europe (USA Facts, 2024). An additional rotating nuclear and conventional force is thus patrolling the sealines of communication and the airspace surrounding Russia: U.S. forces are credibly present in the High North, in the Baltic region, in the Mediterranean, and in the Black Sea, as well as along Russia's Eastern coast in the Pacific (U.S. NDS, 2022, p. 15; CRS, 2024, pp. 22–26). A permanent U.S. military presence of approximately10,000 troops in Poland was also established in 2023. This first ever permanent garrison on NATO's eastern flank includes subunits from an armoured brigade combat group, a combat aviation brigade and forward command elements from the U.S. Army's V Corps (Poland Ministry of Defence, n.d.). Supplemented by over 50 U.S. military facilities inside Russia's northern rim states, in the so-called *Defence Cooperation Agreement*-regime, a credible force projection can rapidly deter Russian aggression.

More importantly is the firm leadership employed by U.S. defence officials in Europe. From the United States' second largest base abroad, in Stuttgart, more than 300 U.S. staff officers work in *The Contact Group Ukraine* (U.S. Department of State, 2023a). From here, U.S. forces lead a coalition of more than 50 Western donor states. As seen from Europe, this is important. Even though Europe has a population threefold the Russian, twice as many troops and eight times Russia's GDP (Stoltenberg, 2023; World Economics, 2024), European states cannot deter Russia without a dedicated U.S. leadership. Regional powers, such as United Kingdom, France, and Poland, possess capable tactical forces. Europe nevertheless depends upon U.S. skills, capabilities, and competencies to accomplish strategic and operational planning and execution of larger multinational campaigns for sustained operations on the continent. As seen from a Russia, however, a credible U.S. military component in Europe is a serious disadvantage.

NATO cohesion

Russia's room for manoeuvre also shrinks due to NATO's cohesion. Having suffered from decades of institutional rivalry and fragmentation, the 2022 invasion has, just like Soviet invasions of Hungary in 1956, Czechoslovakia in 1968, and Afghanistan in 1979, amalgamated the transatlantic security community. The world's mightiest alliance is again rearming its forces under a firm U.S. leadership. This time, however, Western muscle flexing is much closer to Russia's borders than any time since NATO's 1949 inception. The contrast to NATO's tardy response after Russia's Crimean annexation in 2014 is striking: While the Alliance spent more than six months to decide whether to deploy multinational battalions to Poland and the Baltic states out of fear for provoking Russia (Heier, 2019, p. 202), the 2022 response was decisive. Within hours, NATO activated its *NATO Response Force*. Within one week, NATO's decision-making bodies assembled three times, deciding to double the number of multinational battalion groups to the region. The subsequent weeks witnessed the deployment of 140 navy vessels, 130 jet fighters, and 40,000 troops (Clapp, 2022, p. 2; Batchelor, 2022). According to NATO, approximately 300,000 troops are now on high readiness, up from the 40,000 (NATO, 2023). In total, across the Alliance, NATO possessed in 2023 and 2024 more than 20,000 aircraft, 2,150 ships, and almost 3.5 million active troops and other personnel (Statista, 2023a). In contrast, Russian force numbering 4,100 aircrafts, just below 600 vessels, and only 830,000 active troops (Statista, 2023a, 2023b). Europe's GDP in 2024 was also six times Russia's (World Economics, 2024).

Sliding buffer states

The Ukraine war also scared Russia's important rim states into the U.S.-fold. This is a serious blowback for Russia's strategic defences because buffer states are key to keep the United States and other NATO force at an arm's length. As Russia's buffer states, such as Finland and Sweden, are incorporated into the world's mightiest alliance, Western forces can more rapidly and decisively deploy closer to Russia's borders. As seen from Europe, it means that preconditions for a more credible deterrence in the entire northern region, from the Kola Peninsula to Kaliningrad, are more attainable. Not least as Finland, Sweden and Norway alone host over 40 U.S. agreed military facilities and areas (Finland, 2023; Sweden, 2023; Norway 2024).

As seen from Russia, its military force will have to deal with a larger and militarily more potent adversary across all domains: at sea, in the air, on the ground, as well as in space and the cyber domain. Russia thereby runs the risk of being overstretched along a NATO border stretching 2,600 kilometres from the Kola Peninsula in the High North to the Kaliningrad enclave in the Baltic Sea. Finland alone will, with a mobilization force of approximately

260, 000 troops, bind more than half a million Russian soldiers in the event of a crisis (Salonius-Pasternak, 2021, p. 5).

Negotiating from a position of strength with conventional and nuclear forces will therefore be increasingly hard every time a diplomatic crisis needs to be solved with support from Russia's military component. Russia's coercive diplomacy is thus checked more effectively by a modern Nordic infrastructure that can absorb large quantities of U.S. and NATO reinforcements. The strategic depth along Russia's North-Western front is gone, leaving the Kremlin with a considerably shorter warning-time every time a crisis arises on the horizon.

The Ukraine quagmire

Finally, Russia's national pride—the World's fifth largest force of approximately 850,000 standing troops—is severely decimated on Ukrainian fields in Donbas. 16 months into the war, approximately 90 percent of Russia's land force capacity was, according to U.S. intelligence reports, destroyed in a prolonged and highly unexpected war of attrition (Atlamazoglou, 2023; Landay, 2023). Having suffered more than 315,000 casualties in the first 22 months and lost more than two-thirds of its tanks, a large majority of Russia's offensive capabilities—the Battalion Tactical Groups designed to occupy new territories further West—are destroyed (Rommen, 2023). Despite subtle efforts to mobilise reserves and new recruits from a population of approximately 143 million people, three and a half times as large as Ukraine, the Russian army is unlikely to pose a serious threat to European territories anytime soon (UK Defence Intelligence, 2023).

Russian reserves also seem hard to find; not least as over 700,000 potential soldiers fled the country during the partial September 2022 mobilization of reservists (Zakir-Hussain, 2022). According to *The Danish Immigration Service* (2022), the force generation concept has so far failed to bolster the volume needed to fill the gaps. This is because any military organization, whether it Russian or Western, normally spends 15–20 years to educate, train, and build a modern land military component; a competent mechanized combat system with commanders possessing the skills needed to create synergy between mechanised infantry, tanks, artillery, engineers, combat support, and combat service support functions. It takes even more time and resources to neatly integrate this so-called "combined arms structure" into a truly joint force; a modern combat system that produces agility, tempo, surprise, and manoeuvre by seamlessly integrating mutually reinforcing components from all domains (sea, air, land, space, and cyber). Given Russia's cultural, technological, and operative deficiencies displayed in the Ukraine war (Zabrodskyi et al., 2022), it is reasonable to expect a relatively modest military performance compared to U.S. and NATO counterparts in the years to come.

This is, however, not the same as to claim Russia being militarily impotent. Russia's air force and navy, as well as its strategic missile forces, are by and large intact. But as long as the General Staff suffer from a heavily decimated land component, Russia's Armed Forces will remain unbalanced well into the 2030s. It means that Russian ground forces cannot exploit the potential effects caused by strategic bombers or submarines operating with long-range precision guided missiles in the High North or in the Black Sea. Until the army is firmly rebuilt and neatly integrated with the other services, Russia's military may first be seen as an instrument for coercive diplomacy. Not towards NATO forces but towards Western capitals and local communities. This is because the absence of a fully operational land component leaves Russia with little more than an incomplete and fragmented instrument unable to seize geographical control over NATO territories that potentially have been softened by missiles or cyber operations from other services.

The aforementioned empirics suggest that Europe should not expect a calculated large-scale war with Russia any time soon. Europe's security guarantor is back on the continent; NATO's institutional glue has not been stronger since the Cold War; Russia's buffer states in Northern Europe are tied to NATO; and Russia's offensive land forces are decimated for years to come. The Kremlin's freedom of movement is therefore limited, leaving NATO's rival with a few alternative courses of action. How then, may the Russian threat unfold?

Discussing the Russia threat

Being militarily inferior, Russia's course of action can be explored within the battle-avoidant framework of Sun Tzu and Clausewitz. Russia may then prefer to operate below the threshold of war. Targeting soft spots within European communities is obviously less risky than opening new fronts with prospects for a nuclear armageddon against a supreme western adversary. However, if the social contract between Western states and their citizens are weakened, or even splintered, Russia's ambition of redefining Europe's security architecture may have a bigger chance to succeed. This expectation rests on Russia's military problems in Ukraine.

The Ukrainian battlefield

According to two independent journalists Andrei Soldatov and Irina Borogan, Russia's war in Ukraine has been a catastrophe; the Kremlin has not been more vulnerable since the Second World War (Soldatov & Borogan, 2022). Neither conventional nor nuclear threats seem to underscore Russia's leverage against its Western rivals. This statement is underscored by former U.S. Ambassador to NATO and Moscow, CIA-Director Nicholas Burns (2023):

The Ukraine war has exposed Russia's military vulnerabilities, "its military weaknesses laid bare". Similar statements are also echoed from military circles: Russian forces have lost "strategically, operationally and tactically" in Ukraine and beyond (U.S. Department of State, 2023b).

The Kremlin nevertheless needs to prevail. Ukraine is Russia's most important borderland: a territory partly keeping U.S. forces at an arm's length, but also allowing Russia to thrive through linguistic, religious, and historical sentiments. If evicted from Ukraine, the Putin-regime will be accountable for a geopolitical trauma deteriorating Russia's internal stability: Russia's heaviest aviation losses so far comes from Wagner troops advancing towards Moscow during the June 2023 insurrection. Fourteen pilots were killed, while six helicopters and one command aircraft were downed (Horton, 2023). Loosing Ukraine may as such unleash a national trauma: Western liberal ideals, backed by a superior military force, may consolidate its foothold closer to Russia from the Barents Sea to the Black Sea.

A feasible strategy, therefore, is to follow the logic of Sun Tzu: to target Western cohesion; the bonds of trust and confidence that binds Europe together. These are the bonds that also tie long-standing U.S. military commitments and security guaranties to a cohesive community of like-minded states. By targeting transatlantic and European unity, Russian inferiority is compensated though a principle devised by the Chinese general: "breaking the enemy's resistance without fighting". Such an end state may severely undermine the West's long-term solidarity with Ukraine. This is particularly so when it comes to sustain critical deliveries of increasingly sophisticated weapons; technologically advanced firepower with longer range and better precision than Russian forces employ: *F-16* jet fighters, *HIMAR*-artillery, *Scalp* and *Storm Shadow* cruise missiles. Coupled with actionable intelligence from U.S. satellites and SIGINT-sources, the military assistance allows a feeble Ukrainian force to inflict grave casualties on a numerically superior Russian counterpart.

Despite uncertainty as to how sustainable U.S. assistance will be in the years to come, Ukrainian forces have regained half of the territory that were occupied during the 2022-invasion phase (U.S. Department of State, 2023c). Trying to end Western assistance, the Kremlin's most likely alternative therefore seems to be below the threshold of war: the non-kinetic means aimed at degrading the bonds of trust and confidence that keep Western state-leaders together for a steadfast Ukrainian commitment. This is in accordance with Clausewitz' "remarkable trinity": to prevent the citizens' raw emotions and psychic energies from being channelled into the U.S.-led coalition of state-leaders that supports Ukraine. As seen from Russia, if the confidence-based relationship between Western citizens and their political leaders erode, so will the Western cohesiveness and its Ukrainian support.

The peace-time challenge

If the reasoning above is valid, it can be claimed that Russia's high value targets are found located within civilian rather than military circles. But what is a peace-time challenge, and how is it operationalized? A peace-time challenge can be defined as the range of hybrid tools employed by Russian security services and their affiliated non-state actors. As pointed out by former Chair of U.S. National Intelligence Council, Gregory F. Treverton (2021, p. 38), examples are propaganda, fake news, strategic leaks, funding right/left-wing political parties and organizations, organized protest movements, and various cyber tools. During the first year of the Ukraine war, these challenges were particularly prominent in Russia's information domain. According to the European Commission (2023, p. 7), EU citizens and their allies were deliberately targeted with disinformation "reaching an aggregate audience of at least 165 million and generating at least 16 billion views".

Of particular concern are cyber tools and propaganda efforts portrayed on social media. Along these lines of operations, entry costs are low. An extremely large group of European citizens' "hearts and minds" can thus be reached and influenced. A peace-time challenge is also elusive and ambiguous as it operates "outside of and below detection thresholds" (Cullen, 2018, p. 4). But because peace-time challenges emphasize non-military tools, the contenders are opposing social communities rather than armies (Treverton, 2021, p. 37). What is at stake, therefore, is the strategic value of the liberal states' digital infrastructure: the nerve system that keeps sophisticated but very vulnerable public and private services alive and functioning. Examples may be Berlin's undisturbed delivery of electricity for industry and private homes, London's access to drinking water, Prague's availability of internet services, or Rome's functional sewage service. But it could also be a well-functioning railroad system around Warszawa, or a stable airport traffic management system at Schiphol Airport outside Amsterdam. Other examples may be a stable "just-in-time" provision of food to Parisian suburbs, medicine-deliveries to Baltic hospitals, or undisturbed access to raw materials for Germany's industrial production. Soldal describes the cyber dimension in more detail in chapter 5.

As Russian forces are pitted against Western weapons and intelligence in Donbas, the strategic value of Europe's digital nerve system—as an indirect approach to target Europe's cohesion—becomes increasingly attractive. Partly because the Kremlin acknowledges that its conventional forces are unfit to reach political goals outside the Crimea and Donbas territories. Partly because Russia's nuclear forces have reduced credibility when it comes to deter the West from weaponizing Ukraine, and partly because the Kremlin's inferiority leaves Russia with few other alternatives when it comes to end Western cohesion and long-term solidarity with Ukraine.

Europe's first line of defence

Disturbances, malfunctioning, or collapse of Europe's digital nerve system may easily cause anxiety, worries, or outright fear among millions of citizens. This fear may again be translated into civic doubt and mistrust directed towards their national governments. Of particular concern is public agencies' failure to decisively protect vulnerable citizens from malicious cyberattacks. Doubt as to whether local or national authorities have control may grind down the confidence upon which liberal democracies build their legitimate governance on (Wike & Fetterolf, 2018).

Adding to this is economic instability thriving in the aftermath of Russia's invasion. According to the European Commission (2022), economic strain from the Ukraine war has increased citizens' living costs throughout Europe:

> The sharp rise in inflation under the pressure of energy, food and other commodity prices is hitting a global economy that is still struggling with the economic consequences of the pandemic crisis. The EU is among the most exposed economies, due to its geographical proximity to the war and heavy—albeit much diminishing—reliance on imports of fossil fuels.

In 2024, the EU economy continued on a "weaker footing than previously expected" as a technical recession was narrowly avoided in 2023 (ibid., 2024). The economic and social burden may nevertheless transform into domestic resentments. As prices on food, energy, and cost of living accelerate, numerous examples of mass mobilization from frustrated citizens taking to the streets can be found in Italy, Spain, France, and Germany. In 2022, Europe had

> the third most protest events of any world region. Across the continent, people protested the rising costs of living, notably denouncing high energy costs. Protesters included both left-wing activists and far-right political groups, with the latter often calling attention to government inaction on the rising cost of living while also claiming overreach on issues like coronavirus restrictions
>
> *(Hossain & Hallock, 2022, p. 37).*

This may again, according to the World Bank (2023, p. 5), provide fertile ground for social tension. The rise of right wing and left-wing political parties undermine the stable, moderate, centre-oriented, and long-term predictability in European states' governance. In a long-term perspective, this challenge strikes at the core of the Clausewitzian trinity (Clausewitz, 1976, p. 89): by separating the people from its government and armed forces, the raw emotions and psychical energy needed to sustain a superior force may easily crumble. This vulnerability may more over serve as a stepping stone for populism and right/left wing movements. Nurtured by unwanted Russian activities below the threshold of war, such as propaganda, fake news, or strategic leaks on social media, Europe's domestic political landscapes may become more polarised (GEAB, 2023, pp. 14–20). A credible and long-term commitment to Ukrainian reconstruction and development programmes, or costly weapon deliveries, may as such be hard to sustain. Ellingsen describes this challenge in more detail in chapter 3.

Conclusion

This chapter described and discussed Russia's freedom of movement and its most likely course of action towards NATO in the years to come. Two conclusions can be deduced. The first is that Russia, in the years to come, cannot be defined as a military threat. Contrary to *NATO, 2022 Strategic Concept* (p. 4), which claims Russia to be the "most significant and direct threat to Allies' security", Russia should rather be seen as an inferior opponent; a declining power unable to seize and control NATO territories. But Russia can take European cities hostage with hypersonic missiles placed inside submarines and aircraft in the Black Sea, the Baltic Sea, or the North Atlantic. This may have great strategic utility during a stand-off, or a diplomatic crisis, while U.S. and Russian decision-makers try to avoid war. But apart from this, Russia's military capability should instead be seen as an instrument of public intimidation—a mafiosi tool designed to install fear, anxiety, and respect among Western policy makers and citizens rather than a balanced instrument for territorial conquest and geographic control.

The second conclusion builds on the first. Russia's most likely course of action will be of a non-kinetic character, below the threshold of war. That is not to say that outright war is unlikely. Misinterpretations, misjudgements, and human or technical errors may occur, that is, along Russian borders. But by and large, Russia's weakness is most likely advocating strategies aimed to exploit Western vulnerabilities. Civic communities are thereby Europe's first line of defence, not U.S. security guaranties or NATO forces.

The two conclusions highlight a strategic dilemma. Should European authorities prioritize the most likely peace time challenges towards its civic communities? Or should worst-case scenarios, such as preparing for a total war against Russian forces, have primacy? Clearly, NATO member states must do both. But combining the two will nevertheless be demanding. Tough priorities will have to be made across civilian and military domains. Not least in terms of balancing who should be the *supported* and who should be the *supporting* stakeholder? The dilemma grasps the perennial essence codified by the American diplomat and strategist George Kennan. Situated at the American embassy in Moscow in 1948, he described the Russian threat as a "perpetual rhythm of struggle, in and out of war" (Kennan, 1948).

For European states to prevail in such blurred environment, civilian and military stakeholders need to train and exercise below and above the threshold of war. Trimming the crisis management organisation in this transition phase, in and out of war, means building competence in grey zones where inferior opponents prefer to synchronize their coercive actions. Of particular importance is the organizational flexibility needed to respond effectively with civilian and military resources; often through tailor-made and ad hoc–based organizations; headquarters or centres that are neatly designed to address the uniqueness and specific requirements needed for the unique event.

References

Atlamazoglou, S. (2023, February 20). Putin has a problem: 90 percent of his army is fighting (and dying) in Ukraine. *19Fortyfive.com*. www.19fortyfive.com/2023/02/putin-has-a-problem-90-percent-of-his-army-is-fighting-and-dying-in-ukraine/.

Batchelor, T. (2022, March 9). Where are NATO troops stationed and how many are deployed across Europe? *The Independent*. www.independent.co.uk/news/world/europe/nato-troops-russia-ukraine-estonia-map-b2031894.html.

Burns, N. (2023, July 1). *A world transformed and the role of intelligence*. Oxford. www.ditchley.com/sites/default/files/Ditchley%20Annual%20Lecture%202023%20transcript.pdf.

Cawthorne, N. (2020). Introduction. In S. Tzu (Ed.), *The art of war* (pp. 6–12). Arcturus Publishing Ltd.

Clapp, S. (2022). *Russia's war on Ukraine: NATO's response*. European Parliament Research Service. www.europarl.europa.eu/RegData/etudes/ATAG/2022/729380/EPRS_ATA(2022)729380_EN.pdf.

Clausewitz, C. (1976). *On war*. Ed. and trans. by P. Paret & M. Howard. Princeton University Press.

CRS. (2024, January 10). *Great power competition: Implications for defense—issues for congress*. Congressional Research Service. R43838.pdf (fas.org).

Cullen, P. (2018, June 4). *Hybrid threats as a new 'wicked problem' for early warning*. Hybrid COE Strategic Analysis. www.hybridcoe.fi/wp-content/uploads/2020/07/Strategic-Analysis-2018-8-Cullen.pdf.

The Danish Immigration Service. (2022). *Update on military service in Russia*. www.us.dk/media/10558/update-on-military-service-in-russia_tilgaengelig.pdf.

European Commission. (2022). *Autumn 2022 economic forecast: The EU economy at a turning point*. European Commission. (europa.eu).

European Commission. (2023). *Digital services act: Application of the risk management framework to Russian disinformation campaigns*. Publications Office of the European Union. digital services act-KK0923294ENN.pdf.

Finland. (2023, December 18). *Defence cooperation agreement with the United States*. 7ee50234-1dfd-0014-32c5-4ef3549ea301 (um.fi).

Freedman, L. (2013). *Strategy. A history*. Oxford University Press.

GEAB. (2023, September 17). Global economy 2024: Make or break. *Global Europe Anticipation Bulletin*, No. 177, 14–22. Geab-Nr177-en.pdf.

Heier, T. (2019). Britain's joint expeditionary force: A force of friends? In R. Johnson & J. H. Matlary (Eds.), *The United Kingdom's defence after Brexit. Britain's alliances, coalitions, & partnerships* (pp. 189–215). Palgrave Macmillan.

Heuser, B. (2010). *The evolution of strategy. Thinking war from antiquity to the present.* Cambridge University Press.

Horton, J. (2023, June 27). Wagner revolt: How many planes and people did Russia loose? *BBC News.* https://www.bbc.com/news/world-europe-66031403.

Hossain, N., & Hallock, J. (2022). *Food, energy & cost of living protests.* Friedrich-Ebert-Stiftung New York Office. www.library.fes.de/pdf-files/bueros/usa/19895.pdf.

Kennan, G. (1948, May 4). *Policy planning staff memorandum.* www.archive.law.upenn.edu/live/files/9964-kennan-memo-political-warfarepdf.

Landay, J. (2023, December 13). U.S. intelligence assesses Ukraine war has cost Russia 315,000 casualties -source. *Reuters.*

Mearsheimer, J. (2001). *The tragedy of great over politics.* W.W. Norton & Co.

NATO. (2022, June 29). *NATO 2022 strategic concept.* www.nato.int/nato_static_fl2014/assets/pdf/2022/6/pdf/290622-strategic-concept.pdf.

NATO. (2023, July 11). *NATO—news: NATO agrees strong package for Ukraine, boosts deterrence and defence.* NATO—News: NATO agrees strong package for Ukraine, boosts deterrence and defence. https://www.nato.int/cps/en/natohq/news_217059.htm. Accessed: 14 July 2024.

Norway. (2024, February 2). *Norway and USA agree on additional agreed facilities and areas, Oslo. Norway and USA agree on additional agreed facilities and areas under the SDCA.* regjeringen.no.

Paret, P. (1986). Clausewitz. In P. Paret (Ed.), *Makers of modern strategy* (pp. 186–213). Clarendon Press.

Poland Ministry of Defence. (n.d.). *Increasing the US military presence in Poland.* https://www.gov.pl/web/national-defence/increasing-the-us-military-presence-in-poland.

Rommen, R. (2023, December 30). It will likely take Russia up to 10 years to rebuild a highly trained, experienced army, says UK Intel. *Businessinsider.com.*

Salonius-Pasternak, C. (2021). *Defence innovation: New models and procurement implications. The Finnish case.* Policy Paper No. 65. Armament Industry European Research Group. 65-Policy-Paper-Def-Innov-Finland-March-2021.pdf (iris-france.org).

Soldatov, S., & Borogan, I. (2022, April 27). Vicious blame game erupts among Putin's security forces. *The Moscow Times.* www.themoscowtimes.com/2022/04/27/vicious-blame-game-erupts-among-putins-security-forces-a77508.

Statista. (2023a, March 9). Number of military personnel in Russia as of 2023, by type. Russia military personnel by type 2023. *Statista.*

Statista. (2023b, March 30). Comparison of the military capabilities of NATO and Russia as of 2023. NATO Russia military comparison 2023. *Statista.*

Stoltenberg, J. (2023, January 8). *Speech at security conference*, Selen, Sweden. www.nato.int/cps/en/natohq/210454.htm?selectedLocale=en.

Sweden. (2023, December 7). *Defence cooperation agreement with the United States, Stockholm. Agreement on defense cooperation between the Government of the Kingdom of Sweden and the Government of the United States of America.* Agreement on defense cooperation between the Government of the Kingdom of Sweden and the Government of the United States of America. https://www.government.se/contentassets/ad5f87be923e4065b658189a9294f480/agreement-on-defense-cooperation-between-sweden-and-the-united-states-of-america.pdf. Accessed: 14 July 2024

Treverton, G. F. (2021). An American view: Hybrid threats and intelligence. In M. Weissman, M. Nilsson, B. Palmertz, & P. Thunholm (Eds.), *Hybrid warfare* (pp. 36–45). I. B. Taurus.

Tzu, S. (2020). *The art of war.* Arcturus Publishing Ltd.

UK Defence Intelligence. (2023, December 30). *Intelligence update on Ukraine. Ministry of Defence* GB *på X: Latest defence intelligence update on the situation in Ukraine.* Ministry of Defence GB på X: «Latest Defence Intelligence update on the situation in Ukraine. https://x.comDefenceHQ/status/1741026561258573901. Accessed: 14 July 2024.

U.S. Department of State. (2023a, July 23). *Secretary Antony J. Blinken with Fareed Zakaria of GPS, CNN.* United States Department of State.

U.S. Department of State. (2023b, June 2). *Russia's strategic failures and Ukraine's secure future.* www.state.gov/russias-strategic-failure-and-ukraines-secure-future.

U.S. Department of State. (2023c, February 13). *Special online briefing with Ambassador Julianne Smith, U.S. permanent representative to NATO.* United States Department of State.

U.S. NDS [National Defense Strategy]. (2022). *2022 National defense strategy of the United States including the 2022 nuclear posture review and the 2022 missile defense review.* dtic.mil.

USAFacts. (2024, February 2). *Where are US military members stationed, and why? Where are US military members stationed, and why?* usafacts.org.

Waltz, K. (1979). *Theory of international politics.* Addison-Wesley Publishing Co.

Wike, R., & Fetterolf, J. (2018). Liberal democracies' crisis of confidence. *Journal of Democracy, 29*(4), 136–150.

World Bank. (2023, January 12). *Food-security-update.* www.thedocs.worldbank.org/en/doc/40ebbf38f5a6b68bfc11e5273e1405d4-0090012022/related/Food-Security-Update-LXXVI-January-12-2023.pdf.

World Economics.com. (2024). *Europe's combined GDP is far larger than Russia's.* World Economics.

Zabrodskyi et al. (2022, November 30). *Preliminary lessons in conventional warfighting from Russia's invasion of Ukraine: February–July 2022.* Royal United Services Institute. https://rusi.org/explore-our-research/publications/special-resources/preliminary-lessons-conventional-warfighting-russias-invasion-ukraine-february-july-2022.

Zakir-Hussain, M. (2022, October 6). Moscow denies 700,000 have fled since Putin's call-up order—but does not have 'exact figure'. *The Independent.*

9

MUNICIPAL PREPAREDNESS AGAINST HYBRID THREATS

How can military experiences be used?

Jannicke Thinn Fiskvik and Tormod Heier

Introduction

For states that pursue their interests by means of hybrid threats, any vulner-ability within adversary communities represents a valid target. The key point is to combine political, economic, social, and military means in subtle and innovative ways (NOU 2023: 17, p. 115). Of special importance is political leverage; the malign influence imposed on adversary decision-making pro-cesses, their strategies, and adjacent policies; efforts that otherwise would have energised social cohesion, trust, and confidence within an adversary community.

The subtle and unrestricted character of hybrid threats blurs the civil-military boundary in liberal democracies. This is due to hybrid threats' indiscriminate nature. A myriad of public and private agencies may easily be targeted simultaneously, for example through coordinated cyberattacks (see Chapter 5 by Lund). Such a crisis may easily exceed civil resources and com-petence at local levels. For additional efforts, military assistance may be at hand. Communal preparation and response plans should therefore consider how military skills and resources could be exploited more consistently to enhance societal resilience (NOU 2023: 17, p. 26).

Aligning scarce resources across civil-military interfaces is nevertheless difficult—especially in war and war-like situations. This has become more evident over the last decades. Since the 1990s and well into the 2000s, many European states have transformed their armed forces into a more deploy-able and usable expeditionary force. As the Soviet threat disintegrated in 1991, international operations far off own territories became a reality. Not least for thousands of military personnel that for the next two decades were

DOI: 10.4324/9781032617916-12

assigned a politically sensitive role in Central Asia and the Greater Middle East; a role that blurred the traditional civil-military interface at home: counter-terrorist operations inside civic communities; winning the inhabitants' *hearts and minds* through nation- and peace-building efforts, preferably so through corroboration with war lords and power brokers; and finally—before a humiliating retreat from Central Asia in 2021—mentoring Afghan officials and security forces as part of a "good governance-vision" inside tribal communities.

For Russia's European rim states, committing thousands of officers and non-commissioned officers (NCOs) to a politically, culturally, and socially complex environment was a long-term investment. With Russia's 2014-annexation of Crimea and its full-scale invasion in 2022, Europe rebalanced back to territorial tasks. But this time, with a credible U.S. deterrent posture and a cohesive NATO in back-hand. Concurrently, Russia's invasion had catastrophic consequences for its Armed Forces. Two years into the full-scale war, Western intelligence services claim it may take 10 years or more before Russia can rise militarily (Mehta, 2022; Nicholls et al., 2023). In the meantime, posing hybrid threats below the threshold of war seems to be Russia's most likely course of action (see Chapter 8 by Heier).

For mayors, police chiefs, and councillors situated in the districts, often far away from robust governmental agencies, mitigating such threats is not easy. As hybrid threats strike across public and private interfaces, a myriad of actors with different roles and responsibilities are involved. Lacking military ethos, like *unity of command* or *unity of purpose*, a coherent alignment of resources is difficult. This chapter, therefore, analyses the following question: *How can military experiences from international operations, such as from Afghanistan's politicized theatre, be utilised at home, as small and vulnerable municipals prepare for hybrid threats?*

Municipal resources in the periphery tend to be scarce, and thus thinly spread across vast areas. Effective preparations and responses to hybrid threats, therefore, rest on joint efforts from both civilian and military stakeholders. This topic is explored further by Sandbakken and Karlsson in Chapter 11, and by Borch in Chapter 12. More jointness nevertheless requires an updated civil-military framework—a pragmatic approach that seeks to mitigate vulnerabilities and shortcomings in the districts. But which at the same time maintains the sensitive relationship between local mayors and police chiefs, on the one hand, and their military subordinates, on the other.

Second, as civic communities seek resource optimisation, military experiences from foreign operations may be scrutinized. This is because hybrid threats cannot be effectively mitigated without an integrated whole-of-government approach (Christensen et al., 2020). Europe's post-Ukrainian security environment thereby coincides with key elements of the NATO *Comprehensive*

Approach in international operations; a strategy designed to align three mutually reinforcing lines of operations from the civilian and military spheres: security, governance, and reconstruction (Rynning, 2019, p. 27). This resembles the whole-of-government approach that many local emergency response actors must confront in a contemporary hybrid threat scenario at home: To organize themselves effectively so that scarce public and private resources are aligned towards one unifying purpose within crisis management.

To scrutinize the puzzle, the chapter proceeds as follows. First, a theoretical perspective of the "unequal dialogue" between civilian leaders and military servants is outlined. Thereafter, the theory serves as a backdrop to describe how Nordic forces accumulated valuable civil-military experiences from Afghanistan, 2002–2021. On this basis, the strategic utility of such experiences is discussed, but in a context where civic-led municipals at home strive for enhanced resilience against hybrid threats. Finally, four conclusions are made on how military experience can be used to improve civil responses and preparations against hybrid threats.

Perspectives on civilian-military coordination and control

Civil-military relation is a perennial phenomenon encapsulating almost any aspect of civil societies' relationship with their Armed Forces. The scholarly field is just as broad: Valuable insight is given by historians, sociologists, and political scientists. The topics include themes like military coups, civilian control of the military establishment, and soldiers' professional and cultural affinity with their parent society (Brooks, 2019). The key issue is the civilian's political role vis-à-vis its' military subordinates; the timeless dilemma of maintaining civilian control without losing military effectiveness (Feaver, 1996). While there is no question of the existence of civilian control over armed forces in mature democracies, there are nevertheless many ways to frustrate or even evade civilian control (Fiskvik, 2020). Civilian leaders must therefore use every opportunity to guide military efforts into a politically optimal end-state (Cohen, 2012).

Hence, as operational contexts change, so do civil-military relations. In a hybrid threat context, hostile aggressors do not wage conventional wars on a clearly defined battlefield. The enemy's *modus operandi* does not allow the military to defeat their adversaries decisively by brute force only. On the contrary, as wars unfold inside local communities, wars become highly politicized. The blurred interfaces between combatants and non-combatants leave officers and their soldiers in a complex security environment, even the smallest military action may cause profound political consequences (Simpson, 2012, pp. 92–100). It is not politically helpful for any civilian entity to annihilate opponents if the violent course of action alienates local citizens or even the entire international community. You may win the war, but you may also loose the peace—widely defined as the broader political objective.

According to Cohen (2012), therefore, politicians should act more pro-actively to govern their force. Rather than presenting a clear-cut political end state, and thereafter leave the military with planning and execution, the opposite should happen: a detailed regulation of how the organised violence is employed, and how this coercive action is amalgamated with other and more complementary instruments of power, such as diplomatic, economic, or humanitarian levers.

Cohen's model of civil-military relations rests on the "unequal dialogue": a cooperative process where civilian superiors are engaged actively in mili-tary logic. The model is not about civilian leaders dictating their subordi-nates but rather a constructive civil-military dialogue—unequal because the final authority rests with civilians. This is important in both wars and crisis below the threshold of war and on battlefields where tactical and operational manoeuvres unfold within unstable and vulnerable civic communities, and where military action needs to be calibrated to underscore a broader set of political objectives (Cohen, 2012, pp. 227–242). Politicians, supported by their civil servants, should therefore "immerse themselves in the conduct of their wars". Whether you are a parliamentarian, a prime minister, a mayor, or a local police chief, you should "master military briefs as thoroughly as [you do your] civilian ones". In other words: "Both groups must expect a running conversation in which although civilian opinion will not usually dic-tate, it must dominate; and that conversation will cover not only ends and policies, but ways and means" (Cohen, 2012, p. 224).

As military force is used inside civic communities, where lines between war and peace are blurred and political and military activities are intermin-gled (Smith, 2007, pp. 269–271), the notion of a strict division of labour between the civilian and military leadership is challenged. This again, Cohen argues, has grave implications for the Huntingtonian logic, where civilians should abstain from a too detailed interference in strictly military matters (Huntington, 1957). Cohen, on the other hand, maintains that a separa-tion of the two spheres is an arbitrary one. Any civilian stakeholder should be involved meticulously in operative planning and execution; any military activity should be scrutinized through querying, probing, and new sugges-tions on how military skills and resources may underscore political objec-tives. If necessary, advice from senior military advisers should be overruled and those who fail to perform dismissed. At the same time, the civilian side should tolerate disagreements and appreciate blunt counterarguments, even though "the final authority of the civilian leader is unambiguous and unques-tioned" (Cohen, 2012, p. 227).

How can "the unequal dialogue" be used to scrutinise the utility of mili-tary experiences inside a European municipal? The proposition is that while civilian leaders maintain superior authority over the military, officers and NCOs also possess valuable skills and knowledge needed to enhance munici-pals' preparation against hybrid threats. The next section describes how

Nordic forces gained civil-military experience from Afghanistan. Thereafter, the relevancy of this experience is discussed within the context of Cohen's "unequal dialogue".

Civil-military experiences from Afghanistan (2002–2021)

The intervention in Afghanistan involved a comprehensive civilian and military engagement. The International Security Assistance Force (ISAF) focused on humanitarian and stabilization operations to support the central Afghan government in extending its authority to Afghanistan's many provinces (Suhrke, 2011). In this regard, the Provincial Reconstruction Teams (PRTs) played a key role. The PRTs included both civilian and military elements following three lines of operations: security, governance, and development (NATO, 2010).

Embarking on an ambitious state-building project, Afghanistan became a demanding test for the international community. For one, balancing political objectives with military doctrines proved difficult. Moreover, civilian and military assistance had to be more intimately aligned, an issue that blurred their roles and responsibilities. The challenges thereby invoked valuable experiences. Not least for a military profession that, during the Cold War, was accustomed to a total war scenario—a situation that left the military in charge. But political inference became more pronounced as limited wars with limited political objectives gained momentum throughout the 1990s and 2000s.

Here, experiences of Denmark, Finland, Norway, and Sweden are described. The empirics emphasize the quest for better civil-military alignment of resources, in accordance with NATO's *Comprehensive Approach*-strategy in Afghanistan. The description allows us to discuss the utility of military experiences at home, as municipals prepare against hybrid threats.

Denmark

From an initially dispersed participation across Afghanistan, Danish efforts took a decisive turn with the decision to join the British-led PRT Lashkar Gah in 2006. Operating in Southern Afghanistan, Danish forces faced a tribally fragmented population sceptical to foreigners and with little experience of central government (Saikal, 2012). Additionally, there were war lords instrumentalizing tribal rivalries and challenges with heroin production and drug trafficking. Initially, Danish authorities issued a few national caveats and provided a broad mandate for the military engagement (Fiskvik, 2020, p. 215).

Although civilian authorities emphasized the comprehensive approach, the Danish engagement was overwhelmingly military (Rynning, 2019). Situated in one of the toughest parts of the country, there was limited civil-military alignment as Danish NGOs deemed the Danish area of operations too dangerous for humanitarian workers (Fiskvik, 2020, p. 152). Nevertheless, as part of Task Force Helmand, Danish forces were accompanied by stabilization advisors appointed by the Ministry of Foreign Affairs (MFA) (Andersen et al., 2016, pp. 34–36). Jointly they planned and initiated small reconstruction projects for areas where the security situation made the presence of civilian organisations impossible (Danish MFA et al., 2013, p. 22). Lacking clear directions from home, the civil-military coordination on the ground was based on goodwill and compromises between deployed civilian and military personnel (Andersen et al., 2016, p. 29). Experiences from Afghanistan showed that personal relations between civilian advisors and soldiers were of great importance. These relations functioned best when the actors had met beforehand in Denmark with joint education and exercises (Andersen et al., 2016, p. 73).

Many stabilization projects, however, were hindered by the grave security situation. Ultimately, facing insurgents in a high-risk area, the Danish forces prioritized combat operations. In this regard, military advice and force protection weighed heavily in parliamentary debates, and the military contingents were reinforced (Fiskvik, 2020, pp. 146–147). Part of the British-led PRT, Danish forces followed strategic priorities of British forces and adapted to counterinsurgency warfare; the effort shifted from fighting the Taliban across Helmand province to focusing on a permanent and less offensive presence in densely populated areas (Fiskvik, 2020, pp. 158–164). From 2012 onwards, the Danish forces assumed a more withdrawn role, focusing on training and advising Afghan security forces.

As the mission grew more complex and the situation deteriorated, civilian authorities became more involved in planning and operative conduct. They began imposing stricter rules of engagement, in which Danish forces increasingly had to consider both political and ethical consequences when using force (Fiskvik, 2020, p. 152). The Danish Afghanistan strategy from 2008 included a realization that the notion of a process where military forces first secure an area and then followed by development activities did not suit the realities in Afghanistan. Instead, the strategy notes the threat picture "conditions specific demands of an active and flexible interplay between military and civilian efforts" (Danish MFA & Danish MoD, 2008, p. 21). To coordinate efforts, a Danish inter-ministerial Afghanistan working group was established (Fiskvik, 2020, p. 151). In addition, the MFA and the Ministry of Defence (MoD) began jointly to issue Helmand Plans, outlining goals and expected effects. Based on benchmarks and progress evaluations, plans were intended as a tool to ensure political control of the overall effort—signifying a shift in the Danish approach to military affairs (Fiskvik, 2020, p. 152).

In the case of Denmark, the demanding security situation hampered a comprehensive approach. Nevertheless, Afghanistan brought valuable experiences in the use of force under complex conditions, in which Danish forces had to learn counterinsurgency and to interact with a broad range of civilian actors (Rynning, 2019). The experiences in Helmand come with an evolving recognition that civil-military alignment requires trust, a mutual understanding, and joint training before and during operations (Danish MFA et al., 2013).

Finland

Mainly operating with Norwegian and later Swedish forces in Northern Afghanistan, Finnish forces had similarly a low-key approach aiming at gaining local trust (Mustasilta et al., 2022, p. 134). Organized in mobile observation teams, Finnish forces conducted patrols in northern provinces as part of the PRT's aim to make the conditions eligible for reconstruction (Fiskvik, 2020, pp. 183–184). Finnish forces also had to navigate according to guidelines set by the government, such as a geographical restriction to the north and strict rules of engagement, in which they were only to use force in self-defence and in exceptional circumstances (Fiskvik, 2020, p. 183).

The comprehensive approach did not stray far from existing Finnish approach to international crisis management, where Finnish soldiers in previous peacekeeping operations have participated in larger development projects in support of civil society (Fiskvik, 2020, pp. 175–189). This continued in Afghanistan as civilian experts operated alongside Finnish forces in implementing smaller projects between 2004 and 2007. Compared to Norway and Sweden, Finnish forces had more leeway in terms of civic financial assistance to carry out projects (Fiskvik, 2020, p. 200). Moreover, Finland deployed reservists, who's versatile skills from civilian professions were considered a strength in the engagement with civilians (Mustasilta et al., 2022, p. 192). As security deteriorated, however, it became difficult to pursue a fully civil-military approach. Frustrated Finnish aid organizations argued that their line of work became endangered in what they experienced as a prioritization of political and military considerations over development issues (Mustasilta et al., 2022, p. 143). The dialogue between civilian and military spheres stumbled, and as their roles became less distinct, it became important for civilian authorities to have a clear task allocation among the different actors (Fiskvik, 2020, pp. 188–189).

With a more demanding security situation, the original military tasks of stabilizing gravitated towards combat operations. Finnish forces adapted to COIN principles alongside Swedish forces and cooperated in planning and conducting large operations intended to remove insurgents and enable development work (Fiskvik, 2020, p. 201). For the military leadership,

Afghanistan served as an opportunity to test national defence capabilities, develop interoperability, and practice NATO operations (Mustasilta et al., 2022, p. 129).

The Finnish experiences with strategic thinking in a highly politicized security environment are mixed. On the one hand, the decline in social stability inside Afghan communities in Finland's area of operations did not come as a surprise to the military leadership, who adapted the contingents to the conditions on the ground. There were necessary adaptions in rules of engagement, heavier military equipment, and Finnish forces changed to a more traditional military structure of platoons (Fiskvik, 2020, pp. 194–198). In these cases, the dialogue between civilian and military spheres and routines functioned well. Planning the involvement was considered effective and processes for changing the rules of engagement and deploy heavier military equipment went smoothly (Mustasilta et al., 2022, p. 132). On the other hand, when the military leadership requested an increase in the number of troops, the President blocked the proposition and insisted on 195 as the maximum number (Fiskvik, 2020, pp. 183–187). Moreover, deployed personnel perceived that higher officials had a narrow understanding compared to how challenging the situation really was; in this vein, there was a lack of clear strategic guidelines where multiple Finnish objectives, partly vague and at times conflicting, hampered the implementation of a comprehensive approach (Mustasilta et al., 2022, pp. 99–115).

Overall, the understanding of the comprehensive approach and the application of it developed during the ISAF years (Mustasilta et al., 2022, p. 181). There was a cumulative process of awareness of the complexity of operations, including experience with a more active use of force, and importantly an acknowledgement that current conflicts require increased political attention (Fiskvik, 2020, pp. 205–207). Moreover, Finnish forces gained leadership experience in war-like circumstances, as well as new skills in mobility, protection, and intelligence (Mustasilta et al., 2022).

Norway

Norway's contribution to Afghanistan gradually concentrated to the Faryab province in Northern Afghanistan, as the country became lead nation of PRT Meymaneh in 2005. The PRT was led by the military, while civilian efforts were administered by the Norwegian Embassy in Kabul with a small civilian component in Faryab (NOU 2016: 8, p. 125). While acknowledging the comprehensive approach as the only way of addressing the challenges, Norwegian authorities interpreted the coordination of civilian and military efforts to signify a clear separation of civilian and military tasks (NOU 2016: 8, p. 127). Accordingly, with stabilization as the main task, Norwegian forces were to provide security so that civilian actors could do reconstruction work.

Although Northern Afghanistan was mostly calm initially, the Norwegian forces had to navigate between other security problems such as managing the relationship with local warlords and conflicts along ethnic lines (Suhrke, 2011, pp. 89–90).

Aligning civilian and military efforts into a broader comprehensive approach challenged the military. The PRT concept was unfamiliar and included tasks at the margins of core military activities. Furthermore, the PRT had few available funds, and the forces were restricted in doing civilian tasks (Fiskvik, 2020, p. 135). Significantly, the civil-military divide was at odds with ISAF's tight cooperation of civilian and military resources (NOU 2016: 8, pp. 130–131). The official inquiry into Norway's Afghanistan engagement found that in practice, it was difficult strictly to adhere to the civil-military separation (NOU 2016: 8, 2016, pp. 121–151). Co-located in PRT Meymaneh, cooperation evolved between Norwegian civilian advisors and military commanders; without clear instructions for a comprehensive approach, however, this cooperation was largely dependent on personal relationships (NOU 2016: 8, p. 125). Moreover, as the civil-military divide hindered effective coordination with Norwegian-financed NGOs and the civilian component remained small, Norwegian forces coordinated more closely with other international NGOs operating in the province, as well as local authorities when planning operations (NOU 2016: 8, pp. 130–136).

In terms of strategic thinking, Norway lacked a comprehensive strategy. The joint effort of the MFA, the MoD, and the Ministry of Justice and Public Security to concretise political guidelines came short, and few deployed personnel found useful in practice (NOU 2016: 8, p. 132). Recognizing the need for improved coordination, the government established a cross-departmental forum; however, it mainly functioned as an arena for information sharing rather than discussing how best to integrate different elements (NOU 2016: 8, p. 31). Thus, Norwegian civilian and military personnel had to figure out on their own how to implement a comprehensive approach. As the security situation worsened, the government toned down its state-building ambitions, and the PRT was made more robust with heavier equipment and more personnel following military advice (Fiskvik, 2020, pp. 148–154). As military force commanders were left with developing and running operations in the province, the Norwegian forces adapted to ISAF's counterinsurgency strategy and took initiative by seeking out troubled areas in neighbouring districts of the PRT (Fiskvik, 2020, p. 150).

During the engagement, the civil-military dialogue matured as civilians and military gained a better understanding of the mission's complex and demanding nature (NOU 2016: 8). There is an acknowledgement that current crisis management requires short lines of command to ensure both political control and military efficiency. Moreover, at home, an integrated political and military leadership is deemed important to ensure close and persistent political-military coordination (Fiskvik, 2020, pp. 168–169).

Sweden

Assuming command of PRT Mazar-e Sharif in 2006, Sweden became responsible for four northern provinces that varied in stability and economic growth and marred by power rivalry between war lords (SOU 2017: 16, pp. 69–70). Moreover, the PRT organization, which included both military and civilian dimensions, was unfamiliar to the Swedish forces (Fiskvik, 2020, p. 186). With national caveats that limited the scope of Swedish forces to use lethal force, the Swedes had initially a light military footprint, carrying out patrols and stabilization operations to enable development and reconstruction (Fiskvik, 2020, p. 183).

Although Swedish authorities emphasized the comprehensive approach when assuming PRT lead, the civil-military cooperation on the ground was strained. The Swedish International Development Cooperation Agency did not want to interact with military forces, and the Swedish-led PRT did not administer development resources, which limited the coordination with aid and development activities (SOU 2017: 16, p. 74). The aim of establishing trust among the population through civil-military cooperation activities was challenging as the projects created expectations among the Afghans that the military did not have the resources to meet (Fiskvik, 2020, p. 189). By 2010, Swedish forces began questioning the application of the PRT concept as resources were few and with a poorly represented civilian component.

Although civilian authorities back home refrained from referring to counterinsurgency, Swedish forces were influenced by ISAF's counterinsurgency strategy, and officers were enrolled in counterinsurgency courses in Kabul. Adopting COIN principles that signalled a more active approach to ensure security, the Swedish-led PRT planned and conducted several large operations intended to stop and control insurgent activity (Fiskvik, 2020, pp. 190–191). Thus, as the situation deteriorated, the military took on a more active approach and adapted to more demanding conditions. The PRT was reinforced with heavier equipment, and the Chief of Defence removed Swedish national caveats, expanding the possibility to use force in conflict-related situations (SOU 2017: 16, p. 86).

In terms of strategy for the engagement, there were no clear political aims beyond the decision to participate. At the same time, the generals at the armed forces headquarters did not provide much direction either (Ångström, 2020). Consequently, the Swedish force commanders had considerable discretion in interpreting the political overarching mission and defining missions (Fiskvik, 2020, p. 185). Eventually, in 2010, the government published a strategy for Sweden's overall effort in Afghanistan; in essence, the strategy underlined the ISAF military strategy and the comprehensive approach as the solution to an ever-more complex effort (Swedish Government, 2010). At the same time, the strategy outlined a gradual reduction in the military effort and underlined the necessity of keeping civilian and military roles apart. Aiming to improve civil-military coordination, the government established in 2010 a Senior Civilian Representative for coordinating the Swedish efforts and lead the contact with Afghan authorities (Fiskvik, 2020, p. 189). The position was to be an equivalent to the military force commander and had responsibility for overall political guidance, civilian personnel, and development. In 2012, the PRT changed to a civilian command and evolved into a Transition Support Team, with the aim of transition security responsibility to the Afghans (Fiskvik, 2020, p. 196).

The Afghanistan engagement with PRT leadership changed the way Sweden use its military forces (Fiskvik, 2020). The initially perceived peacekeeping operation turned into a complex mission, in which the military was forced to familiarize themselves with a new political context and way of thinking involving a comprehensive approach. The engagement influenced the Swedish Armed Forces in several ways, including valuable experiences, defence reform, and doctrine development that included a comprehensive approach (Fiskvik, 2020, pp. 197–204).

Are Afghan experiences useful at home?

The Afghan war proved demanding for Nordic politicians and their military servants. As years went by, thousands of officers and soldiers gradually socialized into a highly politicized security environment. It was a complex mission that challenged the use of force and various attempts at aligning civilian and military efforts. Partly, it was challenging on the ground where civilian and military actors had to find new solutions but without clearly defined roles and responsibilities. Partly also at home, where political guidance to military subordinates were hesitant and ambiguous. Without clear objectives, the Nordic forces were left with figuring out how to implement a comprehensive approach and arguably developed a stronger political sensitivity. The civil-military interaction endured for almost two decades. Over the years, the two spheres customized their perceptions and expectations. On that basis, we return to Cohen's theoretical framework: while final decisions rest with civilian leaders, civilians should also engage in a detailed dialogue with their military subordinates; not least when it comes to utilize critical skills, knowledge, and competence. Moving from Afghanistan to a European context where civilian leaders prepare for hybrid threats at home, how may the "unequal dialogue" exploit the military experience?

The Nordic cases visualize a clear indication: in the strategic dialogue on ends, ways, and means, civilian leaders should utilize military skills in holistic planning and rigorous execution. Throughout the Afghan war, Western forces adjusted their military logic to a whole-of-government approach. As military efforts were intimately tied to good governance, reconstruction, and development, the violent execution became tamed and politicized into a broader strategic framework. For decades, thousands of officers and NCOs gained practical skills and knowledge. The quest for cross-sectorial, inter-agency synergy became a planning prerogative. Taking political guidelines into account, such as strict rules of engagement inside tribal communities, contextual sensitivity became a constant priority in planning and execution. Contextual sensitivity became important not least to cope with a fragmented cooperative landscape where local authorities, international aid workers, and civilian advisors were incapable or even reluctant to align their resources.

On the one hand, comprehensive planning-experience may be a critical asset for civilian leaders preparing for hybrid threats. This is because local communities often are scarcely populated, and hence suffering from endemic resource scarcity. By integrating experienced liaison officers, NCOs, officers' advisory groups, or even military veterans with relevant experience into their local crisis management groups, administrative councillors, mayors, and police chiefs are energized by professional expertise. Building on politically sensitive experience from tribal communities, military personnel can assist in comprehensive threat assessments inside civic communities. As municipals seek mitigate threats, military experts can be used to enhance leadership, elaborate complex threat assessments, and produce executive orders. The overall purpose would be to enhance a coherent unity of effort: partly by identifying mutually reinforcing instruments of power in the wider society; partly by organizing critical capabilities in an effectively and orderly manner; and finally, systematically identifying critical vulnerabilities that should be secured for effective response measures (NATO, 2019, pp. 3–5).

The operational benefits are two-fold. First, civil-military integration is customized to a specifically tailored context inside the local community. This is crucial to energize the vibrant link between municipals' strategies and their practical output vis-à-vis local citizens that are strained by fear, anxiety, and apprehension. Second, by integrating military skills, competence, and drills into the civilian chain-of-command, the civilian side may more easily seize the initiative. This is because experienced staff members can streamline, and thereby energize, decision-making processes with rigour and structure. Seizing the initiative is often seen as an important step, partly to reassure citizens but also to prevent further damage from malign actors.

On the other hand, an "unequal dialogue" that invites militaries into a civilian-led operation may also cause friction. Three problems may arise. First, officers with extensive combat experience from abroad may easily be seen as too intrusive or too influential in a civilian-led planning process. Unless personal bonds of mutual trust and confidence are thoroughly cultivated for a longer period, and unless the military experts behave benevolently, a sense of institutional rivalry, personal scepticism, or even professional prejudice may arise amid a crisis. This again may undermine the municipal's "unity of effort" or "unity of command"—in a critical phase where resources need to be properly aligned at the right place, at the right time. As demonstrated by the Nordic cases, if political objectives are not firmly anchored and contextualized into a broader a whole-of-government approach, civil-military friction will arise.

Second, friction may evolve if the civilian chain-of-command is challenged by a numerically superior military support element. Should a local crisis intensify to a level where civilian and military roles become blurred, an integrated civil-military planning process may erode or even collapse. This is because mayors or police chiefs may feel military professionals exceed their supporting role by encroaching more authority than admissible or necessary. This follows the military's resourceful organisation: Equipped with heavy material, such as dedicated transport companies, engineers, signal units, field hospitals, military police, or logistics, the sheer size and volume of skilled and experienced manpower may easily outnumber the civilian planning and execution capacity. As noted in the Nordic cases, all four states eventually ended with a clear civil-military divide. Simultaneously, the civilian side increasingly recognized the need for pragmatism and flexibility to address threats that cut across the civil-military interface. Nurturing a healthy civil-military dialogue arguably requires close coordination and training as to familiarize different elements across diverging cultures.

Finally, as militaries are incorporated into the "unequal dialogue", civilians should know that military experience is not necessarily compatible to any European context. As staff officers and NCOs are intimately tied to civilian headquarters, the execution of response measures may become unnecessary proactive. As pointed out in the Nordic cases, the military gravitated towards traditional combat tasks due to increased resistance from Taliban. Seizing the initiative to gain momentum, and hence influence the opponent's calculus, became an operational necessity. In a European crisis management context, a too proactive posture with armed Gendarmerie or Home Guard soldiers may unintentionally stir local anxiety rather than reassurance. Military support may as such undermine broader political objectives.

Conclusion

As small and vulnerable municipals prepare for hybrid threats, this chapter discussed how military experiences from abroad could be utilized at home. To conclude, two empirical and two theoretical findings are deduced.

First, the civilian side could be more forward leaning and take braver steps to exploit the military experiences situated inside their local community. If civilian crisis-management units are invigorated with relevant military expertise, civilian authorities may gradually develop a common frame of reference; an operational approach where local civilian leaders can balance ends and means more effectively, determine viable courses of actions, and thus orchestrate and direct the local community's instruments of power in an even more coherent manner. A poorly applied operational approach may easily lead to the opposite: A deteriorating relationship between citizens and the state because communal response measures are deemed ineffective and inadequate to meet inhabitants' expectations and needs in a dire situation. Such events may furthermore undermine legitimacy, trust and confidence.

Second, this is not the same as claiming that military expertise and personnel should dominate the civilian chain-of-command, or that civilian side should cede roles and responsibilities over to the military. Mayors, police chiefs, and regional councillors are still the superiors. But as scarcely resourced districts prepare for hybrid threats, pragmatism and local adjustments must prevail over institutional traditions originating from a Cold War era where civil-military boundaries were clearer.

Third, there is no clearcut division between civilian and military roles and responsibilities. If military resources are exploited more systematically by civilians, the two spheres will gradually coincide. This is consistent with Churchill's perennial dictum from Great Britain's existential war in 1941, claiming that "policy and strategy becomes one". This conclusion validates Cohen's theoretical proposition, where civilians must be more engaged in an operative sphere that traditionally has been dominated by militaries.

Fourth, while Cohen focuses on major wars and civil-military relationships at the national level, hybrid threats are more likely to materialize where critical vulnerabilities are most prominent. This is inside small, transparent, and scarcely resourced municipals. Accordingly, an interesting future research avenue is to examine how a closer integration in terms of organization and governance can be facilitated in a hybrid threats context. In this vein, Cohens' theory may be developed further to encapsulate regional and local levels in a state's chain-of-command.

References

Andersen, S. B., Vistisen, N. K., & Schøning, A. S. (2016). *Afghanistan. Erfaringsopsamling 2001–2014 DEL III: Danske erfaringer med stabiliseringsprosjekter og CIMIC [Afghanistan. Collection of experiences 2001–2014 PART III: Danish experiences with stabilization projects and CIMIC].* Forsvarsakademiet. https://www.diis.dk/files/media/publications/afghanistan-lessons/reports/dk_afghan_erfaring_del_iii-final-enkeltsidet.pdf.

Ångström, J. (2020). Contribution warfare: Sweden's lessons from the war in Afghanistan. *The US Army War College Quarterly: Parameters, 50*(4). https://doi.org/10.55540/0031-1723.2688.

Brooks, R. A. (2019). Integrating the civil–military relations subfield. *Annual Review of Political Science*, 22(1), 379–398. https://doi.org/10.1146/annurev-polisci-060518-025407.

Christensen, T., Lægreid, P., & Røvik, K. A. (2020). *Organization theory and the public sector: Instrument, culture and myth*. Routledge.

Cohen, E. A. (2012). *Supreme command: Soldiers, statesmen and leadership in wartime*. Simon and Schuster.

Danish MFA, & Danish MoD. (2008). *Den danske indsats i Afghanistan 2008–2012 [The Danish effort in Afghanistan 2008–2012]*. https://www.fmn.dk/globalassets/fmn/dokumenter/strategi/afghanistan/-dansk_indsats_afghanistan-2008-2012-.pdf.

Danish MFA, Danish MoD, & Danish MoJ. (2013). *Denmark's integrated stabilisation engagement in fragile and conflict-affected areas of the world*. https://amg.um.dk/-/media/country-sites/amg-en/policies-and-strategies/stability-and-protection/stabiliseringspolitik_uk_web.ashx

Feaver, P. D. (1996). The civil-military problematique: Huntington, Janowitz, and the question of civilian control. *Armed Forces & Society*, 23(2), 149–178. https://doi.org/10.1177/0095327X9602300203.

Fiskvik, J. (2020). *Expeditionary warfare and changing patterns of civil-military relations. The politics of war in the Nordic countries, 2001–2014* [PhD thesis]. Norwegian University of Science and Technology.

Huntington, S. P. (1957). *The soldier and the state: The theory and politics of civil–military relations*. Harvard University Press.

Mehta, A. (2022, May 20). Russia's military is now a 'wounded bear.' Can it revive itself? *Breaking Defense*. https://breakingdefense.sites.breakingmedia.com/2022/05/russias-military-is-now-a-wounded-bear-can-it-revive-itself/.

Mustasilta, K., Karjalainen, T., Stewart, T. R., & Salo, M. (2022). *Finland in Afghanistan 2001–2021* (72; FIIA Report). https://www.fiia.fi/wp-content/uploads/2022/12/report72_finland-in-afghanistan-2001-2021.pdf.

NATO. (2010). Provincial reconstruction teams look at way forward in Afghanistan. *News*. https://www.nato.int/cps/en/natolive/news_62256.htm?selectedLocale=en.

NATO. (2019). *Allied joint doctrine for the planning of operations (AJP-5)*. NATO Standardization Office. https://www.coemed.org/files/stanags/01_AJP/AJP-5_EDA_V2_E_2526.pdf.

Nicholls, D., Daly, E., & England, J. (2023, May 26). Russian army needs a decade to rebuild—and NATO can take advantage. *The Telegraph*. https://www.telegraph.co.uk/world-news/2023/05/26/russia-ukraine-nato-army-rebuild-decade-watch-defence/.

NOU 2016: 8. (2016). *A good ally: Norway in Afghanistan 2001–2014*. Norwegian MFA & MoD. https://www.regjeringen.no/contentassets/09faceca099c4b8bac85ca8495e12d2d/en-gb/pdfs/nou201620160008000engpdfs.pdf.

NOU 2023: 17. (2023). *Nå er det alvor—Rustet for en usikker fremtid [Now it is serious—Prepared for an uncertain future]*. Ministry of Justice and Public Security. https://www.regjeringen.no/no/dokumenter/nou-2023-17/id2982767/.

Rynning, S. (2019). Denmark's lessons. *The US Army War College Quarterly: Parameters*, 49(4). https://doi.org/10.55540/0031-1723.3117.

Saikal, A. (2012). *Modern Afghanistan*. Bloomsbury Publishing. https://www.bloomsbury.com/uk/modern-afghanistan-9781780761220/.

Simpson, E. (2012). *War from the ground up: Twenty-first century combat as politics* (1st edition). Oxford University Press.

Smith, R. (2007). *The utility of force: The art of war in the modern world*. Alfred A. Knopf.

SOU 2017: 16. (2017). *Sveriges samlade engagemang i Afghanistan under perioden 2002–2014 [Sweden's overall engagement in Afghanistan during the period 2002–2014].* Regeringen och Regeringskansliet. https://www.regeringen.se/contentassets/257a87e121a14684b4fb7e4488131827/sveriges-samlade-engagemang-i-afghanistan-under-perioden-20022014-sou-2017.16.pdf.

Suhrke, A. (2011). *When more is less: The international project in Afghanistan.* Hurst & Company.

Swedish Government. (2010). *Strategi för Sveriges stöd till det internationella engagemanget i Afghanistan [Strategy for Sweden's support to the international engagement in Afghanistan].* https://www.regeringen.se/contentassets/6284170ece4f493cad8960d2369bbcf6/strategi-for-sveriges-stod-till-det-internationella-engagemanget-i-afghanistan.

10

HYBRID THREATS AND THE "NEW" TOTAL DEFENCE

The case of Sweden

Joakim Berndtsson

Introduction

To meet the challenges of complex, "non-traditional" threats and "trans-boundary" crises (e.g. Ansell et al., 2010; Boin, 2019), political leaders, national defence, and crisis management organisations are seeking ways to strengthen capabilities, preparedness, and resilience. While strategies vary, they often emphasise intergroup and interorganisational integration and collaboration across public-private, civilian-military, or national-international boundaries (see also Chapter 11 by Karlsson and Sandbakken). Concepts such as the "whole force" (UK), the "total force" (the United States), or the "defence team" (Canada) capture some aspects of this move towards collaborative and integrated defence arrangements (Goldenberg et al., 2019; Louth and Taylor, 2018; Berndtsson et al., 2023). In other contexts, notably in the Nordic and Baltic Sea regions, the perceived transformation and escalation of threats have seen a revival of national defence concepts such as "total defence" or, as in the case of Finland, "comprehensive security" (e.g. Valtonen and Branders, 2021, see also Chapter 13 by Kasearu, Truusa and Tooding).

The concept of total defence has been described as a "whole of society approach to national security intended to deter a potential enemy by raising the cost of aggression and lowering the chances of its success" (Wither, 2020, p. 62). In the past, total defence arrangements were primarily designed to meet the challenges of "total war". Today, these and similar concepts are being re-invented and adapted to address threats and crises across multiple sectors and levels of society, and across the peace-war spectrum. In the process, governments, militaries, and crisis management organisations face the

DOI: 10.4324/9781032617916-13

challenge of explaining the meaning and nature of hybrid or transboundary threats, as well as to organise robust and integrated whole-of-society and whole-of-government responses. This chapter focuses on the case of Sweden, and in particular the framing of emerging threats to mobilise support for, and participation in, the "new" total defence (see also Chapter 13 by Kasearu, Truusa and Tooding).

Swedish total defence can be traced back at least to the 1940s (Larsson, S., 2021; Angstrom and Ljungkvist, 2023). Thus, the concept is familiar terrain for many Swedes. What is currently emerging, however, is quite different; society has undergone substantial changes in the last decades, as have global and regional security landscapes. At the same time, definitions and interpretations of threats have changed, and concepts such as "hybrid threats" or "non-linear" warfare have gained considerable traction in political and military discourse. While the meanings of such concepts remain elusive (Rinelli and Duyvesteyn, 2018, pp. 19–23; Chapter 1 by Borch and Heier; Chapter 2 by Bjørge and Høiby), security policies in Sweden and elsewhere are characterised by "entangled" security logics (e.g. territorial and societal) that shape both understandings of, and responses to, perceived threats across the peace-war spectrum (e.g., Wrange, 2022). Broadening definitions of threats and security has meant increasing demands for capabilities and readiness but also a "securitization" of a growing number of issues (Larsson, O. 2021; Stiglund, 2021). Still, the fundamental idea of total defence remains; the concept requires active participation, planning, and preparation by nearly all parts of society—from the armed forces and crisis management institutions to municipalities, private businesses, and individuals. Thus, state actors face the dual challenge of explaining and communicating the nature and urgency of potential threats and the need for preparedness to an array of audiences, and to construct the "new" total defence as a credible response.

The aim of this chapter is to contribute to our understanding of how threat perceptions evolve and how they are mobilised to explain and legitimise decisions on defence organisation. To achieve this, the chapter asks: *how have understandings of hybrid threats by key political actors in Sweden developed in the last decade, and how have they been communicated in the process to re-invent total defence?* Through a constructivist lens, the analysis focuses on the framing (meaning making) of threats to national security by political actors, using ideas about hybrid threats and transboundary crises as a backdrop. Empirically, the analysis centres on the case of Sweden and, more specifically, on the framing of threats by political actors and in particular the Swedish Defence Commission (SDC). In the process, notions of societal and national security, service, duty, preparedness, and readiness—all central ideas for total defence—are bestowed with meaning. The ambition here is not to explain specific decisions, nor to investigate the level of political or popular support for specific policies, but rather to identify key shifts in the construction of meaning of threats and responses. Doing so will pave the way for a better understanding of ideas and issues that underpin the ongoing re-invention of total defence in Sweden and elsewhere.

Theory

Frame analyses have frequently been used to analyse threat perception and policy framing in the context of defence and security (e.g. Mörth, 2000; Desrosiers, 2012; Björnehed and Erikson, 2018; Wikman, 2021; Coetzee et al., 2023). Frame theory—originally associated with the work of sociologist Erving Goffman (1974)—commonly adopt a constructivist approach to social phenomena. Analyses frequently centre on sense-making processes in communication by different actors and how issues and events are linked to metaphors, symbols, norms, and narratives to create meaning or to mobilise support (Björnehed and Erikson, 2018). Framing is a process whereby aspects of a perceived reality are selected and made meaningful or salient in communication to promote a particular problem definition, causal interpretation, moral assessment, or solution (e.g. Entman, 1993). Frames—as expressed in speech, text, or images—thus seek to impose a certain logic or interpretation on an audience and define the terms of understanding an issue. Frames may shape public opinion and influence policymaking, and they may legitimate certain decisions and actions (Mörth, 2000, pp. 173–174). Framers commonly seek to align the public's pre-existing expectations, ideas, and perceptions with their views and goals (Desrosiers, 2012, p. 5).

From previous research on crisis management, we know that communication is a key challenge for authorities facing an ongoing or imminent threat. As Ansell et al. (2010) observe, public leaders confronted with a (transboundary) crisis need to be able to communicate an accurate and reliable account of events (meaning making) and their strategies for dealing with a situation (also Boin, 2019). The same line of thinking can be applied to the work needed to prepare and organise for *potential* crises or threats; here too, public leaders need to explain the nature and meaning of threats to various audiences, including political allies and opponents, civil and military defence organisations, and the public. Thus, state actors need to create an authoritative and credible narrative (or frame) where the hybrid nature of threats to society—as well as their remedies—is clearly explained. In Sweden, the main response is the re-invention of total defence.

Total defence is a whole-of-society *and* whole-of-government approach to national security. As such, it is a way to prepare and organise for potential threats and crisis that are essentially understood transboundary in nature. Total defence is not a formal organisation but involves myriad actors on multiple levels and across societal sectors. We might think of it as a complex "security network" (Whelan, 2017), organised, operated, and governed on multiple levels and through both military and civilian centres of authority, and shaped by different "strategic logics" (Angstrom and Ljungkvist, 2023). Because of the sheer number of actors and interests involved, establishing a common understanding of threats is vital. Only by (successfully) convincing different audiences of the meaning and gravity of potential threats can enough political, institutional, and popular support be generated to support decisions on, and engagement in, total defence and (war) preparedness. Thus, investigating *how* threats are framed to mobilise support for the ongoing re-invention of total defence emerges as an important undertaking.

Data and analytical approach

The Swedish Defence Commission (SDC) is described as a "forum for consultations between the Government and representatives of the political parties represented in Parliament" (Swedish Defence Commission website, April 20, 2023). The body comprises members of the parties in Parliament, representatives and experts from several ministries and the Prime Minister's Office, the Swedish Armed Forces, the Swedish Defence Materiel Administration, and the Swedish Civil Contingencies Agency, MSB. The stated ambition of the SDC is "to reach a broad consensus on Sweden's defence and security policy", and the government frequently refers to SDC reports and analyses when presenting proposals to Parliament (ibid.). The Commission is also tasked with contributing to public debate on security issues, and their analyses and findings are frequently discussed in the media. Taken together, the SDC is an influential body when it comes to constructing meaning around security issues in Sweden.

The SDC has released several key reports where key challenges and threats—as well as necessary responses—are identified and discussed in some detail. For the present study, five reports are of particular interest. First, the reports from 2013 and 2014 are relevant as they signal a shift in threat perception, fuelled primarily by the Russian annexation of Crimea in early 2014. Next, the reports from 2017 and 2019 are important as they deal with the need to adapt the total defence concept to tackle both conventional and emerging threats to Swedish security. Finally, the report from June 2023 is important because it covers several issues and events from the past few years, notably the war in Ukraine and Sweden's application to join NATO. The reports follow a similar structure, with one or more sections devoted to national, regional, and global security, as well as the state of Swedish security policy and readiness.

In addition to key sections of the reports, some key government decisions and actions are included to provide context on total defence development since 2015. The analysis below is presented chronologically, starting with the shift from 2013 to 2014 and moving forwards. The focus is on identifying dominant ideas or frames in threat and security perceptions and how these are linked to the need to re-establish and develop total defence capabilities.

In practice, this means systematically scanning the texts for clusters of salient messages, key themes, and recurring tropes that help illustrate how communicators or framing actors (the SDC) define and describe problems (threats), their causes, and their remedies or solutions (see, e.g. Entman, 1993). A simple search for key words and phrases will serve to further illuminate changes and dominant ideas, while the closer reading of key passages moves us towards a more detailed understanding of framing processes.

Framing threats and the "new" total defence

Since the beginning of the Cold War, Swedish defence and security policies and organisation have been closely tied to the country's position as neutral and, later, militarily non-aligned (Andrén, 1972; Kronvall and Petersson, 2012). Originally, total defence was built around the idea of war-time neutrality and self-sufficiency, and included civilian organisations, military defence, and the active participation of the civilian population (e.g., Westberg, 2021, pp. 206–208; Larsson, O., 2021). From the early 1990s, however, Sweden followed a wider European pattern of military downsizing, specialisation, and transnationalisation (King, 2011). The policy aimed at war-time neutrality was abandoned, but the country remained militarily non-aligned. The mission of the Swedish Armed Forces (SAF) shifted towards international deployments and expeditionary capabilities, and many of the previous national and total defence plans and structures were dismantled (Angstrom and Ljungkvist, 2023, pp. 16–18). In this period, Sweden became an active and increasingly integrated NATO partner country (Ydén et al., 2019; Petersson, 2018).

Recently, however, there has been a marked territorial (re-)turn, with a renewed focus on national defence. Following the annexation of Crimea in 2014, the Swedish Government decided in 2015 to resume its coordinated total defence planning (Swedish Government Decision, December 10, 2015). In addition, in May 2022, and following Russia's invasion of Ukraine in February the same year, Sweden applied to become a NATO member.[1] Thus, the country is currently seeking to re-invent the total defence concept for a new time while at the same time adjusting both military and civil defence to conform to NATO standards, norms, and requirements. How have these shifts in regional and international security been framed by the SDC, and how have such framings helped pave the way for substantial shifts in Swedish security thinking and organising?

2013–2014: globalisation and the quest for societal security

In the 2013 report, there are few signs of direct, conventional military threats to Sweden. Instead, the main point of departure is the double-edged sword of globalisation and attendant developments in areas such as technology, finance, science, and security politics (SDC, 2013, pp. 15–16). Globalisation is, the Commission observes, both a source of development and a potential cause of disruption and escalating threats:

> Because the interaction between countries increases, there is also a more pronounced spread of social change and their effects. Taken together, globalisation makes it more difficult to predict developments. . . . Interdependencies and factors such as technological developments means that interruptions to energy supplies, the financial system or world trade can have considerable and escalating effects.
>
> *(SDC, 2013, p. 21)*[2]

From this perspective, rapid and often unpredictable changes brought about by globalisation are a key concern. These threats are also connected to issues of climate change and the control and supply of natural resources (SDC, 2013, pp. 24–29). In addition, international terrorism, the spread of (mainly internal) armed conflicts, developments in the cyber domain, and issues surrounding nuclear, biological, and chemical weapons are outlined and discussed (ibid., pp. 30–44). While some of these threats are well-defined in terms of their sources, most are described as diffuse, complex, and difficult to predict.

Discussing consequences of this development for Swedish defence and security policy, the Commission observes:

> The preconditions for Sweden's defence and security policy have changed fundamentally since the end of the Cold War. Globalisation has created and will continue to create new possibilities and challenges. The global financial centre of gravity has pivoted towards Asia. The area of security policy changes due to technological and demographic developments, increased demand for strategic resources, dependence and development of flows, the consequences of migration, urbanisation and climate change.
>
> *(SDC, 2013, p. 213)*

Again, globalisation is the central frame through which most other security issues are understood. The Commission promotes a wide understanding of security and observes that responses to potential threats need to be similarly wide-ranging and coordinated across sectors and levels of Swedish and international society (SDC, 2013, p. 215). In terms of military defence, the Commission concludes that "a military attack directly aimed at Sweden remains unlikely for the foreseeable future" (SDC, 2013, p. 221). Instead, the Commission favours continued Swedish participation in UN-led operations and increasing defence cooperation in the Nordics and within EU frameworks (ibid., pp. 227–228). While potential military threats in the region are discussed—notably in relation to the conflict in Georgia in 2008—such threats are not (yet) at the forefront.

In 2014, the pendulum begins to swing back towards territorial defence. As before, the SDC uses globalisation as a point of entry for the strategic outlook, but new geopolitical realities and potential military threats are placed centre stage. The 2014 report is a direct consequence of events in the region; in the instructions to the Commission, the Minister for Defence specifically calls for an update of the previous analysis in view of the Russian annexation of Crimea (ibid., preamble). While the broad conception of security and potential sources of threats remains, more focus is placed on Swedish societal security and military defence. In addition, the Commission underlines, security is a concern for all of society:

> To strengthen societal security is a duty for several actors; individuals, companies, organisations, municipalities, county administrative boards, government and parliament. We are collectively responsible for our security.
>
> *(SDC, 2014, p. 11)*

While similar formulations exist in the 2013 report, the explicit reference to individual and collective responsibilities in the first few paragraphs signals a shift in SDC framing. The dominant frame of globalisation is supplemented with a more "territorial" view of Swedish security. This is evident in the wording on the risk of military attack: the SDC maintains that an attack remains unlikely, but the phrase "for the foreseeable future" has been taken out (SDC, 2014, p. 35). Simultaneously, diffuse military threats have largely been overshadowed by a clearly identified source of increased tension: Russia.

When it comes to remedies, the report mentions total defence (13 times compared to 1 in the 2013 report), but mainly with reference to existing legislation and not as a cornerstone of societal and national security. In terms of the military, the Commission emphasises the need to increase war-fighting capabilities and to improve recruitment and retention (SDC, 2014, Ch. 8–9). Further, the Commission notes the need to deepen bi- and multilateral defence cooperation and recommends that the government should sign the Host Nation Support Agreement with NATO (ibid., pp. 37–38).[3] Discussing civil defence, the report underscores the need to identify strategic civilian assets and targets and to increase readiness as well as support capabilities and coordination between authorities and organisations (SDC, 2014, pp. 104–106). In addition, the SDC identifies a need for information and education about crisis management and defence structures in Sweden. In this context, the report also identifies a need to develop society's resilience against information and influence campaigns (SDC, 2014, p. 107).

While many parts of the central narrative remain in place in both reports, there are changes in terms of emphasis and tone when it comes to the nature and meaning of threats. These changes are not surprising, but they do provide some insight into the framing process, whereby national and territorial defence regain prominence for the first time in several decades, and where the need to engage more parts of society in defence and crisis management is placed at the forefront. Ultimately, this reframing helped pave the way for several subsequent changes, such as the 2015 decision to resume total defence planning and organising (Swedish Government Decision, December 10, 2015).

2017–2019: War preparedness and total defence

In the two following SDC reports, the emphasis on territorial and national defence becomes even more apparent, and total defence planning and organising is placed centre stage. In the 2017 report (titled "Resilience"), the understanding of threats as hybrid is clearly conveyed:

> The threats to our security can be antagonistic as well as non-antagonistic. Typical antagonistic threats to our security can be war, influence campaigns, and information warfare, network attacks in the cyber domain, terrorism, and the use of weapons of mass destruction. In addition, there are several challenges and threats to our security that may be both antagonistic and non-antagonistic, such as organised crime, disruptions vital systems and flows, state failures, financial crises, political and religious extremism, threats to our democracy and legal system, threats to human freedoms and rights, social exclusion, migration flows, threats to our values, climate change, natural disasters, and pandemics.
>
> *(SDC, 2017, p. 15)*

In the face of this wide range of threats, the SDC reiterates that "security work is a task for the whole of society" (SDC, 2017, p. 16). Globalisation still forms an important backdrop to the Commission's understanding of the wider security landscape and potential sources of threats, but much more emphasis is placed on war preparedness, on the idea of rebuilding total defence capabilities, and on the need to increase awareness and knowledge among the Swedish population:

> The lack of planning for a heightened state of alert and war has meant that the public's knowledge of total defence is very limited today. The favourable security situation that has dominated in Europe and the western world has also meant that ideas of heightened states of alert and war appear foreign to many people. Therefore, awareness and knowledge about how

society can be affected by and armed attack and how one should act if war comes needs to be improved. This is an important issue because total defence capabilities to a large extent depends on the population's willingness to defend [the country].

(SDC, 2017, p. 45)

Significantly, the 2017 report has also dropped the word "unlikely" from its assessment of the risk of war, concluding instead that "an armed attack against Sweden cannot be ruled out" (SDC, 2017, p. 61). Further, the SDC observes how the nature of war and warfare has changed and points to antagonistic tactics used in the "grey zone" between war and peace (ibid., p. 65). The "grey zone" concept was not used in the previous two reports, but here it serves both to define the changing nature of threats and to underscore the need for (war) preparedness across multiple sectors (ibid., pp. 65–69, 75–77).

Following this framing of potential threats, total defence appears as the key remedy. The report delves into several aspects of civil defence, including organisation and governance, cyber security, shelters, law and order, supply chain security, and the roles of volunteer organisations and businesses (SDC, 2017). Additionally, the idea of "psychological defence" re-emerges. The concept is defined as consisting of three parts: ensuring defence willingness among the population; maintaining the ability to communicate official information; and identifying and countering influence campaigns (ibid., p. 106). Discussing the importance of psychological defence, the SDC continues:

Even in peacetime, Sweden must have such individual and societal readiness and capability in psychological defence that the shock of being threatened or being attacked does not passivate decision makers or the population. . . . [L]eading public representatives must have such insight and ability as to be able to engage and communicate the importance of total defence in their communication with the public.

(SDC, 2017, p. 109)

According to the SDC, responsibility rests not only with public officials but also with individuals. This adds to previous statements about the importance of "defence willingness" to meet increasingly complex threats, as well as the (legal and moral) duty to participate in total defence activities. Threat perceptions and framings not only inform the perceived need to rethink the roles and mandates of public institutions, businesses, and civil society organisations, but also emphasise the importance of the public. Even though the population is not the primary audience of the SDC, the Commission's status buttresses the framing of both problems and solutions, thus paving the way for political discussion and decisions.

As we move to 2019, the focus shifts to Swedish military defence. The framing of the nature of threats and the risk of war (conventional and hybrid) largely follows the same pattern as in 2017. However, the SDC also observes a "return" of geopolitics in international politics and places even more emphasis on the potential threats posed by Russia ("Russia" or "Russian" are used 384 times in the 2019 report, compared to 98 times in 2017). In terms of Swedish security, building a credible total defence is again described as a key part of the solution: "By making clear that an attack against Sweden comes at a high cost, total defence, together with political, diplomatic, and economic means, is a war deterrent" (SDC, 2019, p. 107). In addition, total defence organising is also part of the wider effort towards building societal security:

> Assessments of what the total defence should be able to do . . . also includes an important psychological component. To plan for a serious situation strengthens total defence credibility internally vis-à-vis the population and externally, internationally, vis-à-vis external actors. . . . Defence willing-ness and public support are the basis of building a credible defence. People and decision makers must be made aware of the demands of war. *Crisis awareness* needs to be complemented with *war awareness.*
>
> (SDC, 2019, p. 108, emphasis added)

The trope "war awareness" is key here; it helps underline the need to involve the wider public in total defence and war preparedness work and serves as a "reality check" for both political leaders and individuals.

Together, the 2017 and 2019 reports show an increasing emphasis on the risk of war. From this vantage point, total defence is framed as a vital part of Sweden's response to both "conventional" threats *and* a complex set of "hybrid" threats across the peace-war spectrum (see also Angstrom and Ljungkvist, 2023, pp. 18–20). Improving total defence capabilities, the reports conclude, strengthens peacetime crisis management capabilities and readiness. In addition, the need to re-establish links between (total) defence and the population is emphasised. As we move to the final report included in this analysis, several events and decisions potentially influence the threat perceptions of the SDC, including the Covid-19 pandemic, the invasion of Ukraine in 2022, and the Swedish Government's decision in May 2022 to apply for NATO membership.

2023: The "reality of war" and the road to NATO

In June 2023, the SDC submitted their report titled "Serious Times" to the Minister for Defence. The introductory chapter deals with Swedish security policy, and unsurprisingly, the point of departure is the ongoing war in Ukraine:

> Because of Russia, a large-scale war rages in Europe. The effects of war are not limited to Ukraine. Therefore, the preconditions for Swedish security policy have changed fundamentally. . . . Russia's full-scale invasion of Ukraine, along with China's escalating territorial claims, demonstrate that conflicts over territory [fought with] military means are, again, a reality. As a consequence, Swedish security policy is currently undergoing the greatest changes in modern times.
>
> *(SDC, 2023, p. 11)*

Russia is clearly identified as the main source of the threat of war ("Russia" or "Russian" is mentioned 784 times; "China" 397 times). It is also apparent that this new reality demands dramatic changes. Even though the threat of war has been placed centre stage, this is not the only challenge to Swedish security. The SDC also observes that "[t]he antagonistic threat against Sweden is broad and is becoming ever more complex" (SDC, 2023, p. 11). Threats include information manipulation, disinformation, influence campaigns, cyber attacks, illegal intelligence gathering, terrorism and sabotage, threats to critical infrastructure, and exploitation of economic dependencies (ibid.). Thus, the war frame is accompanied by a wide range of hybrid and transboundary threats to Sweden and international security. This ultimately leads the SDC to conclude: "Sweden is best defended within NATO. Sweden's NATO membership increases security for both Sweden and NATO" (ibid.).

After the invasion of Ukraine, the "war awareness" trope of 2019 is placed centre stage to explain fundamental changes to Swedish security policy and practice. At the same time, other, equally transboundary and arguably serious challenges are pushed to the background. The 2023 report does touch upon issues of climate change, pandemics, and natural disasters, but these are clearly not seen as equally critical—at least not at present. In their discussion of overarching security policy assessments, the SDC conclude:

> The Defence Commission believe that Sweden's total defence must be developed to be able to effectively meet a military attack. Historically in Sweden and internationally, it has proven impossible to restore acceptable total defence capabilities only when a crisis or war strikes. The Defence Commission want to stress that it is not the most likely turn of events that is most important when dimensioning the total defence, but the events that would have severe consequences should they occur.
>
> *(SDC, 2023, p. 23)*

From this perspective, speeding up total defence development is a key priority. Although international defence cooperation and collective security have been discussed at length in previous reports, Sweden's NATO membership represents a break with long-standing traditions of neutrality and non-alignment.

In the report, the SDC explicitly connects total defence to Sweden's role as a NATO member:

> Sweden's contributions to NATO will increase the collective defence of the Alliance. Sweden should be able to withstand an armed attack and defend its own territory within the framework of collective defence. Thus, the Commission asserts that Sweden has a responsibility to continue to strengthen its national defence, including societal resilience, through a cohesive total defence.
>
> *(SDC, 2023, p. 199)*

Connecting the need to reinforce total defence to the country's credibility and responsibilities vis-à-vis the Alliance lends further weight (and urgency) to established narratives about the need for "fundamental changes" to meet a growing set of conventional and hybrid threats.

In sum, the analysis of SDC framings from 2013 and onwards reveal an increasing focus on both "traditional" and "new" threats, and the construction of total defence as a key response. This is in line with previous findings on the recent development of Swedish total defence strategy as based on both a "narrow" and "broad" conception of security (Angstrom and Ljungkvist, 2023). From this perspective, the SDC gradually outlines a vision of total defence as capable of dealing with threats across the peace-war spectrum. This not only requires the strengthening of military capabilities or "war awareness" among the public but also demands total defence engagement from the whole of Swedish society, where "every public agency, private business and even every single individual is ready to resist and defend against hybrid attacks" (ibid., 2023, p. 20). Whether the ongoing re-invention of Swedish total defence will deliver on its promise, and to what extent various audiences (including the public) fully accept framings of responsibilities and duties, remains to be seen.

Conclusion

The aim of this chapter has been to contribute to our understanding of *how* perceptions and understandings of threats develop, and how they are communicated in the context of (total) defence organising and planning. While the analysis is limited in terms of empirical scope, some interesting conclusions can be drawn about changes in Swedish security thinking and practice in a period shaped by ideas of "hybrid" and "transboundary" threats.

First, the analysis shows a gradual shift towards territorial, national, and societal security and defence, and an attendant re-emergence of war as a dominant narrative. Over time, it is also clear that Russia is seen as the main source of such threats. As the assessment of the risk of war shifts from "unlikely in the foreseeable future" to "unlikely" to "cannot be ruled out", so the narrative gradually moves towards war preparedness and total defence. Second, as "war awareness" emerges as a priority, understandings of both conventional and hybrid threats grow more complex. The hybrid nature of external, antagonistic threats necessitates responses that involve not only the military but also wider society. After the decision in 2015 to resume total defence planning and organising, increasing emphasis is placed on the need for popular (and psychological) resilience, as well as the need for a widespread willingness to defend the country. In the process, individuals' duties and responsibilities are explicitly linked to the country's readiness and its ability to meet a wide range of perceived threats. Finally, as the "war frame" grows more prominent, other threats (e.g. climate change) become less visible. Facing the threat of (hybrid and conventional) war, previous defence policies and modes of cooperation are no longer seen as sufficient. Total defence, the SDC argue, needs to be coupled with a Swedish NATO membership. Thus, the responsibility to continue the development of civil and military defence capabilities is linked to the hybrid nature or increasing complexity of threats, as well as to the future role of Sweden as a credible NATO member.

Framing processes provide insight into the ways in which hybrid threats are bestowed with meaning by different actors across time and space, and how they are linked to social and historical ideas about national and international security. The ways in which key political actors frame threats, their sources, and their remedies help us understand political decisions on defence policy and organisation. In addition, identifying key elements and ideas in these processes also helps us understand how they evolve, how they are communicated to different audiences, and how they can serve different purposes. While the empirical focus of this chapter has been on the Swedish process of re-invention of total defence, future research can further increase our understanding of framing processes through comparative analyses. Expanding the empirical focus and developing theoretical frameworks will further our knowledge of perceptions, communication, and meaning making in (total) defence and security policy processes, both within and across national settings.

Funding

Work for this chapter has been supported by the Total Defence for the 21st-century project at the School of Global Studies (Swedish Research

Council, project no. VR2021–06292) and the Modern Military Profession Project at the Swedish Centre Studies of Armed Forces and Society (CSMS) and the Institute for Management of Innovation and Technology (Swedish Armed Forces Technical Development Grant no. AT.9225360).

Notes

1 Sweden became a NATO member on March 7, 2024.
2 Original in Swedish; all translations by the author unless otherwise stated.
3 The 2014 SDC report was delivered on May 15, 2014; the Host Nation Support agreement was signed in September the same year and entered into force on July 1, 2016.

References

Andrén, N. (1972). Sweden's Security Policy. *Cooperation and Conflict*, 7(1), 127–153.
Angstrom, J., & Ljungkvist, K. (2023). Unpacking the Varying Strategic Logics of Total Defence. *Journal of Strategic Studies*, 1–25.
Ansell, C., Boin, A., & Keller, A. (2010). Managing Transboundary Crises: Identifying the Building Blocks of an Effective Response System. *Journal of Contingencies and Crisis Management*, 18, 195–207.
Berndtsson, J., Goldenberg, I., & von Hlatky, S. (2023). *Total Defence Forces in the 21st Century*. Montreal: McGill Queens University Press.
Björnehed, E., & Erikson, J. (2018). Making the Most of the Frame: Developing the Analytical Potential of Frame Analysis. *Policy Studies*, 39(2), 109–126.
Boin, A. (2019, Summer). The Transboundary Crisis: Why We Are Unprepared and the Road Ahead. *Journal of Contingencies and Crisis Management*, 27(1), 94–99.
Coetzee, W., Larsson, S., & Berndtsson, J. (2023). Branding 'Progressive' Security: The Case of Sweden. *Cooperation and Conflict*, 59(1), 86–106.
Desrosiers, M. (2012). Reframing Frame Analysis: Key Contributions to Conflict Studies. *Ethnopolitics*, 11(1), 1–23.
Entman, R. (1993). Framing: Toward Clarification of a Fractured Paradigm. *Journal of Communication*, 43(4), 51–58.
Goffman, E. (1974). *Frame Analysis: An Essay on the Organization of Experience*. Cambridge, MA, USA: Harvard University Press.
Goldenberg, I., Andres, M., Österberg, J., James-Yates, S., Johansson, E., & Pearce, S. (2019). Integrated Defence Workforces: Challenges and Enablers of Military–Civilian Personnel Collaboration. *Journal of Military Studies*, 8, 28–45.
King, A. (2011). *The Transformation of Europe's Armed Forces: From the Rhine to Afghanistan*. Cambridge and New York: Cambridge University Press.
Kronvall, O., & Petersson, M. (2012). *Svensk säkerhetspolitik i supermakternas skugga 1945–1991 [Swedish Security Policy in the Shadow of the Superpowers 1945–1991]* (2nd edition). Stockholm, Sweden: Santérus Academic Press.
Larsson, O. L. (2021). The Connections between Crisis and War Preparedness in Sweden. *Security Dialogue*, 52(4), 306–24.
Larsson, S. (2021). Swedish Total Defence and the Emergence of Societal Security. In S. Larsson & M. Rhinard (Eds.), *Nordic Societal Security: Convergence and Divergence*. London and New York: Routledge, pp. 45–68.
Louth, J., & Taylor, T. (2018). *British Defence in the 21st Century* (1st edition). Routledge. https://doi.org/10.4324/9781315202389.

Mörth, U. (2000). Competing Frames in the European Commission—The Case of the Defence Industry and Equipment Issue. *Journal of European Public Policy*, 7(2), 173–189.

Petersson, M. (2018). The Allied Partner: Sweden and NATO Through the Realist-Idealist Lens. In A. Cottey (Ed.), *The European Neutrals and Nato: Non-Alignment, Partnership, Membership?* London: Palgrave Macmillan.

Rinelli, S., & Duyvesteyn, I. (2018). The Missing Link: Civil-Military Cooperation and Hybrid Wars. In E. Cusumano & M. Corbe (Eds.), *A Civil-Military Response to Hybrid Threats*. Cham: Springer International Publishing, pp. 17–41.

Stiglund, J. (2021). Threats, Risks, and the (Re)turn to Territorial Security Policies in Sweden. In S. Larsson & M. Rhinard (Eds.), *Nordic Societal Security: Convergence and Divergence*. London and New York: Routledge, pp. 199–223.

Swedish Defence Commission. (2013). Vägval i en globaliserad värld [Paths in a Globalised World]. Ds. 2013:33.

Swedish Defence Commission. (2014). Försvaret av Sverige: Starkare försvar för en osäker tid [The Defence of Sweden: Stronger Defence for Uncertain Times]. Ds. 2014:20.

Swedish Defence Commission. (2017). "Motståndskraft: Inriktningen av totalförsvaret och utformningen av det civila försvaret 2021–2015 [Resilience: The Total Defence Concept and the Development of Civil Defence 2021–2025]. Ds 2017:66.

Swedish Defence Commission. (2019). Värnkraft: Inriktningen av säkerhetspolitiken och utformningen av det militära försvaret 2021–2025 [Defensive Power—Sweden's Security Policy and the Development of its Military Defence 2021–2025]. Ds 2019:8.

Swedish Defence Commission. (2023). Allvarstid: Försvarsberedningens säkerhetspolitiska rapport 2023 [Serious Times: The Defence Commission's Security Policy Report 2023]. Ds. 2023:19.

Swedish Defence Commission. Updated 20 April 2023. https://www.government.se/government-of-sweden/ministry-of-defence/defence-commission/ [English Version] Accessed 5 June 2023.

Swedish Government, Decision. (2015, December 10). *Regeringen beslutar om återupptagen totalförsvarsplanering [Government Decision on Resumed Total Defence Planning]*. https://www.regeringen.se/artiklar/2015/12/regeringen-beslutar-om-aterupptagen-totalforsvarsplanering/ Accessed 25 January 2021.

Valtonen, V., & Branders, M. (2021). Tracing the Finnish Comprehensive Security Model. In S. Larsson & M. Rhinard (Eds.), *Nordic Societal Security: Convergence and Divergence*. London and New York: Routledge, pp. 91–108.

Westberg, J. (2021). *Svenska säkerhetsstrategier: teori och praktik [Swedish Security Strategies: Theory and Practice]* (2nd edition). Lund: Studentlitteratur.

Whelan, C. (2017). Managing Dynamic Security Networks: Towards the Strategic Managing of Cooperation, Coordination and Collaboration. *Security Journal*, 30(1), 310–327.

Wikman, L., (2021). *Don't Mention the War: The Forging of a Domestic Foreign Policy Consensus on the Entry, Expansion and Exit of Swedish Military Contributions to Afghanistan*. Diss. Uppsala: Uppsala University, 2021.

Wither, J. K. (2020). Back to the Future? Nordic Total Defence Concepts *Defence Studies*, 20 (1), 61–81.

Wrange, J. (2022). Entangled Security Logics: From the Decision-Makers' Discourses to the Decision-Takers' Interpretations of Civil Defence. *European Security (London, England)*, 31(4), 576–96.

Ydén, K., Berndtsson, J., & Petersson, M. (2019). Sweden and the Issue of NATO Membership: Exploring a Public Opinion Paradox. *Defence studies*, 19, 1–18.

11

INFORMATION SHARING IN COMPLEX CRISES

The case of joint domain-specific centers in Norway

Line Djernæs Sandbakken and Ørjan Nordhus Karlsson

Introduction

Transboundary crises cross political, geographical, and functional barriers and could simultaneously impact multiple sectors and regions within a nation (Ansell et al., 2010). As with the other Nordic countries, Norway's approach to managing such crises entail collaborative and coordinated efforts among various actors, including governments, civil society organizations, and private companies (Larsson & Rhinard, 2021). A concerted response is crucial when problems are interconnected and complex, requiring crisis managers to navigate multiple and often conflicting perspectives. In addition, it is not always clear what the main problem is, thus making it challenging to identify which actors need to be involved (Ansell et al., 2010; Boin, 2019). Consequently, transboundary crises often present "wicked problems" for crisis managers (Rittel & Webber, 1973).

A hybrid threat is one example of a transboundary crisis that governments must deal with. Hybrid threats are coordinated actions designed to create confusion and dilemmas across sectorial lines of responsibility, that is, exploit sectorial and administrative gray areas (Cullen & Wegge, 2021). The objectives differ, but undermining decision making, creating instability, influencing public opinion, and disrupting democratic processes are well-known aims (Hybrid COE et al., 2020). Hybrid threat actors often aim at staying below the threshold of armed conflict, and the concerted use of different instruments of power, such as disinformation, economic pressure, diplomatic sanctions, and/or outright sabotage of critical infrastructure, makes it challenging to measure the overall effect and complexity, in turn creating potential ambiguity on how to counter the threats (Hoffman, 2018).

DOI: 10.4324/9781032617916-14

As summarized by the Finnish Hybrid Center of Excellence: "Hybrid threat actors seek to undermine and harm the integrity and functioning of democracies" (Jungwirth et al., 2023). As such, hybrid attacks can affect several sectors and administrative levels simultaneously, potentially creating a complex, transboundary crisis. In order to counter and mitigate this kind of cross-sectorial multi-level threat, information sharing and a common situational awareness is key. See also chapters by Cullen, Berndtsson, Borch and Heier, and by Bjørge and Høiby in this volume for in-depth discussions on the phenomena of hybrid and ways to counteract such threats.

In Norway, ministerial rule is a key trait of the public administrative system, which means that there is a lowered threshold for individual ministers to intervene in tasks and priorities of line agencies (Larsson & Rhinard, 2021: 15). This tendency forms part of the backdrop that further emphasizes the need for strong cross-sectorial collaboration in both crisis management and information sharing, especially with regard to hybrid threats.

That said, most security services have experience with the handling of hybrid threats, highlighted in the open security assessment for the Norwegian Police Security Service (PST), Military Intelligence (NIS), and the National Security Agency (NSM).

However, collaboration between different intelligence and security agencies comes with its own set of challenges. This has been highlighted after major incidents such as 9/11 2001, the London 7/7-bombings in 2005, and more recently the multiple terror attacks in Paris in 2015 and Brussels 2016 (Argomaniz, 2009; Ateş & Erkan, 2021; Boer, 2015; Bures, 2016; Thomson, 2021). For Norway, a recent example is how the fourth crisis management principle of "cooperation" was included after July 22, 2011, terrorist attacks (Lagreid & Rykkja, 2015).

To deal with these kinds of complex problems, some security services opt to collaborate through joint domain-specialized centers.[1] However, the mandate and operational modus might differ across sectors and countries, as highlighted by Van der Veer et al. (2019) in their study of six so-called "fusion centers" working across organizations to target transboundary crises such as terrorism. Other ways of collaborating for handling transboundary crises might be through the set-up of ad-hoc organizations (Mintzberg, 1983), or specific tailor-made partnerships (see Borch, this volume).

In this chapter, we will examine how three different Norwegian joint domain-specific centers manage transboundary crises such as hybrid threats: the Joint Cyber Coordination Center (FCKS), the National Cyber Security Center (NCSC), and the National Intelligence and Security Center (NESS).[2] These three centers all belong to the same policy area of national security, deal with transboundary crises and hybrid threats, and focus strongly on information sharing. More specifically, we will explore how meta-organizational design impacts information sharing. Key variables are boundary spanning roles, the different designs of the centers, and their mode of governance. We believe our findings can be of interest to similar centers in other countries as well.

The rationale for establishing joint domain-specialized centers for coordination and collaboration is often to attain a broader situational understanding and facilitate an effective response to the challenges that may arise (Borch & Andreassen, 2020). Well-known challenges for these organizational designs are different goals and ways of working, which may hamper the information sharing between organizations (Owen et al., 2013). We will study the design of three domain specific centers, posing the following research question: *How can the organizational design of domain-specific centers facilitate information sharing in transboundary crises?*

Theory

Transboundary crises and information sharing

Covid-19 and the war in Ukraine have led to an increased focus on crisis management in many organizations. The awareness of hybrid threats is more acute. These threats have both an international and domestic dimensions, and a hybrid campaign can lead to significant sectorial implications within a country (Jungwirth et al., 2023). A key aim of a hybrid threat actor is to create instability and confusion. Thus, information sharing across sectors and agencies is vital to establish a common situational awareness. Poor or ineffective collaboration between different sectors may have a negative impact on the crisis management effort. As such, it is important in advance to understand how these kinds of threats affect society (Nilsson et al., 2021).

Inter-organizational collaboration has been seen as a panacea for responding to transboundary crises (Alford & Head, 2017; Ansell & Gash, 2008; Boer, 2015). While there are benefits in collaborating across sectors and levels of governments to manage hybrid threats, there are challenges as well, as highlighted in the introduction.

Information sharing is key to ensure efficient collaboration during a transboundary crisis (Ansell et al., 2010; Boin, 2019; Christensen et al., 2016). Since organizations have different goals, values, structures, and ways of working, communicating effectively across organizational borders/sectors could be challenging (Owen et al., 2013).

Effective horizontal and vertical coordination in crises facilitates for information sharing. When this fails it could impact the national crisis management system's ability to handle a crisis effectively (Boin et al., 2019). Domain-specific centers are seen as one answer to efficient information sharing.

To achieve shared understanding in a collaborative setting, it is important both to agree on what the information means, as well as sharing the information in such a way that it is understood by all partners. The former can be achieved through written communication, while the latter is better transmitted face to face (Dennis et al., 2008). Thus, sharing complex information is best achieved through physical interaction, as it allows for more rich communication (Valaker et al., 2018).

Information sharing is also about establishing situation awareness, which is an important part of achieving successful inter-organizational collaboration (O'Brien et al., 2020). In a crisis, organizations may have different goals and priorities. This in turn makes it difficult for various actors to attain a similar understanding of a situation or threat. At the same time, sectorial or agency-specific situation awareness can both overlap and complement that of other actors (Stanton et al., 2006).

Security services, both nationally and internationally, often encounter problems that can hinder effective collaboration. Legal and jurisdictional frameworks may be different, hampering what kind of information can be shared. There may also be cultural and language barriers, which can lead to misunderstandings. Moreover, these agencies sometimes have competing priorities and objectives driven by national or agency interests (Argomaniz, 2009; Ateş & Erkan, 2021; Boer, 2015; Bures, 2016; Thomson, 2021; Jansen et al., 2023).

Inter-organizational collaboration could also be seen as challenging crisis management principles such "sectorial responsibility", making joint situational awareness more difficult to attain. This is true for Norway even though the fourth crisis management principle of "Collaboration" was introduced after the terror attacks on July 22, 2011 (NOU 2012:14, 2012; Lagreid & Rykkja, 2015).

An uncertainty of who is the lead actor/agency might lead to underlap or overlap of responsibility, meaning either that actors wait for someone else to take responsibility or that several organizations take lead at the same time (Boin et al., 2020; Christensen et al., 2016). Thus, when these mechanisms are not working sufficiently, challenges in tackling transboundary crisis might arise (Alford & Head, 2017).

The design of meta-organizations

To theoretically identify joint domain-specialized centers and how they are structured, we use the concept of meta-organizations. There is a debate in the literature between meta-organizations as formal organizations that organize other formal organizations (Ahrne & Brunsson, 2005) and meta-organizations as more loosely bounded networks of organizations that come together based on shared common goals (Gulati et al., 2012). In this article, we base our understanding on the latter definition, as meta-organizational design in this manner is usually not bound by formal authority through employment, even if individual workers may have such links to their home organization. Informal authority and trust gained through long-term collaboration are central traits (Gulati et al., 2012). However, the different design of the joint domain-specific centers in our study may influence how the footprint of the various "mother"-organizations materializes within each center.

Boundary spanners have a key role in securing and sharing information as well as building trust within meta-organizations. Boundary spanning roles are identified as central for achieving inter-organizational collaboration in several studies (Brown et al., 2021; Curnin & Owen, 2014; Kapucu, 2006; Kalkman, 2020). Additionally, the link between the center and its parent organization is important for securing resources, linking personnel and activities, as well as securing decisions (Albers et al., 2016). This link is especially important for joint domain-specific centers, as network participants must balance the tasks and responsibilities of their parent organization with the needs of the network.

The literature identifies a range of different boundary spanning roles, and one person often holds several roles at the same time. Considering the data material in this chapter, we will focus on the role of the liaison/representative, the communicator and the organizational expert (Ancona & Caldwell, 1992; Curnin & Owen, 2014).

As liaison, the representative retains close contact with the parent organization.

The role of the communicator highlights the importance of vertical and horizontal coordination, and the ability to bring important information between the center and the parent organization, and to other partners when relevant. The role of the organizational expert refers to having broad knowledge of the priorities and capabilities of own organization, but also of the other organizations being connected or affiliated with the center (Curnin & Owen, 2014). Through these roles, the boundary spanners can facilitate inter-organizational collaboration and alleviate some of the challenges that may arise.

The way meta-organizations such as domain-specific centers are designed and might also impact on information sharing. For instance, being co-located as opposed to distributed when collaborating might mitigate silo-thinking and can lead to increased understanding and communication across organizational borders (Hærem et al., 2022). Another approach is to collaborate within a common, overarching structure. This could facilitate information sharing among members and compound long-term intelligence and analysis, as done by the EU's Hybrid Fusion Cell (Cullen, 2021).

Establishing permanent meta-organizations such as joint domain-specific centers to foster collaboration across organizations is one way of dealing with the collaboration challenges discussed earlier. The same kind of design could be applied to an ad hoc center to deal with a time-specific problem/crisis.

Methodology

This is a qualitative research study based on semi-structured interviews and document studies where we compare three different joint domain-specific centers in the Norwegian context. These centers were chosen because they all work with topics related to national security. In addition, the center structure is a unique form of collaboration for handling transboundary crises and therefore provides an interesting arena for research. A comparison opens for exploration of differences and similarities, including the conditions they appear under (Bennett, 2005). The Norwegian context is interesting because of its small size, high degree of political trust among the population, low degree of political conflict, and the ministerial rule that leads to a sectorized public sector. As discussed earlier, the latter leads to an increased need for cross-sectorial collaboration. The centers have chosen different meta-organizational designs, in turn providing us with an opportunity to compare these designs with regard to different modes of information sharing and the role of boundary spanners.

The data collection focused on document studies of publicly available documents such as White Papers, legislative proposals from the government, and annual reports where domain-specific centers have been mentioned and discussed (Creswell & Creswell, 2018). The interviews were conducted in winter 2022 and spring 2023. Altogether, we did seven semi-structured interviews with key personnel working in or closely connected to these centers. All the informants either held leadership positions in their respective organizations or were appointed to participate by their leaders, thus being able to provide hands-on information on the daily work within the centers as well as collaboration with partners.[3] The informants represented the Norwegian National Security Authority (NSM), the National Crime Investigation Service (Kripos), the Ministry of Defence, the Police Directorate, the Directorate for Civil Protection (DSB), and the Crisis Support Unit (KSE). The Police Security Service (PST) and the Norwegian Intelligence Service (E) declined participation. As such, we have not interviewed representatives from all organizations in the centers. However, since the focus is on the centers, not specific participants, this seems negligible. The main objective was to gain insight into the formal structures, informal relations, and collaborative mechanisms within and between the centers as well as the connection with the parent organizations (Brinkmann & Kvale, 2015; Tjora, 2017). We chose to specifically ask for publicly available information. In that respect, we formulated a research question that would make it possible to collect, analyze, and publish this material.

We have used reflexive thematic analysis to identify themes in the data material (Braun & Clarke, 2022). Through an inductive approach, coding and themes were identified through the content of the data. In addition, we have used different sources to triangulate data, for example relating interviews with key personnel from different organizations and to public documents such as White papers and Official Norwegian reports (NOU).

Empirical description

In this section, a description of the three domain-specific centers will be given, with key points and characteristics summed up in Table 11.1. Table 11.1 also provides a comparison across central context variables (network model,

TABLE 11.1 Comparison of the domain-specific centers

Name of center	The Joint cyber coordination center (FCKS)	The National cyber security center (NCSC)	The national intelligence and security center (NESS)
Established	2017	2019	2022
Type of crisis/ delineation	Transboundary crisis within specific sector.	Transboundary crisis within specific sector. Outreach to public/private/ academic sector	Cross-sectorial transboundary crisis.
Key stakeholders	National security authority (NSM), National Criminal Investigation Service (KRIPOS), Norwegian Intelligence Service (NIS), and Norwegian Police Security Service (PST)	Competent authority: National Security Authority (NSM). Multitude of public sector authorities and private organizations. Academia. About 30 organizations in all	NSM, PST, NIS, and Kripos
Mandate/ Main task	Information sharing. Crisis and emergency planning. Competence development. Development of incident management/collaboration with partner— hybrid included	Main coordination and information sharing hub during a cyber/digital attack (also relating to an overall hybrid attack)	Enhance national ability to detect and understand hybrid threats. Support government/ministerial decision making. Information sharing
Membership	Appointed	Members apply	Appointed
Network model	Shared governance	Lead agency (NSM)	Shared governance
Design of center	Co-located	Distributed	Co-located
Main boundary spanning roles	Communicator/ liaison	Organizational expert/liaison	Communicator/ liaison

Source: created by the authors

boundary spanning roles, and design of center). For clarity, the table is presented first, but will also be referred to in the discussion.

A brief outline of the Norwegian centers

The joint cyber coordination center (FCKS), the national cyber security center (NCSC), and the National Intelligence and Security Center (NESS) have had an evolutionary development, from informal collaborations between key actors to formal establishments through government funding (Ministry of Justice and Public Security, 2017a, 2017b, 2018a, 2018b).

A key difference between the three centers relates to membership. With regard to NCSC, government agencies and public and private sector actors are free to join based on their perceived need. As for NESS and FCKS, participation is by invitation only.

The joint cyber coordination center (FCKS) was established in 2017 as a permanent, co-located center focusing on analysis, protection, and overview of cyber space. The center is mandated by the government (Ministry of Justice and Public Security, 2017a, 2017b). The center does not have any principal legal authority; thus, the establishment did not entail any changes in the institutional framework related to roles, responsibilities, or legal basis.

The national cybersecurity center (NCSC) was established in 2019. About 30 organizations from public and private sectors, academia and the military are part of the center (NSM, 2023). The center is coordinated by NSM. Initially the plan was to gather the members on a regular basis at the physical office in Oslo. However, due to Covid-19, most of the meetings have been digital, and the trend of virtual meetings, at least partial, seems to continue.

The National Intelligence and Security Center (NESS) was established at the end of 2022. The collaborating partners in the co-located center are NSM, PST, NIS, and Kripos. Its predecessor NEST was established in April 2022 in collaboration between PST and Kripos, when the government shortly after Russia's invasion of Ukraine granted an extra 100 million NOK to combat foreign intelligence, hybrid threats, and increased presence in Northern Norway (Ministry of Justice and Public Security, 2022).

Although hybrid threats are not a new phenomenon, the war in Ukraine in February 2022 has been a catalyst for establishing a domain-specific center to target hybrid threats. Several informants (#4, 5, 6) admit that they were surprised when the news of NEST was publicized by the Minister of Justice and Public Security in April 2022. One reason for this was that an inter-organizational group was looking at the center's mandate at the time. Their task was assessing different collaboration structures best fitted to counter/mitigate hybrid threats. NEST was established before the working group had concluded their work. However, some informants (#2, 7) argued that it was better to get something started and at the same time continue in parallel the analysis of the future make-up of the center. Late in 2022, NESS (the new center) was established. As exemplified with NESS, there has been an evolution from informal collaboration between partners to increasingly more formal structures.

Discussion

The aim of this chapter is to examine how information sharing takes place in three joint domain-specific centers, with specific focus on the structures of three meta-organizations, and the roles of the boundary spanners working within the centers. The task environment itself also plays an important part. Transboundary crises, such as hybrid threats, add complexity to the tasks each center is supposed to deliver on (Table 11.1). Based on the rich data from interviews with key informants, a total of 15 quotes have been selected to reflect the aforementioned context variables. In the following, the quotes will be referred to by the numerical (#) signifier in Table 11.2.

Information sharing in transboundary crises

As highlighted in the introduction and theory, information sharing in a transboundary crisis is key both to attain the same situational understanding and to ensure both horizontal and vertical collaboration (Alford & Head, 2017; Ansell & Gash, 2008; Boin et al., 2019; O'Brien et al., 2020). In this context, the informants drew attention to three major issues:

(I) The necessity for a distributed, situational awareness (#1), (II) active collaboration (#2& 3), and (III) how difference in the organization's mandates can hamper the first two activities (#3 & 4). Seen together, these challenges entail the main reasons why these centers were established in the first place.

With regard to information sharing in a complex crisis, the centers differ. NESS and FCKS *share* information between the participants as part of the task of producing common analytical- and informational products (Table 11.1). As for NCSC, the center *distributes* information to its many members to update/inform/warn on a specific or developing situation. The information shared in FCKS and NESS is of more sensitive character, a fact reflected in the restriction on participating members. Overall, it seems clear that a smaller milieu of known organizations makes it easier to strengthen personal relations (#12) as opposed to a larger center with a distributed second tier of participants (#13).

Formal mandate also matters. For example, intelligence services are mandated by the Security Act on how and what kind of information they can

TABLE 11.2 Domain-specific centers: Quotes aligned by context variables.

Category	Quote (informant#)	Remarks	#
Information exchange	*One aims to have as similar an understanding as possible of the threat and risk landscape in cybersecurity across the various organizations and ensure that information flows smoothly across them. (#1)*		1
Information exchange	*We are here to facilitate collaboration and, in some cases, to be able to compile analytical assessments. When you look at things across all four organizations, you may get something more. 4x1 is sometimes more than just 4. (#1)*	Specifically for NESS and FCKS	2
Information exchange	*All security services must, in one way or another, deal with hybrid threats. But where does one's mandate begin and the other's end? It creates an inherent uncertainty, doesn't it, making it difficult to take countermeasures. (#2)*	Informant discussing constraints of sharing info.	3
Information exchange	*What has been very challenging with hybrid (threats) and, yes, in general, is what mandate one has when working with information. (#7)*	Informant elaborating on restricted information and sectorial issues	4
Boundary spanning	*It is, in a way, one of the core tasks, that of good information sharing and being a focal point within the cyber field among the various professional communities. (#1)*	Referring to the role of liaison	5
Boundary spanning	*There are different needs among the individual chiefs. The Police Director has a need for periodic reporting, either weekly or biweekly, providing a summary that goes to the Ministry of Justice, demonstrating the ongoing work in this area. On the other hand, the Intelligence Services may desire a more comprehensive and finalized analytical product that delves deeper into assessments and conclusions. (#2)*	The term "Chiefs" here refers to the different heads (Directors) of the various collaborative services.	6
Boundary spanning	*It can be challenging in a center to determine what can be shared when other services are also involved (. . .) There may be differences in the cultural emphasis placed on what is considered most important. This can vary based on the organization one comes from and how individuals perceive and prioritize information. (#4)*		7
Boundary spanning	*They produce common products to a greater extent, and therefore, they do more than just being liaisons. The problem area they work on is more demanding and necessitates a common product to a greater extent than is the case for cyber. (#5)*	"They" referring to NESS.	8

(Continued)

TABLE 11.2 (Continued)

Category	Quote (informant#)	Remarks	#
Boundary spanning	*How do you know and how do you identify whether it's one sector authority or another that should handle an incident? And often, there's an assumption that someone else will handle it if it's not clearly defined, and you may risk it falling through the cracks.* (#3)	Informant discussing the ambiguity of hybrid threats and (who's) lead agency	9
Boundary spanning	*In a hybrid context, some are mostly concerned with detection and attribution, while there are also those who are concerned with the consequences. And that's us.* (#6)		10
Boundary spanning	*What is important is that the analysts and coordinators in FCKS have close dialogue with their line environments. We work every day to ensure that we don't develop this, yes, the Stockholm syndrome.* (#1)		11
Design of center	*One advantage is that all services are co-located. (. . .) My experience is that it is easier to share information and have a lower threshold for sharing information with individuals you know, even if they are in a different organization.* (#1)		12
Design of center	*It is true that we have not been able to achieve the same level of trust building in one-on-one interactions. It is more challenging to encourage people to interact and share in a digital space where you may not have complete visibility of who is behind the screen. Building trust in such environments requires additional effort and strategies to foster open communication and collaboration.* (#3)	Informant also referring to challenges in relation to corona-pandemic.	13
Design of Center	*The center structure can contribute to building trust and culture, as well as sharing expertise. This comes into play automatically when people begin to work together over time, both on a personal and organizational level.* (#4)		14
Design of Center	*The disadvantage of that structure is the sector principle itself because you end up with some gray areas that don't fit perfectly. In other words, it's not a square box you fit inside or fall outside of.* (#3)		15

Source: created by the authors

share. Other organizations, such as the police or other civil agencies, operate differently and may not have access to certain information. Boer (2015) touches upon this in his article of information sharing between intelligence services in the EU, especially regarding problems related to oversight on ownership and integrity of data. The networked governance model of the FCKS and NESS can be seen as a way of mitigating this, since all organizations are legally autonomous, and they collaborate on an equal standing. Still, there is a perceptive worry that gray areas still exist, leading to what Christensen et al. (2016) calls underlap.

Information sharing and the structuring of the centers

The structuring of the joint domain-specific centers will in the context of this article be divided into two parts: the modes of governance (Boin et al., 2014), and whether the centers are distributed or co-located (Hærem et al., 2022).

In general, informants from or with knowledge of co-located centers put emphasis how this meta-organizational design is positive for information sharing (#12 and 14). Developing trust, getting to know members from other organizations, and a common culture are specified as positives in this regard (Gulati et al., 2012).

However, some highlight the sector principle as hindering collaboration (Boin et al., 2020; Lagreid & Rykkja, 2015), creating uncertainty in what kind of information that can be shared, leading to possible gray areas (#15). This concern was also raised regarding mandate (#3), as exemplified earlier. Here, the role of boundary spanners becomes essential (next section).

As for the only distributed center (NCSC), building trust and collaboration are highlighted as areas where more effort is needed (#13). However, the mandate and task of NCSC compared to FCKS and NESS are different (Table 11.1). Creating an information network where distribution of information "down the line" is the main task and is less demanding than compiling and analyzing different information from security organizations and producing a concerted report or give governmental advice (Van der Veer et al., 2019).

This fact is further illustrated by the choice of governance model. NCSC employs a lead agency-model, whereas the collaborative environment of FCKS and NESS is reflected in their shared governance approach and close, daily collaboration (Boin et al., 2014). In this regard, it could be argued that both the design (co-located and distributed) and governance model (lead agency and shared governance) seem a right fit to the center's specific tasks (Table 11.1).

However, in order to avoid what an informant called "The Stockholm syndrome" (#11), and to deal with uncertainties of possible gray areas (#7), the role of boundary spanners is key.

Information sharing through boundary-spanning roles

Boundary spanners are identified as important for successful inter-organizational collaboration (Brown et al., 2021; Curnin & Owen, 2014; Kapucu, 2006; Kalkman, 2020). The role of the representative/liaison (Curnin & Owen, 2014) is identified in all the three centers (#5). In FCKS and NESS, the members meet on daily basis and will therefore have more flexibility in how they execute the liaison role as opposed to NCSC, which meets more infrequently. FCKS and NESS constantly exchange information they see as relevant for the partners. However, even if these centers have a similar design, there may still be differences in how information is shared and utilized. One informant argued that FCKS can be described as an arena where common challenges are discussed among permanent liaisons from each organization, while decisions and reports are made by each individual organization. NESS, on the other hand, seems to a larger extent to produce reports made by the center as a whole (# 8). The informant reflected that one reason for this differentiation may be that cybersecurity can be seen as more limited in scope, while hybrid threats are so broad and complex that they require closer collaboration when producing analyses. This differentiation will need further looking into.

The boundary spanning role of the communicator is central for horizontal and vertical coordination of information (Curnin & Owen, 2014; Ancona & Caldwell, 1992). Particularly for NESS and FCKS, one of the most important tasks is to produce relevant reports to decision makers, which require precise information sharing and agreement on the most important focus areas. However, the mother-organizations also have their own specific expectations to NESS. This in turn poses a problem for the center, with regard to both its priorities and its output. What the Chief of Police wants is not necessarily what the Head of military intelligence is looking for, as exemplified by an informant (#6). This illustrates some of the complexities with inter-organizational collaboration, such as competing goals and priorities as well as cultural barriers (Argomaniz, 2009; Ateş & Erkan, 2021; Boer, 2015; Bures, 2016; Thomson, 2021; Jansen et al., 2023).

For NCSC, the role of the communicator is at the core of the center's mandate, as it gathers and shares information with partners from a broad range of sectors and levels of government. Through close communication, the centers can avoid underlap or overlap (Boin et al., 2020; Christensen et al., 2016), since there are many actors working on closely related topics (#9). This might be easier in a co-located context; however, the NCSC has become quite efficient in sharing information with its members through electronic means.

Moreover, the role of the communicator is also important for reaching out to partners outside the meta-organization. An informant who was involved in the initial discussions on the design and output of NESS pointed out the complexity of hybrid threats. Working with hybrid threats, it's not a given that all the necessary information will be available when only four organizations are members (#10). Thus, it would be important for the core members to include views and information from partners from outside the center as this may present a broader scope of transboundary threats.

The role of the organizational expert is important in the center structure because the members must represent their parent organization's priorities and capabilities within the center, but also due to the organization experts' knowledge of the partner organizations. This in turn makes it easier to share relevant information (Curnin & Owen, 2014).

One informant explained that working closely together in a co-located context enables the analysts to know better what kind of information the others need. In this way, they can move past the challenge of different cultural backgrounds and priorities (#7). This way of working is especially evident in the co-located settings. At the same time, keeping close relations with the parent organization is important for maintaining one's own ties and identity (#11), especially since all members in the meta-organizations are employed in their parent organizations and not in the center (Gulati et al., 2012).

Summing up, the balance between being an organizational expert of the parent organization as well as building relations and trust through close cooperation with the partners is both a central and challenging balancing act. This seems more important for the distributed centers, such as NSCS, as the outreach to different organizations is so much larger than in FSKC and NESS.

Conclusion

This study has explored how the organizational design of three joint domain-specific centers facilitates for information sharing in transboundary crises in the Norwegian context. FCKS, NESS, and NCSC all have information sharing as one of their core activities, but the way they have operationalized this task differs. This also points back to the flexibility of meta-organizational design.

For NCSC, the main task lies in the distribution of information to its members. The information is of less sensitive character than for FCKS and NESS, and it's not expected that the members participate in creating new products (although information shared by members to NSM, the network's lead agency, can certainly be included in the information packages). As a lead agency model, a distributed network and a boundary-spanning role that focuses on organizational aspects allow for expedient information sharing to many organizations in transboundary crises.

For FCKS and NESS, the sensitivity of the information, the expectations of shared/common products, and a problem structure that could lead to gray areas (or underlap) make for a different organizational design. A co-located, collaborative shared governance model with few members allows for confidence building and information sharing, which again may lead to more effective handling of the threat. They are also consensus-based meta-organizations, where the boundary-spanning role of communicator, both horizontal in the center and vertical to parent organization, is key.

However, as stated by several of the informants, producing concerted analytical reports and updates is a time-consuming process. The challenge, especially for NESS dealing with hybrid threats, is to deliver timely information to different decision-makers trying to deal with a fast-moving, transboundary crises both within their own sector and distributed. In such a scenario, the lead agency model of NCSC would have allowed for greater speed. The trade-off might have been that the participating organizations in both FCKS and NESS would be less willing to share their information.

An alternative organizational design to the lead agency model and shared governance would be for the parent organizations to give NESS and FCKS a kind of pragmatic autonomy as discussed with Van der Veer et al. (2019) in relation to fusion centers. In such a scenario, the boundary-spanning role of communicator would be more focused on the horizontal effort, between the participants in the center.

Transboundary crises in the form of hybrid threats are a global phenomenon. The findings in this study may therefore be of relevance to other countries, particularly to those administrative systems like that in Norway, with a high level of trust and comparative low level of conflict within the political system.

When designing cross-sectorial collaborative entities such as joint domain-specific centers, the flexibility in the meta-structure means that the organization can be tailored according to the specific needs of the country. Our research has shown that different organizational designs allow for information sharing in different ways, and that working through joint domain-specific centers may reinforce cross-sectorial collaboration.

All the three centers in this study are relatively new, and the participants in NESS are still finding their role. In this respect, further research into the role of boundary spanners in joint domain-specific centers would be of interest.

In such an endeavor, the methodological approach of observation and shadowing could allow for data gathering on the different roles that are present in a work environment when managing and/or preparing for a transboundary crisis.

Notes

1 This organizational design is also used in other sectors as well, for instance search and rescue authorities.
2 See also Table 11.1.
3 All informants have been anonymized.

References

Ahrne, G., & Brunsson, N. (2005). Organizations and meta-organizations. *Scandinavian Journal of Management*, 21(4), 429–449. https://doi.org/10.1016/j.scaman.2005.09.005

Albers, S., Wohlgezogen, F., & Zajac, E. J. (2016). Strategic alliance structures: An organization design perspective. *Journal of Management*, 42(3), 582–614.

Alford, J., & Head, B. W. (2017). Wicked and less wicked problems: A typology and a contingency framework. *Policy and Society*, 36(3), 397–413.

Ancona, D. G., & Caldwell, D. F. (1992). Bridging the boundary: External activity and performance in organizational teams. *Administrative Science Quarterly*, 37(4), 634–665.

Ansell, C., Boin, A., & Keller, A. (2010). Managing transboundary crises: Identifying the building blocks of an effective response system. *Journal of Contingencies and Crisis Management*, 18(4), 195–207. https://doi.org/10.1111/j.1468-5973.2010.00620.x

Ansell, C., & Gash, A. (2008). Collaborative governance in theory and practice. *Journal of Public Administration Research and Theory*, 18(4), 543–571.

Argomaniz, J. (2009). Post-9/11 institutionalisation of European Union counterterrorism: emergence, acceleration and inertia. *European Security*, 18(2), 151–172.

AteŞ, A., & Erkan, A. (2021). Governing the European intelligence: Multilateral intelligence cooperation in the European Union. *International Journal of Politic and Security (Online)*, 3(3), 230–250. https://doi.org/10.53451/ijps.900302

Bennett, A. (2005). *Case studies and theory development in the social sciences*. MIT Press.

Boer, M. D. (2015). Counter-terrorism, security and intelligence in the EU: Governance challenges for collection, exchange and analysis. *Intelligence and National Security*, 30(2–3), 402–419. https://doi.org/10.1080/02684527.2014.988444

Boin, A. (2019). The transboundary crisis: Why we are unprepared and the road ahead. *Journal of Contingencies and Crisis Management*, 27(1), 94–99.

Boin, A., Busuioc, M., & Groenleer, M. (2014). Building European Union capacity to manage transboundary crises: Network or lead-agency model? *Regulation & Governance*, 8(4), 418–436.

Boin, A., Ekengren, M., & Rhinard, M. (2020). Hiding in plain sight: Conceptualizing the creeping crisis. *Risk, Hazards & Crisis in Public Policy*, 11(2), 116–138. https://doi.org/10.1002/rhc3.12193

Boin, A., Kuipers, S., & de Jongh, T. (2019). A toxic cloud of smoke: Communication and coordination in a transboundary crisis. In P. Laegreid & L. Rykkja (Eds.), *Societal security and crisis management*, 133–150.

Borch, O. J., & Andreassen, N. (2020). *Beredskapsorganisasjon og kriseledelse*. Fagbokforlaget.

Braun, V., & Clarke, V. (2022). *Thematic analysis. A practical guide*. SAGE Publications Ltd.

Brinkmann, S., & Kvale, S. (2015). *InterViews. Learning the craft of qualitative research interviewing* (3rd ed.). Sage Publications Ltd.

Brown, O., Power, N., & Conchie, S. M. (2021). Communication and coordination across event phases: A multi-team system emergency response. *Journal of Occupational and Organizational Psychology*, 94(3), 591–615.

Bures, O. (2016). Intelligence sharing and the fight against terrorism in the EU: Lessons learned from Europol. *European View*, 15(1), 57–66. https://doi.org/10.1007/s12290-016-0393-7

Christensen, T., Lægreid, P., & Rykkja, L. H. (2016). Organizing for crisis management: Building governance capacity and legitimacy. *Public Administration Review*, 76(6), 887–897.

Creswell, J. W., & Creswell, J. D. (2018). *Research design: Qualitative, quantitative & mixed methods approaches* (5th ed.). Sage Publications Ltd.

Cullen, P. (2021). A perspective on EU hybrid threat early warning efforts. In M. Weissmann, N. Nilsson, B. Palmertz, & P. Thunholm (Eds.), *Hybrid warfare. Security and asymmetric conflict in international relations* (pp. 47–57). I.B. Tauris Bloomsbury Publishing.

Cullen, P., & Wegge, N. (2021). Hybrid warfare. In S. Stenslie, L. Haugom, & B. H. Vaage (Eds.), *Intelligence analysis in the digital age*. Routledge. https://doi. org/10.4324/9781003168157

Curnin, S., & Owen, C. (2014). Spanning organizational boundaries in emergency management. *International Journal of Public Administration, 37*(5), 259–270.

Dennis, A. R., Fuller, R. M., & Valacich, J. S. (2008). Media, tasks and communication processes. A theory of media synchronicity. *MIS Quarterly, 32*(3), 575–600.

Gulati, R., Puranam, P., & Tushman, M. (2012). Meta-organization design: Rethinking design in interorganizational and community contexts. *Strategic Management Journal, 33*(6), 571–586.

Hoffman, F. G. (2018). Examining complex forms of conflict Gray Zone and hybrid challenges. *PRISM, 7*(4), 30–47.

Hybrid CoE, Giannopoulos, G., Smith, H., & Theocharidou, M. (2020). *The landscape of hybrid threats: A conceptual model*. European Commission, ISPRA, PUBSY No. 123305.

Hærem, T., Valaker, S., Lofquist, E. A., & Bakken, B. T. (2022). Multiteam system handling time-sensitive targets: Developing situation awareness in distributed and co-located settings. *Frontiers in Psychology, 13*, 1–13.

Jansen, P. T., Eriksen, C. C., Hoven, S., Løberg, A., Ravndal, J. A., Rolfheim-Bye, C. L., & Skoglund, G. (2023). *Rapport fra 25. juni-utvalget. Evaluering av PST og politiet.* https://www.politiet.no/globalassets/04-aktuelt-tall-og-fakta/evaluering-srapporter/evaluering_25juni2022.pdf

Jungwirth, R., Smith, H., Willkomm, E., Savolainen, J., Alonso Villota, M., Lebrun, M., Aho, A., & Giannopoulos, G. (2023). *Hybrid threats: A comprehensive resilience ecosystem*. Publications Office of the European Union. https://doi. org/10.2760/37899, JRC129019

Kalkman, J. P. (2020). Boundary spanners in crisis management. *International Journal of Emergency Services, 9*(2), 233–244.

Kapucu, N. (2006). Interagency communication networks during emergencies: Boundary spanners in multiagency coordination. *The American Review of Public Administration, 36*(2), 207–225.

Lagreid, P., & Rykkja, L. H. (2015). Organizing for "wicked problems"–analyzing coordination arrangements in two policy areas: Internal security and the welfare administration. *International Journal of Public Sector Management, 28*(6), 475–493.

Larsson, S., & Rhinard, M. (Eds.). (2021). *Nordic societal security. Convergence and divergence*. Routledge.

Ministry of Justice and Public Security. (2017a). *Meld. St. 10 (2016–2017) Risiko i et trygt samfunn—Samfunnssikkerhet. (Meld. St. 10 (2016–2017))*. Oslo. www. regjeringen.no

Ministry of Justice and Public Security. (2017b). *Meld. St. 38 (2016–2017) IKT-sikkerhet—Et felles ansvar*. Oslo. https://www.regjeringen.no/no/dokumenter/ meld.-st.-38-20162017/id2555996/

Ministry of Justice and Public Security. (2018a). *NOU 2018: 14 IKT-sikkerhet i alle ledd—Organisering og regulering av nasjonal IKT-sikkerhet*. Oslo. https://www. regjeringen.no/no/dokumenter/nou-2018-14/id2621037/

Ministry of Justice and Public Security. (2018b). *Prop. 1 S (2018–2019). Proposisjon til Stortinget (forslag til Stortingsvedtak) for budsjettåret 2019.* https://www. regjeringen.no/contentassets/f4d34526e4914a5793a9b4c668cf70b7/no/pdfs/ prp201820190001_jddddpdfs.pdf

Ministry of Justice and Public Security. (2022). *Regjeringen vil styrke sikkerheten, særlig i nord.* Retrieved 25 June. https://www.regjeringen.no/no/aktuelt/ regjeringen-vil-styrke-sikkerheten-sarlig-i-nord/id2907492/

Mintzberg, H. (1983). *Structure in fives: Designing effective organizations.* Prentice-Hall.

Nilsson, N., Weissmann, M., Palmertz, B., Thunholm, P., & Häggström, H. (2021). Security challenges in the grey zone: Hybrid threats and hybrid warfare. In M. Weissmann, N. Nilsson, B. Palmertz, & P. Thunholm (Eds.), *Hybrid warfare. Security and assymetric conflict in international relations* (pp. 47–57). I.B. Tauris Bloomsbury Publishing.

NOU 2012: 14. (2012). *Rapport fra 22. juli-kommisjonen.* Oslo. www.regjeringen.no

NSM. (2023). *Nasjonalt cybersikkerhetssenter (NCSC).* Retrieved 29 June. https:// nsm.no/fagomrader/digital-sikkerhet/nasjonalt-cybersikkerhetssenter/

O'Brien, A., Read, G. J. M., & Salmon, P. M. (2020). Situation awareness in multi-agency emergency response: Models, methods and applications. *International Journal of Disaster Risk Reduction, 48,* 1–11.

Owen, C., Bearman, C., Brooks, B., Chapman, J., Paton, D., & Hossain, L. (2013). Developing a research framework for complex multi–team coordination in emergency management. *International Journal of Emergency Management, 9*(1), 1–17.

Rittel, H. W. J., & Webber, M. M. (1973). Dilemmas in a general theory of planning. *Policy Sciences, 4*(2), 155–169. https://doi.org/10.1007/BF01405730

Stanton, N. A., Stewart, R., Harris, D., Houghton, R. J., Baber, C., McMaster, R., Salmon, P., Hoyle, G., Walker, G., & Young, M. S. (2006). Distributed situation awareness in dynamic systems: Theoretical development and application of an ergonomics methodology. *Ergonomics, 49*(12–13), 1288–1311.

Thomson, J. (2021). *Conflict and cooperation in intelligence and security organisations: An institutional costs approach.* Routledge.

Tjora, A. (2017). *Kvalitative forskningsmetoder i praksis* (3rd ed.). Gyldendal akademisk.

Valaker, S., Hærem, T., & Bakken, B. T. (2018). Connecting the dots in counterterrorism: The consequences of communication setting for shared situation awareness and team performance. *Journal of Contingencies and Crisis Management, 26*(4), 425–439.

Van der Veer, R., Bos, W., & van der Heide, L. (2019). *Fusion centers in six European countries: Emergence, roles and challenges.* The International Center for Counter-Terrorism.

12

HYBRID THREAT INTELLIGENCE THROUGH VIRTUAL NETWORKS

The role of network entrepreneurs

Odd Jarl Borch

Introduction

Intelligence gathering and assessment are vital to making sense and creating a shared threat picture (Rühle & Roberts, 2021). However, detection of activities, distribution of information about suspicious actions, and connecting of the "dots" of information may prove challenging. Notaker (2023) highlights how the complexity related to the number of aggressors' building blocks, the ambiguous nature of the links between them, and surveillance limitations due to laws and regulations may increase the number of intelligence "blind spots".

To investigate hybrid threats, we need to cover a broad range of areas with high-quality data access. Hybrid threats may be regarded as extremely intelligence intensive as one has to cover a broad range of arenas, instruments, and aggressors (Häggström, 2021). Keeping an intelligence capacity as a constant team of responders or establishing a permanent organization is important not the least at strategic, national level, as illuminated in by Sandbakken and Karlsson (chapter 11). Building intelligence organizations covering all parts of society at all levels may, however, prove time-consuming, inadequate, and costly. This study illuminates how the mobilization of intelligence resources through ad-hoc, tailor-made networks may represent a prudent alternative. An open-source, grassroot approach is regarded as of special relevance for intelligence at operational and tactical levels, where a vigilant population may represent a critical asset (Treverton, 2021).

This paper focuses on the role and relationship of the network entrepreneur in mobilizing and running adapted grassroot networks for intelligence gathering. A network entrepreneur is a person who establishes links to and

DOI: 10.4324/9781032617916-15

between different resources and who provides the necessary governance mechanisms. Tailor-made, ad hoc, and easy adjustable network organizations have been named virtual networks because of their flexibility and elasticity as to tasks, staffing, and cross-over between organizations and geographical regions (Davidow & Malone, 1992). In this chapter, a virtual organizational form is presented as a fast and flexible solution for integrating actors from a wide range of civilian-military, private-government, voluntary organizations as well as the public. How the relationship between a varied set of actors is developed, and how these resources may be mobilized, interlinked, and governed by the network entrepreneur are illuminated.

Theory

Hybrid threats and intelligence

The composite, transboundary, often hidden, and diffuse nature of crises such as hybrid threats call for fast detection, information gathering, and sensemaking as a platform for appropriate response action (Boin 2019; Cullen & Wegge, 2021; Weissmann et al., 2021). However, access to adequate information and understanding of the crisis pattern may prove challenging. Galeotti (2022) illuminates how antagonist countries and groups may achieve their goals in a competitive and conflict-ridden situation through a "whole of society" attacker approach. Such an approach may include tools such as disinformation, falsifying narratives, and destabilization through conflict stirring, crime, and sabotage. As shown in several chapters in this book, these actions may be synchronized and appear in different habits and combinations over time. There may be areas where we do not expect these threats, we may not see the relations between different incidents, and we may not be able to attribute the threats to certain groups or nations, nor reveal their short- and long-term motives. A special challenge is present if institutional barriers restrict intelligence activity, as seen in many Western countries. Such "blind spots" related to cultural, legal, political, or economic aspects may severely hamper intelligence gathering and avoid the detection of threats in time for mobilization (Notaker, 2023).

Compared with conventional conflicts, hybrid threats may be regarded as more intelligence intensive as they must cover a broad range of arenas, instruments, aggressors, and have a longer time span (Häggström, 2021).

The term multilateral intelligence has been launched to highlight the need for collection, protection, and analysis of both publicly available and secret information including a broad range of sources (McGruddy, 2013). In the NATO operational planning schemes, the comprehensive approach framework has an emphasis on broader civilian-military cooperation and integration in complex conflicts. This doctrine implies a need for joint knowledge development processes providing a more holistic understanding of the operational arena in time and space, with information from a wider and more differentiated range of sources.

Treverton (2021) claims that the challenges of hybrid threat intelligence call for thinking creatively employing new media and networks. Crowd sourcing represents an important intelligence avenue. It integrates the formalized intelligence system with grassroot organizations, social media, the victims of hybrid threats, and concerned citizens (Brabham, 2013). These sources may add to the more formal systems of intelligence for the early detection of weak signals, providing the necessary breadth of the threat picture, and improving the validation process. With increased understanding of the threat picture, it may provide an apparatus for horizontal scanning, providing indications of hybrid threats as well as reflections on the logic behind an aggressor's action (Gill et al., 2021).

Virtual networks

Early warning intelligence from various sources, multilateral data collection, and collective sensemaking may prove critical for providing the necessary mobilization of information resources and adequate hybrid threat response (Cullen & Wegge, 2021). In hybrid threat defense, the speed of action is important. Sensemaking is a retrospective development process of plausible images where people extract meaning from the present information, to understand ambiguous, misleading, or bewildering issues or events (Weick et al., 2005; Brown et al., 2015). Involving individuals with different backgrounds in collecting and interpreting the data as well as information sharing activities may create a rich information space (Daily & Starbird, 2015). Within the communities, significant persons such as municipality officials, politicians, business leaders, journalists, and concerned citizens may assemble local data and reflect on the information value. Internet and social media may provide both input of information and an arena for collective reflections. The term digital volunteers have been used to describe actors in the communities linking up to provide crowdsourced information (Chernobrov, 2018). These digital volunteers may contribute through collecting, verifying, and mapping updated information about a critical incident, among others using various digital channels. Even though false information and rumors may circulate in a situation of hybrid threats, hampering sensemaking, these crowd feeding sources may prove valuable weak signals, especially before a broader range of data sources can be mobilized and data triangulation performed.

One important challenge is how these resources are recognized and linked up to a hub where the information can be critically evaluated and put together in a sensemaking process.

However, mobilizing resources from other organizations and parts of the society may be regarded as a challenging task due to sensitivity, security issues, and general risk avoidance. Intelligence officers may be reluctant to move beyond the prevailing form of alliances and the "strategic comfort zone" of the agency (Forte et al., 2000; Greenwood & Hinings, 1993). There may also be fear of giving away information that may be used against you. Creating a new network based on a patchwork of organizations and single people may be challenging, both in finding the right persons and in gaining the necessary trust-based links to the information gate keepers. Thus, the first phase of developing the virtual network may be to build awareness, create a platform of trust, and open channels for a broad range of input from the population as well as stakeholders.

The network entrepreneur

Single people may represent a critical force within the intelligence community. Not the least within the process of designing intelligence networks. The concept network entrepreneur underlines the constructive capabilities of the person building new structures. The network entrepreneur must detect and unite various sources, reveal threat and information needs, build social trust relations, and develop flexible collaboration structures (Larson, 1992; Zhao & Aram, 1995).

An extra challenge as to hybrid threat intelligence is that the sources may have a different sense of rationality linked to their thinking about threats and information gathering (Weber, 1978; Habermas, 1985). There must be governance mechanisms according to the motivation, strategic importance of the relation, and the risks involved for all parties (Borch, 1992; Kapucu & Hu, 2022). The governance mechanisms should safeguard the interests of its members and reduce the thresholds for participation. To achieve openness and trust, the network entrepreneur must manage several arenas of interpersonal action. This includes not only the instrumental systems based on formal positions and authority but also social-emotional and political-strategic elements, providing the necessary motivation and trust between the source and the network architect (Weber, 1978; Habermas, 1985).

The governance mechanisms that serve as a starting point for the network entrepreneur may build upon a formal institution, including formal contracts and positions that may provide both professional trust and a formal authority. Second, the administrative coordination mechanisms may have to be developed providing joint arenas for interaction and dialogue and predictable role behaviors. Third, to achieve greater flexibility and acceptance, a relational contract based on personal-cultural relations, common values, norms, and emotions may allow for higher-order trust and more rich information, an active contribution to the sensemaking process, and a predictable, future commitment (Larson, 1992; Larson & Starr, 1993). The delicate and subtle aspects of relational contracts depend on the network entrepreneur's understanding of the social fabric in the community. Relations are rooted in the informants' life sphere, expressed feelings and needs, friendship, and artistic expression. These socio-cultural and subjective-emotional relations of a relational contract may add to both the motivation and engagement in the collaboration (Borch, 1994; McLaughlin & Read, 2014). Through a broad set of contracts, you may also achieve an increased level of systemic complexity by adding the number of transaction partners, the range of information flows, reducing uncertainty about future behavior, and protecting each participant from opportunistic behavior from the others.

Methodology

This research built upon an exploratory, longitudinal study examining a range of incidents in the Norwegian county of Finnmark over a period of six years. Secondary data collected from written sources represented the starting point of this study. Government evaluations and policy documents gave further details as to threats and response measures. Newspaper and other media articles were an important corrective and platform to reveal the processes at the time from persons outside the process. Third, interviews were performed with key personnel involved in the process of gathering data.

The analysis of data was characterized by an interactive dialogue between theory, collected data from the actors, written documents, and own experience using data triangulation (Borch & Arthur, 1995; Yin, 2009). The theoretical framework was confronted with the empirical findings, developing a more detailed model of the process of developing a virtual network.

To increase validity and open up other, more critical perspectives and views of the process, transcripts of interviews, written documents, and analyses were evaluated by independent researchers in the field. They contributed by pointing out weaknesses in reliability and biased opinions and giving innovative ideas for presentation.

Data and analysis

Context: The county

Finnmark is the northernmost and largest county in Norway by area, about the size of Switzerland. It is also a sparsely populated region with around 75,000 inhabitants. Finnmark has the only Norwegian border to Russia. The region also borders to the Northern part of Finland and Sweden. The sea areas outside the county include both the Norwegian Sea and the Barents Sea, reaching up to the Svalbard archipelago. The sea areas are important fishing grounds and have significant oil and gas reserves. There is a strict quota regime on fish resources in the Barents Sea, where Norway and Russia are the dominant stakeholders. In this sea area, the Russian Navy has most of its second strike capacity, and sea-based nuclear deterrent capability is operating out of the Kola Peninsula from its base a few miles from the Norwegian border. The region is multi-cultural. About one thousand inhabitants (1.4 % of the population) have Russian background.

Especially after the Cold War, the people of the Arctic region experienced a broad range of academic, cultural, and commercial exchange with Russia. This also included the development of joint political fora, such as the Barents Euro-Arctic Council, Arctic Council, and Arctic Coast Guard Forum. Within maritime preparedness, partnership agreements on search and rescue and oil spill response went back to the Soviet area. All collaboration, except border control, the maritime SAR partnership, and the joint fish stock governance, ended abruptly with the Russian full-scale war against Ukraine in February 2022.

After Russia's annexation of Krim in 2014 and the following sanctions, Norway experienced a range of destructive incidents from outside aggressors. Norway was in 2023 on the Microsoft list of the top ten countries that had suffered most cyberattacks, including phising, malware, and digital espionage. Russian military started jamming the GPS-satellite navigation signals over the Finnmark region in 2023 almost daily, causing navigation problems for air passenger and medivac traffic (Forsvarets Forum, 2024). In 2015, Russia, without further notice, sent over five thousand refugees over the border crossing point at the small town of Kirkenes, causing severe strains on the receiving institutions on the Norwegian side. Other issues have been so-called memorial diplomacy with Russian liberation memorials located in sensitive regions used for political purposes, conflicts about open harbours for the Russian fishing fleet, and tension about the local effect of sanctions against Russia.

Finnmark police district is responsible for the safety and security of the population in Finnmark. The police district has approximately 600 employees. The chief of police is located in the border town of Kirkenes. The police district is responsible for the border station between Norway and Russia at Storskog and has authority over the border guard troops from the Norwegian military along the whole Norwegian-Russian border.

The background of the network entrepreneur

The network entrepreneur has a broad background from the police and is also born in this region. She is now serving as the head of police in Finnmark. Her family has roots in the indigenous Sami population. In her youth, she was active in snowmobile speed races and won medals in regional competitions, a very popular sport in Finnmark. After law education, she served with the police district in various positions before she became regional chief of police. She has received several awards and prizes for her contribution to a secure society, among others from the Norwegian Army and the Finnish border guard. After managing the mass immigration crisis in Eastern Finnmark in 2015, she was chosen by the newspapers in the region as *the Person of the Year*. In March 2023, she received the reward *the Preparedness prize 2022* from the Norwegian Total Defense Forum for her concern for society safety and preparedness, for having a clear and strong voice, and for her stamina in turbulent times.

Building virtual intelligence networks

In a sparsely populated area with long distances and a varied, multi-cultural population, with many immigrants, building an intelligence network is a challenging task. Both access to information and trust in the receiver are extremely important. The network entrepreneur started out emphasizing an open dialogue with all regions and groups. She frequently visited local communities and expressed her opinions in the newspapers and other media. She took part in both local and national debates, underlining the challenges of the region. She was clear about the dilemmas emerging from the increased tension that sanctions against Russia caused, the importance of building a resilient population in the region, and the difficulties of having the national authorities understand the local challenges.

Developing her broad regional network further through informal dialogue and being a media person provided a platform for relation building and to bring together resources without too much bureaucracy. The administrative capacities of the local police helped in creating meeting places in local communities and with municipality leadership. Informal networks and trust developed served together with the formal position as regional chief of police to facilitate information input.

During the mass immigration across the Russian border in 2015, she profited from these efforts in the handling of approximately 5,000 refugees arriving at the border post for two months. The mass immigration happened suddenly with no early warning.

In the mass immigration case in 2015, we had to react quickly based on weak signals of what may come and make sense of a chaotic situation. We had to bring together a broad range of people to find out how to take care of the immigrants with food and shelter, check their identity and give them a medical check, and inform all levels. It was necessary to develop close contacts with a broad range of resources and stakeholders such as the municipality leaders, hospitals, transport companies, volunteer organizations, etc.

We had to coordinate activity, create a joint situational awareness about what was happening each day on the border, and inform both internally and externally. This included the higher levels of the county administration, the national police directorate, intelligence services, and the immigration authorities.

(Regional newspapers, Northern Norway, 2023)

To face the imminent crisis, different ad hoc, cross-sectoral management teams with representatives from several institutions were developed and authorized to work closely together.

> We had to develop an ad hoc management team from the different actors for the creation of information packages, decision-making and implementation. This went through several steps.
>
> *(Interview by author, June 6, 2023)*

The mobilization role at higher levels was, however, a challenge. The network entrepreneur tried to inform through several channels and to receive more resources. However, there was a lack of understanding at higher levels of the severeness of the crisis. There were also challenges about who should fund the response measures introduced at local levels. Third, there were discussions about the legal aspects and the rights of the immigrants, among other, as to their refugee status.

> We received limited response from regional and national levels for a long time. Normally, the county governor should serve as regional coordinator of response and resource acquisition. But they did not respond fast enough. We also had limited response from national levels, especially the immigration authorities. . . . It is important to create joint situational awareness in all channels and sectors. We maybe did not inform well and make it formal enough.
>
> *(Interview by author, June 6, 2023)*

The discussion above shows the importance of creating trust-based contacts through an open and frank relation to communities and local institutions. Highlighting the potential threats through her frequent media presence increased the awareness in the population and gave the network entrepreneur a central position as an information exchange hub. The efforts to build relations over years became an asset at local level when mobilization to a major crisis was necessary.

The incident also shows the challenges of creating the necessary situational awareness and mobilizing networks across all sectors and levels. This may have to do with formal-legal mechanisms and the prepared principles of crises managed at the lowest possible level, the unwritten rules of not crossing sector borders, as well as budget limitations for unforeseen expenses within the public administration. Thus, dealing with the national level may rely more on the formal and administrative governance mechanisms than the informal networks to achieve the necessary connections.

After the full-scale Russian invasion of Ukraine, the Norwegian government acknowledged the need for strengthening the intelligence capacities of the region.

The Ministry of Justice gave the Police Security Service (PST) increased funding to establish more regional capacity to support the police district and its chiefs of police. Finnmark police district received significant additional capacity. The transboundary and diffuse nature of hybrid threats also provided challenges at national intelligence levels and ended in a new joint hybrid threat intelligence network between the Police Security Service, the Norwegian Intelligence Service, the National Cyber Security Center (NSM), and the Police. The objective was to detect and analyse hybrid threats and vulnerabilities and provide decision support among others, through joint situational awareness reports.

> The reports from the national level have now become an asset for the regional police. The center works all right, and we contribute to their analyses through the regional office of the Police Security Service.
> *(Interview by author, June 6, 2023)*

The foundation of a formal joint venture between the intelligence and security agencies showed the importance of broad networks to meet hybrid threats, with both vertical and horizontal integration.

The level of trust between the local and national level was, however, a challenge. In the region, there had for decades after the Cold War been close collaboration with Russia. Over the years, several high-level meetings and arrangements had taken place. Good local relations had developed with cultural co-arrangements, frequent visits over the border, and trade relations. This changed dramatically after February 2022, with the Russian full-scale invasion, in Ukraine. This abrupt stop in cross-border relations was, however, difficult for the community to grasp. It took time to understand the changes in the threat picture after 30 years of collaboration. Many were reluctant to see Russia as an aggressor in a regional context.

The Police Security Service experienced skepticism hampering cooperation with the local population (NRK, 2022a). In February 2020, the Intelligence Service's assessment of current security challenges warned that Russia tried to stir conflict between national authorities and local interest groups. In response, the mayor of the border town claimed that the foreign intelligence service in their open threat assessment tried to make the community collaboration with Russia suspect through their statements. He emphasized the need for stricter control of the Norwegian intelligence services and their activities.

> It is surprising that (the director of the Intelligence Service) says this: it means that all of us that wants a good and close collaboration with Russia is regarded as a tool for Kremlin. I find this very regrettable.
>
> *(Klassekampen, 2020)*

This was a severe attack on the authority of the national intelligence resources. It showed the difficult balance of increasing vigilance among the population on the one side, and not stirring antipathy and a feeling that all cross-border contact was suspect on the other side. The network entrepreneur as a regional chief of police was frank about this in a newspaper chronicle:

> They (the intelligence services) are doing their job, telling what they experience and focus on what we must be aware of. I have served at the border for almost 20 years and seen the change in rhetoric. Even though we live here in in the border region we should not be naïve. We may think that we know Russia better than those in the south. But we cannot run our own private foreign policy not being in line with the national one. We have to believe that the national authorities have the best competence and know things we are not aware of . . .
>
> *(Haetta, 2020)*

The same type of conflicts emerged in a discussion on what was claimed to be influence operations from Russian institutions. Researchers from the regional university, warned against the memorial diplomacy and the cynical history-based propaganda strategy that lay behind. They claimed that we had not been able to register, prepare for and protest this type of offensive (Myklebost & Markussen, 2023; Myklebost, 2023). This stirred significant criticism towards both the researchers and the newspaper in the border town. The researchers were severely criticized by a local author for creating myths about the region and creating doubts about the personal integrity of the many who has worked with the Russians on WW2 history (Olsen, 2023).

The region was vulnerable for this kind of conflicts around the narrative, both of WW2 narratives and the threat picture of Russia. NRK—The National Public Broadcaster claimed that there was a risk of severe tension in this region as parts of the Russian population in the region supported Russia's war in Ukraine (NRK, 2022b).

The cases above shows the range of challenges that were present when it comes to sensing the incidents and create an understanding of the situation, extracting real threats from democratic expressions of different opinions. Integrating different agencies, industries, and the community level to provide the necessary situational awareness was a challenging task with rising tensions in the population. In an interview with the national broadcasting company, the network entrepreneur expressed that the dependency of open borders in the region have done something about the people's view of Russia. She confirmed the skepticism against the security services in the region.

> It is not naivety but dependency that have made us neglect breach of human rights in Russia, for example against journalists . . . People is reluctant to talk to officials from the Police Security Agency. One may think that Russian intelligence may find out about the contact.
>
> *(NRK, 2022a).*

The network entrepreneur in an interview with the leading national business newspaper October 16, 2022, claimed that we as a nation had not given sufficient notice to hybrid threats.

> We have no tradition of thinking this way. My region may have become a test bed for Russia where hybrid threats are evaluated for their use in hybrid warfare. Russia is good at their job. We must respond as best we can. The best thing would be if we could use my region as a laboratory for testing out new response solutions, to make us capable of locating the hybrid threats. More knowledge would be available if one dares to make more efforts in this region.
>
> We can now see that the people are more vigilant and report from incidents. Earlier one wrote about an incident on Facebook. Now they send messages to the police, and that is good.
>
> *(Dagens Næringsliv, 2022)*

In the neighbouring police district, the Chief of Police warned industry leaders against increased security threats. Foreign spies are using honey traps and performing influence operations. She was talking about significant increase in intelligence threats after the full-scale war against Ukraine in 2022.

> We now must be much more aware, notice suspicious activity and put the incidents together. Recent incidents have been Russian citizens using drones over restricted areas and simulated photo sessions close to military installations.
>
> *(High North News, 2023)*

In February 2024, the police district launched a brochure to the population and a dialogue with the tourist industry to inform them about suspicious behaviour that should be reported to the police. In a radio interview, the network entrepreneur claimed that new weapons from other countries may be tested in her region, such as mass immigration, fake news, and illegal use of drones close to military installations (NRK, 2024).

The network entrepreneur and the governance mechanisms

This study shows how building intelligence networks for hybrid threats depend on a broad range of trust relations, and sensitivity as to differences in local interests, values, and emotions in a period of increased tension. Warning against threats and efforts towards educating the population may be regarded as a provocation. The network entrepreneur of this study had the position as chief of police in a county bordering to Russia. This provided her with a formal position and authority, as well as an administrative platform through police liaisons in each municipality. However, relational contracts with the people in her home district built over a long time served as an especially important asset.

The network entrepreneur emphasized from her start as chief of police the importance of being frank and open in here dialogue with her constituency. The culture of this region is very much about being outspoken and expressing directly what one means about a case. The network entrepreneur followed suit, and interviews and speeches were frank about the challenges present. This also included critics of the national authorities, if necessary.

When the network entrepreneur was elected the county person of the year by the largest newspapers in the region, she gave an interview where she reflected about the dilemma of being open about information, and the challenges of reaching out to the national levels in a severe crisis, such as the mass immigration across the Russian border. In an interview, she claimed:

> It was extremely tough months facing the migration crisis. All employees stood in it and managed to provide a high-quality job. Extra challenging it was to be open about the many immigrants not being refugees crossing the border. You must balance your message and not giving the impression that you have prejudices. You must be objective and neutral. This was emotionally challenging.
>
> It took a long time before we had help from national authorities, but this was understandable as this was a crisis they did not recognize and did not know how to respond.
>
> *(IFinnmark, 2015)*

The frank and outspoken strategy also meant that there were critics, both at national levels and sometimes from local interest groups. As a response to recommendations from the chief of police that all harbors should be closed for all Russian vessels, including fish trawlers, local stakeholders responded negatively. In one article in regional newspapers, local and regional leaders of the labour union, local politicians, the local business association, and two leading businesspeople expressed "concern about an isolationist policy that leads to closed borders, with no empathy for the people that live in the border areas". The network entrepreneur, however, continued to present openly the challenges:

> We need to have a joint understanding of the threat situation to avoid unnecessary fear. The Police Security Service (PST) tells us that we must expect a more aggressive Russia. In the North, we have important values to protect. We encourage the industry to build a common situational awareness together with the police and discuss how the society may oversee the threat. You may invite the police to help in your threat analyses.
>
> *(Regional newspaper, 2023)*

The network entrepreneur showed through several threat incidents that she had built relations that served as fast responding and direct channels of communication. When the Russian started jamming the GPS satellite positioning in the region in 2017, the pilots of the passenger planes contacted her immediately directly through SMS. She was then able to use her formal position to open a dialogue with both the aviation and Ekom authorities and contribute to the mobilization of the Ministry of Foreign Affairs. She also went to the media informing about the challenges present.

The discussion above shows how a combination of a trust-building, open and frank dialogue with different groups of the population, a formal position that gives both authority and links to all levels of government, and a well-functioning and fine-grained, administrative platform may represent the combination of governance mechanisms for an effective information network to work in the local communities. The data presented also shows the strain of "standing in the storm" in a situation where the population is moving from an extended period of strong socio-cultural and commercial relations with a seemingly friendly neighbour, towards national introduced sanctions and restrictions on exchange, and a gradually and partly hidden increase in aggressive action from a neighbor.

Conclusion

This study emphasizes the challenges of information gathering, sensemaking, and creating joint situational awareness in a society where the threat picture suddenly changes. It shows the challenges of building information channels in a region with strong cultural, economic, and political tensions. Information about threats from a neighbouring country may be met by skepticism and critics about not respecting the people-to-people collaboration. The society may meet such allegations with skepticism and critique and refuse to contribute.

In a hybrid threat context, this study illuminates the need to both motivate and inform the population to increase vigilance and provide the necessary trust in authorities. This may be challenging if there is a tension between groups, regions, and towards the national government. In many countries, there may be strong political voices in rural areas against the national government and conflict of interests between groups. These tensions may be strengthened by the aggressor building dividing narratives, stirring up conflict areas, and stimulating demonstrations against the national response measures, including national security agencies.

This study emphasizes the need for a broad range of sources to provide the necessary intelligence related to hybrid threats. It highlights the development of virtual context-adapted and tailor-made networks for fast and adequate response to hybrid threats, and the role of single persons and organizations providing the necessary foundation for information gathering. The study shows the critical importance of a network entrepreneur in developing a network based on trust. A broad set of governance mechanisms may be activated both to motivate, provide the necessary formal backing and create both professional and personal trust within the network. To achieve this, the role of distinct types of trust should be highlighted. The formal position of the network entrepreneur illuminating the capacity behind the efforts may provide structural trust, showing high degree of professionalism may be an additional trust facilitator. More important is the network entrepreneur showing strength as well as alignment with the virtues and values of the society creating personal, norm-based trust.

This study has its limitations in looking at a single network entrepreneur in a specific region only. The data is collected mostly from local and secondary sources. Further studies should make comparative studies across regions, illuminating the common characteristics of network entrepreneurs and the build-up of several types of networks with adjoining governance mechanisms. How to bridge regional and personal tensions and create trust between different actors in information networks in turbulent times may represent a prudent avenue of research.

References

Boin, A. 2019. The transboundary crisis: Why we are unprepared and the road ahead. *Journal of Contingencies and Crisis Management,* 27, no. 1 (Summer): 94–99.

Borch, O. J. 1992. Small firms and the governance of interorganizational exchange. *Scandinavian Journal of Management,* 8, 321334.

Borch, O. J. 1994. The process of relational contracting. Developing trustbased strategic alliances among small business enterprises. In P. Shrivastava, J. Dutton, & A. Huff (eds.). *Advances in Strategic Management.* Greenwich, Connecticut: JAI Press Inc.

Borch, O. J., & Arthur, M. B. 1995. Strategic networks among small firms: Implications for strategy research methodology. *Journal of Management Studies,* 32, no. 4: 419–441.

Brabham, D. C. 2013. *Crowdsourcing.* Boston: MIT Press Essential Knowledge.

Brown, A. D., Colville, I., & Pye, A. 2015. Making sense of sensemaking in organization studies. *Organization Studies,* 36, no. 2: 265–277.

Chernobrov, D. 2018. Digital volunteer networks and humanitarian crisis reporting. *Digital Journalism,* 6, no. 7: 928–944.

Cullen, P., & Wegge, N. 2021. Hybrid warfare. In S. Stenslie, L. Haugom, & B. H. Vaage (eds.). *Intelligence Analysis in the Digital Age*. London: Routledge.

Dagens Næringsliv. 2022. Politimesteren i Finnmark: Nord-Norge er et laboratorium, hvor Russland eksperimenterer. Interview National Business Newspaper, *Dagens Naeringsliv*, October 10, 2022. https://www.dn.no/teknologi/ellen-katrine-hatta/finnmark/droner/politimesteren-i-finnmark-nord-norge-er-et-laboratorium-hvor-russland-eksperimenterer/2-1-1335355?zephr_sso_ott=pxmliN

Daily, D., & Starbird, K. 2015. "It's Raining Dispersants." Collective Sensemaking of Complex Information in Crisis Contexts. CSCW'15 Companion, March 14–18, 2015, Vancouver, BC, Canada.

Davidow, W. H., & Malone, M. S. 1992. *The Virtual Corporation. Structuring and Revitalizing the Corporation for the 21st Century*. New York: Harper Collins Publishers.

Forsvarets Forum. 2024. Voldsom økning i GPS-forstyrrelser. Feature, February 26, 2024. Oslo.

Forte, M., Hoffman, J. J., Lamont, B. T., & Brockmann, E. N. 2000. Organizational form and environment: An analysis of between-form and within-form responses to organizational changes. *Strategic Management Journal*, 21: 753–773.

Galeotti, M. 2022. *The Weaponizing of Everything. A Field Guide to the New Way of War*. New Haven, USA: Yale University Press.

Gill, M., Heap, B., & Hansen, P. 2021. *Strategic Communications—Hybrid Threats Toolkit. Applying the Principles of NATO Strategic Communications to Understand and Counter Grey Zone Threats*. Riga, Latvia: NATO Strategic Communications Centre of Excellence. ISBN: 978-9934-564-38-3.

Greenwood, R., & Hinings, C. R. 1993. Understanding strategic change: The contribution of archetypes. *Academy of Management Journal*, 36: 1052–1081.

Habermas, J. 1985. *The Theory of Communicative Action: Volume 2: Lifeword and System: A Critique of Functionalist Reason*. Boston: Beacon Press.

Haetta, E. K. 2020. Falske motsetninger på grensen mot Russland. Chronicle, *IFinnmark*, February 25, 2020. https://www.ifinnmark.no/falske-motsetninger-pa-grensen-mot-russland/o/5-81-1126024

Häggström, H. 2021. Hybrid threats and new challenges for multilateral intelligence cooperation. In M. Weissmann, N. Nilsson, B. Palmertz, & P. Thunholm (eds.). *Hybrid Warfare: Security and Asymmetric Conflict in International Relations*. London: I.B. Tauris, Bloomsbury.

High North News. 2023. Politimesteren i Nordland advarer: Etterretningstrusselen fra Russland har aldri vært større. Interview, *High North News*, September 29, 2023. https://www.highnorthnews.com/en/nordland-chief-police-russian-intelligence-threat-has-never-been-greater

IFinnmark. 2015. Ellen Katrine er årets finnmarking i 2015: Tøft å stå i immigrasjonsbølgen. Interview, December 30, 2015. https://www.ifinnmark.no/alta/nyheter/sor-varanger/ellen-katrine-er-arets-finnmarking-i-2015-toft-a-sta-i-immigrasjonsbolgen/s/5-81-178757

Kapucu, N., & Hu, Q. 2022. An old puzzle and unprecedented challenges: Coordination in response to the COVID-19 pandemic in the US. *Public Performance & Management* Review, 45, no. 4: 773–798.

Klassekampen. 2020. Mistenkeliggjør mitt arbeid. *Feature*, February 12, 2020. https://klassekampen.no/utgave/2020-02-12/mistenkeliggjor-mitt-arbeid

Larson, A. 1992. Network dyads in entrepreneurial settings: A study of the governance of exchange relationships. *Administrative Science Quarterly*, 37, no. 1: 76–104.

Larson, A., & Starr, J. A. 1993. A network model of organization formation. *Entrepreneurship Theory and Practice* (Winter): 5–15.

McGruddy, J. 2013. Multilateral intelligence collaboration and international oversight. *Journal of Strategic Security*, 6, no. 3 (Suppl.): 214–220.

McLaughlin, J., & Raed, E. 2014. Ian Macneil and relational contract theory: Evidence of impact. *Journal of Management History*, 20, no. 1: 44–61. https://doi.org/10.1108/JMH-05-2012-0042.

Myklebost, K. A. 2023. Memory diplomacy of the borderland. Russian-Norwegian patriotic memory tours, 2011–2019. *Nordisk Østforum*, 37: 130–155.

Myklebost, K. A., & Markussen, J. A. 2023. Norge under russisk minnepolitisk press. Chronicle, *Nordlys*, September 8, 2023. https://www.nordnorskdebatt.no/norge-under-russisk-minnepolitisk-press/o/5-124-264790

NRK. 2022a. Finnmarkinger skeptisk til PST. Interview PST-Official, NRK-Norwegian National Broadcaster, July 12, 2022. https://www.nrk.no/tromsogfinnmark/finnmarkinger-skeptisk-til-pst-1.16035086#:~:text=Finnmarkinger%20skeptisk%20til%20PST%20Politiets%20sikkerhetstjeneste%20har%20i,russisk%20etterretning%2C%20som%20er%20sv%C3%A6rt%20aktive%20i%20%C3%98st-Finnmark.

NRK. 2022b. Russere i Finnmark støtter krigføringen: Skaper strid i det russiske miljøet. Feature NRK-Norwegian National Broadcaster, March 26, 2022. https://www.nrk.no/tromsogfinnmark/russere-i-finnmark-stotter-krigforingen_-skaper-strid-i-det-russiske-miljoet-1.15904220

NRK. 2022c. PST om Finnmark: Det er en skepsis mot å snakke med oss. Interview, NRK-Norwegian National Broadcaster, July 22, 2022. https://www.nrk.no/norge/pst-om-finnmark_-_-det-er-en-skepsis-mot-a-snakke-med-oss-1.16012973

NRK. 2024. Politiet til finnmarkingene: Slik avslører du russisk spionasje. Feature, NRK-Norwegian National Broadcaster, February 17, 2024. https://www.nrk.no/tromsogfinnmark/politiet-laerer-folk-i-finnmark-a-takle-trusselen-fra-russland-1.16767767.

Notaker, H. 2023. In the blind spot: Influence operations and sub-threshold situational awareness in Norway. Journal of Strategic Studies, 46, no. 3: 595–623. https://doi.org/10.1080/01402390.2022.2039634.

Olsen, P. K. 2023. Tvilsom forskning på Norge – Russland. Chronicle, *Nordlys*, September 13, 2023. https://www.nordnorskdebatt.no/tvilsom-forskning-pa-norge-russland/o/5-124-265090

Regional Newspaper. 2023. Speech at the Regional Conference 2023, Regional Newspaper, May 4, 2023. https://www.e-pages.dk/finnmarkdagblad/24698/?gatoken=dXNlcl9pZD0xNDYxMjE5OS0zM2JiLTQ2MzQtODYwMC0xODkxYTQ3YmJmMDMmdXNlcl9pZF90eXBlPWN1c3RvbQ%3D%3D&token=YV91c2VyX2lkPTY4NGQ0MWFiLWVlM2ItNDdhMC04ZDFjLTRhZDc3ZTJhNGMxYSZhX3VzZXJfaWQ9PTE0NjEyMTk5LTMzYmItNDYzNC04NjAwLTE4OTE4OTY4ZHNiYmYwMw%3D%3D

Regional Newspapers, Northern Norway. 2023. Interview Chief of Police Finnmark, June 6, 2023.

Rühle, M., & Roberts, C. 2021. Enlarging NATO's Toolbox to Counter Hybrid Threats. *NATO Review*, March 19, 2021. https://www.nato.int/docu/review/articles/2021/03/19/enlarging-natos-toolbox-to-counter-hybrid-threats/index.html

Treverton, G. F. 2021. Hybrid threat and intelligence. In M. Weissmann, N. Nilsson, B. Palmertz, & P. Thunholm (eds.). *Hybrid Warfare. Security and Asymmetric Conflict in International Relations*. London: I.B. Taurus.

Weber, M. 1922/1978. *Economy and Society*. Berkeley: University of California Press.

Weick, K. E., Sutcliffe, K. M., & Obstfeld, D. 2005. Organizing and the process of sensemaking. *Organization Science*, 16: 409–421.

Weissmann, M., Nilsson, N., Palmertz, B., & Thunholm, P. (eds.). 2021. *Hybrid Warfare. Security and Asymmetric Conflict in International Relations*. London: I.B. Taurus.

Yin, R. K. 2009. *Case Study Research. Design and Methods* (4th ed.). Thousand Oaks: Sage Publications.

Zhao, L., & Aram, J. D. 1995. Networking and growth of young technology-intensive ventures in China. *Journal of Business Venturing*, 10: 349–370.

PART IV

Knowledge and resilience

13

HYBRID THREATS AND COMPREHENSIVE DEFENCE IN SMALL DIVERSE SOCIETIES

The case of Estonia

Kairi Kasearu, Tiia-Triin Truusa and Liina-Mai Tooding

Introduction

The security environment of constantly changing and evolving hybrid threats, aimed at exploiting a country's vulnerabilities and seeking to undermine fundamental democratic values and liberties (Joint Communication to the European Parliament and the Council. Joint Framework on Countering Hybrid Threats a European Union Response, 2016; see also chapter 2 by Bjørge and Høiby, and by chapter 3 by Ellingsen in this volume), underscores importance of fostering societal resilience and cohesion. In recognition of this, our chapter asks the question: *how is ethnic diversity associated with the will to resists threats to the state's welfare and independence?*

Front-line countries, with comparably minor territories with small yet diverse populations, face the important question of ensuring internal social cohesion in case of an adverse event, be it crisis or trans-boundary crisis spanning multiple domains of societal life, hybrid threats, or war. These events place weighty demands on society through the need to pool resources and engage in a joint effort. The "rally 'round the flag" effect, bringing people together for a cause, is short lived and not sustainable in a protracted crisis or indeed in the near constant presence of multifarious hybrid threats. This reinforces that societal resilience and cohesion form the bedrock of comprehensive defence, which in its essence is a complex network of actors across multiple societal domains to ensure national security and well-being (see Berndtsson, this volume).

Reckoning with hybrid threats and the targeted destabilization of the European security architecture evidenced in Estonia during the Bronze Soldier crisis in 2007 (Ehala, 2009; Juurvee & Mattiisen, 2020) and continuing with various Russian activities and attacks have led to a steady decrease of the

DOI: 10.4324/9781032617916-17

sense of ontological security of European nations. Ontological security refers to an actor's confidence of expectations (like adherence to international law and order) and allows for certainty of action and the sustainment of identity (Mitzen & Schweller, 2011). Arguably states that are geopolitically situated in a near-peer threat environment such as these in the Nordic-Baltic region have developed defence and security models that reflect this ontological insecurity. Most of these countries employ a broad security concept reliant on the whole-of-society approach. Countries in the Nordic-Baltic region all use some form of total defence paradigm, a defence posture mostly adopted by nations bordering on hegemonic powers (Fiala, 2019). Largely for historic reasons, Scandinavian nations tend to use the term total defence and the Baltic States are more inclined to use comprehensive defence, however at the heart of it lies the need to mitigate the smallness of these nations and to raise awareness, readiness, and resilience of the society, the will to defend and resist.

Some countries like Finland and Estonia have adhered to total defence principles throughout the post–Cold War re-workings of the Western European military systems (Szymański, 2020). Sweden (Objectives for Swedish Total Defence 2021–2025—Government Bill 'Totalförsvaret 2021–2025'—Government.Se, 2020) and Lithuania (National Threat Assessment, 2020) have reestablished total defence thinking, with contemporized modifications, aspiring to meet today's security environment. The comprehensive defence as a total defence concept denotes the mental, physical, economic, and other potential of government structures, local governments, defence forces, and the entire nation to manage a situation of crisis and to preserve the nation (Veebel & Ploom, 2018; Wither, 2020; Lillemäe et al., 2023). Thus, this model of defence denotes a shift from state-provided security to a society provided security approach (Chandler, 2012), accentuating that all members of society are producers of security and defence (Truusa, 2022).

The aim of this chapter is to analyze and elaborate based on the Estonian example on how ethnic diversity is associated with the will to resist. First, we will frame the will to defend within the comprehensive defence paradigm and explore the theoretical points of departure concerning social cohesion. In the empirical section, we will focus on how different ethnic groups perceive threats and how threat perception is associated with the will to resist. The variation in association between threat perception, institutional trust, and the will to defend one's country across different social groups can be interpreted as an indicator of societal cohesion. We will sum up the limitations of the study and make concluding remarks.

Theory

In the comprehensive defence context, the will to defend plays an important role. It is a phenomenon that has been extensively studied as part of the

European/World Value Survey and that has been one of the focuses of study in the majority of Nordic-Baltic states. The will to defend, which means the willingness of the population to protect their country within the limits of their abilities, is mainly used as a substantive term in strategic documents that determine the development directions of the defence field, such as the Estonian Security Policy Foundations and the National Defence Development Plan. Although the will to defend is often talked about in a military sense—to defend one's country with a weapon—the term holds a broader field of meaning, especially in countries employing comprehensive defence principles, including activities that—directly and indirectly—contribute to the country's defence capability. Although the terms will to defend and will to fight are sometimes used as synonyms, the meaning and connotations differ across contexts and disciplines (Silm et al., submitted). In this chapter, we concentrate on the will to defend and resist as a constituent part of societal cohesion, which in turn is paramount for countering and dealing with hybrid threats.

Although the Nordic-Baltic region has diverging historical tracks with the Baltic states having joined the free European community only after the dissolution of the Soviet Union, these nations are, nonetheless, similar in that due to economic and socio-demographic processes, the societies of the Nordic-Baltic countries have become quite diverse. This has also been boosted by migration waves from the global south and the war in Ukraine. However, diversity and the acceptance of diversity as one of the hallmarks of a democratic society make for an ideal venue for employing hybrid tactics against a nation.

In the context of intensified geopolitical risks and the decrease of ontological security, the frontline small states struggle with existential questions, at the same time having to balance the demand for societal cohesion that belies the comprehensive defence concept and the democratic value base for accepting diversity and personal freedoms of opinion and beliefs.

Societal cohesion is the underlying principle for a comprehensive defence model to be effective. For example, in the Estonian policy documents, social cohesion is seen as one of the tools for reducing security risks. Increasing the cohesion of Estonian society means that in a crisis or armed conflict, the society would not polarize, and inhabitants would not only trust and support the state's actions in resolving the situation but also understand their active roles within the nation's security and defence structure.

The question of how ethnic diversity is associated with social cohesion has been a focus of interest for several decades. In 2006, Robert Putnam stated that ethnic diversity may negatively impact social cohesion, which led to formulating a constrict proposition—ethnic diversity lowers the quantity and quality of interpersonal contacts (Putnam, 2007). However, Van der Meer and Tolsma (2014: 474), in their review, conclude that "people in the ethnically heterogeneous environments are less likely to trust their neighbors or to have contact with them; this does not spill over to generalized trust, to informal help and voluntary work, or other forms of pro-social behavior and attitudes, at least not in Europe." However, empirical findings are highly diverse, depending on how diversity and social cohesion are conceptualized and operationalized. Ethnic diversity can be conceptualized and operationalized broadly and narrowly; the narrow approach focuses on ethnic composition and diversity as fragmentation. The broader approach considers the concentration and polarization of ethnic groups (Dinesen et al., 2020). There is even more vagueness with social cohesion (Friedkin, 2004; Larsen, 2013; Schiefer & van der Noll, 2017).

Social cohesion has been applied and elaborated on individual and group levels (Friedkin, 2004) and societal levels (*Discussion Document—Resilient Social Contracts | United Nations Development Programme, 2018*). At the individual level, social cohesion is understood as a person's attraction or attachment to a group and is measured by individuals' membership attitudes and behaviors. Group-level cohesiveness considers how group members interact and influence each other's membership attitudes and behaviors, how they form social networks, and how the social processes in networks affect individuals' attitudes and behaviors. In the current chapter, we follow the broader approach: social cohesion as the formal and informal ties holding members of a society (individuals, groups, and institutions) together horizontally (across citizens, between groups) and vertically (between citizens/groups and the state) and that involve trust, belonging, identity, and participation (*Discussion Document—Resilient Social Contracts | United Nations Development Programme, 2018*). Lowe et al. (2020) emphasize that, in general, social cohesion is high when the majority in society voluntarily follow rules and regulations and adhere to social contracts and do not consider each other as cultural "strangers" but as "similar" and believe that the same norms and expectations govern everyone. Thus, social cohesion can be seen as a causal system; the individuals and groups create social cohesion, but at the same time, it determines individuals' and groups' attitudes and behaviors. In the case of ethnic diversity and social cohesion, the main question is the interplay between the group and societal level cohesion. In the case of polarized ethnic diversity, the high ethnic-based intergroup cohesion might jeopardize societal cohesion and thus create a fertile ground for hybrid threats. It means that societal resilience and resistance might be weaker and the risk of being harmed by hybrid threats higher.

Nevertheless, the mechanism may work another way around as well. The hybrid threats, for example, disinformation operations and interior conflicts, may hinder social cohesion while magnifying intragroup cohesion and intergroup confrontation. Thus, social cohesion, in case of intensified situations of hybrid threats, is, on the one hand, the buffer and pillar of resistance, but on the other hand, it is the main target of hybrid threats.

Our analytical model relies on societal cohesion and focuses more precisely on resistance, which comprises three sub-dimensions: will to fight, will to defend, and intention to leave the country, institutional trust, and the threat perception. The pattern of associations between these three dimensions: resistance, institutional trust, and threat perception indicate the cohesion of the society. The different perceptions of threat across ethnic groups, institutional trust, and the readiness to defend the country are indicators of societal cohesion—the broader gap between the perceptions of ethnic groups denotes the lower level of societal cohesion.

Empirical evidence

Our analysis uses data from the "Estonian Public Opinion on National Defence" (*National Defence and Society | Kaitseministeerium, n.d.*) polls of Estonia that have been conducted under the aegis of the Ministry of Defence since 2000. We utilize the data from 2000 through 2021. The sample is representative of Estonia's permanent residents aged 15 or older and is formed based on the population proportional model. The basis of the model is the regions and the size of the settlement (the number of inhabitants). The source address selection within each region is made randomly on the basis of the address lists of the Population Register. From 2000 to 2020, an address-based random sample with the so-called young man's method was applied. 2021 was an online panel. Depending on the year, the sample size varies between 950 and 1,250 respondents. Since November 2011, in addition to the primary sample, an additional sample of 200 respondents was also formed to better represent the Russian-speaking population in the sample and to compare the different groups of population. However, in the general overview of the distributions of opinions, the national composition of the population is considered, and the data are weighted by increasing the number of Estonians and reducing the over-represented Russian speakers' influence of the respondents. The surveys were conducted as personal interviews, using paper forms until March 2014, and since October 2014 using tablets.

Resistance index

For the purpose of this study, a resistance index was created, and it measures the intent of the individual to resist the aggressor to the best of their ability, in other words to be an active producer of security and defence. The resistance index is combined of the measures of an active will to defend (personal contribution to military or resistance activities) and passive will to defend (support the government's decision for the nation to fight back), and the decision to stay in Estonia in case of an attack. The items used were: (1) "If

Estonia is attacked, are you ready to participate in defence efforts to the best of your abilities and skills?" (1—yes, definitely . . . 4—definitely no); (2) "If Estonia is attacked by any country, should we, in any case, provide armed resistance, regardless of the attacker?" (1—yes, definitely . . . 4—definitely no); (3) "If Estonia is attacked would you try to leave Estonia?" (1—yes, definitely . . . 4—definitely no).

The resistance rate shows how much the value of three items tends towards an affirmative answer on the first two characteristics (an attack on Estonia must be resisted, I will or likely will participate in the resistance) and on the negative side on the third characteristic (will not leave or probably will not leave Estonia). "Cannot say" and the absence of an answer are not distinguishable in the datasets; they are counted as non-affirmative answers (not leaving on the third characteristic, i.e., a positive answer). Count results are 1 (0 and 1 considered together), 2, and 3. Thus, the resistance rate 3 means "high resistance"—an individual is convinced that Estonia must resist, he/she will participate in resistance and does not leave the country. Score 2 denotes an "average resistance"—means that from three items at least two have been marked positively, for instance an individual stays in Estonia or has no opinion, thinks that Estonia should resist and be protected (passive will to defend) but is not ready to participate in defence (active will to defend). Score 1 marks "low resistance"—the individual agrees only with one of the statements or does not agree with statements 1 and 2 and intends to leave Estonia. Thus, resistance rate varies from 1 to 3, from low resistance to high resistance.

Trust in institutions

In addition to the resistance rate, we considered it important to gauge trust in the defence institution as trust in institutions is an important tenant for societal cohesion. This was measured through the question "To what extent do you trust the following . . . institutions?" Estonian Defence Forces and Estonian Defence League.[1] Answers ranged from 1—do not trust at all . . . 4—I have full trust.

Threat perceptions

Threat perceptions were measured with six statements: armed conflict in East-Ukraine; Russia's attempts to restore its influence in neighboring countries; immigration of war refugees and illegal immigrants to Europe; a large-scale military attack by some foreign country; cyberattacks against national information systems, institutions, companies, or citizens; intervention by a foreign country to influence Estonian politics or economy for its own benefit (scale: 1—very likely . . . 4 not likely at all).

Analysis and results

Institutional trust as an indicator of societal cohesion

In the case of the Estonian- and Russian-speaking population, the largest minority of approximately 33% (*Rahvaarv | Statistikaamet, n.d.*), a different attitude towards the institutions that ensure national defence is evident. While for Estonians, both the Estonian Defence Forces (EDF) and the Estonian Defence League (EDL) are undeniably institutions that ensure the safety of society, and their trust is on the same level as the police, border guards, and rescue service; the Russian-speaking respondents' trust in state defence institutions tends to remain on the same level as their trust in political institutions. Likewise, this group of respondents has low trust in NATO. However, non-Estonians are a heterogeneous group; those non-Estonians who trust the Parliament and the government also express greater trust in the EDF, and it is another way that low trust toward political institutions is associated with low trust towards national defence institutions. It has given reason to conclude that for some Russian-speaking residents, the defence of the Estonian state is not a matter of ensuring the safety of society but expresses a view on the political course of the Estonian state—belonging to the so-called collective West. As a rule, such respondents consider the threat posed by Russia to the Estonian state to be non-existent and therefore do not understand the necessity of developing national defence. The starting position of the rating trends in 2000 was not exceptionally high. Only half of the population trusted the EDF, but mistrust prevailed in the case of the EDL (see Figure 13.1). In the decade after regaining independence, the EDF was still being built up, and prejudices from the Soviet period still influenced attitudes toward the military. However, the increase in trust in the EDF and the EDL has become steady since 2002, when Estonia's accession to NATO gained a clear perspective among Estonians and non-Estonians.

On the one hand, the process of becoming a NATO member state forced the Estonian national defence institutions to exert themselves more; on the other hand, it also signaled to the public that our armed forces are recognized as meeting Western standards. Thus, the trust of Estonians in the EDF has remained stable at the level of 90% for a long time, but also, among non-Estonians, trust has been earned by more than half of the respondents. The trustworthiness of the EDL among Estonians has consistently reached over 80 percent, while the trust of non-Estonians in this organization has remained modest—presumably not so much because of distrust but because of the higher proportion of "can't say" answers due to fewer contacts that the Russian-speaking population has with the EDL.

Looking at the time trend of institutional trust ratings of non-Estonians, it can be seen that in periods when more severe problems have arisen in the relations between the Western countries and Russia; the trust has declined (see Figure 13.1). As a result of the Bronze Night in 2007, the reliability of the EDF in the estimations of non-Estonians dropped from a high of 75 percent to 60 percent. The Russia-Georgia War in August 2008 led to a drop in trustworthiness to 48 percent. The subsequent decline occurred during the Arab Spring in 2011–2012 and then in 2014, when Crimea was annexed, and the military confrontation in Donbas began. In the case of all these events, Russia's and Western countries' approaches to the nature and causes of conflict have radically differed. Public opinion, known to be most influenced and shaped by the information field in which the subject of the opinion lives, clearly mirrors these processes. Thus, indications of media-tization are visible as the Russian-speaking majority tends to follow more the information available through the Russia-driven mass media. This makes them also more vulnerable for influence operations, something that Mahda and Semenenko explore in Chapter 6, Hordiichuk et al. in Chapter 7, as well as Akrap and Kamenetskyi in Chapter 14. Figure 13.1 clearly demonstrates the polarization trend, which has two sources: on the one hand, native Esto-nians are demonstrating very high trust towards defence institutions, and over the years, the trust towards EDL has increased to the same level with trust of EDF.

In general, half of the non-Estonian population trusts defence institutions, but we can see a tendency towards a slight decrease. Thus, the polarization is occurring due to the very high trust of Estonians and slight decrease of trust level among non-Estonians. It is of note that previous studies have shown that at the individual level, the likelihood of resisting is positively associated with the feeling of pride to live in Estonia and high trust towards defence institutions (Berglund et al., 2022).

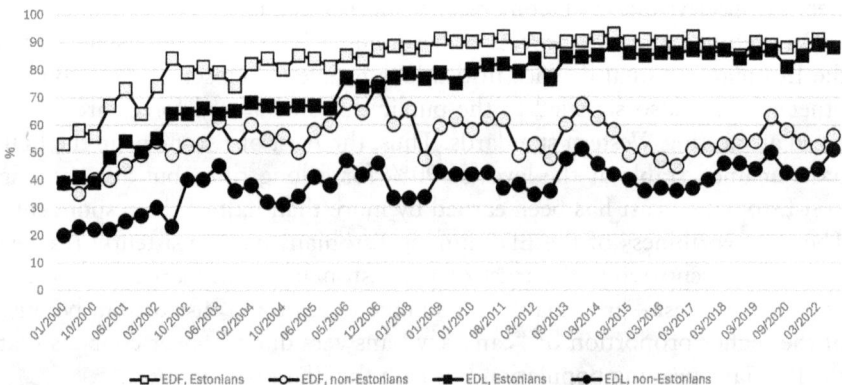

FIGURE 13.1   Trust in Estonian Defence Forces (EDF) and Estonian Defence League (EDL), %

Source: created by the authors

Resistance

The mean rate of the resistance index of Estonians and non-Estonians has been relatively stable over the years. The difference between Estonians and non-Estonians in the inclination towards a stronger will to resists tends to increase rather than decrease over time (except for the end of the period). At the beginning of the economic depression of the first decade of 2000 and after its end, and with the deepening of the Ukraine-Russia conflict, we see a sharp increase in the chances of leaning towards a higher resistance rate among Estonians compared to other nationalities. Perhaps, reflecting the decrease in ontological security and "rally 'round the flag'" effect. By the end of the period, the gap had narrowed but remained double. Looking at the individual components of the resistance index, a similar pattern emerges. In most years, the relative empirical probability of Estonians supporting state to resist an attack is several times higher than that of representatives of other nationalities. However, there are also years with a similar level of attitudes. Refusing to answer or hesitating to answer is statistically more characteristic of representatives of other nationalities than Estonians (except for the middle years of the period, where the difference does not emerge).

In the personal contribution (defence readiness/will to fight), we see a 1.5–3.5 times statistical difference in the second half of the period towards a higher defence readiness of Estonians compared to other nationalities. However, the difference decreases during the COVID period, but after the Russian invasion to Ukraine in February 2022, the differences increased again. In 2022, 91% of Estonians agreed that in case of an attack, the Estonian state has to resist and 76% were ready to individually contribute, among non-Estonians the percentages were respectively 61% and 44%. A year later, 89% of Estonians and 71% on non-Estonians support the state's decision to resist, and 72% of Estonians and 48% of non-Estonians are ready to contribute. Thus, the Russian large-scale attack to Ukraine directly and sharply affected Estonians' readiness; for non-Estonians, it took longer, and we can see the increase in 2023.

The third component of the resistance index—leaving Estonia—varies less among Estonians and non-Estonians. The risk of leaving Estonia is lower for Estonians than that of representatives of other nationalities, but the discrepancy based on nationality is the least important compared to the other two characteristics.

Resistance and threat perceptions

There must be a sense of danger and threat for the readiness to resist to arise, a sense of ontological insecurity. For resistance to occur, there must be a motive and an opponent. Confirmation of the need to defend Estonia, personal readiness to participate in the defence, and the intention to stay in

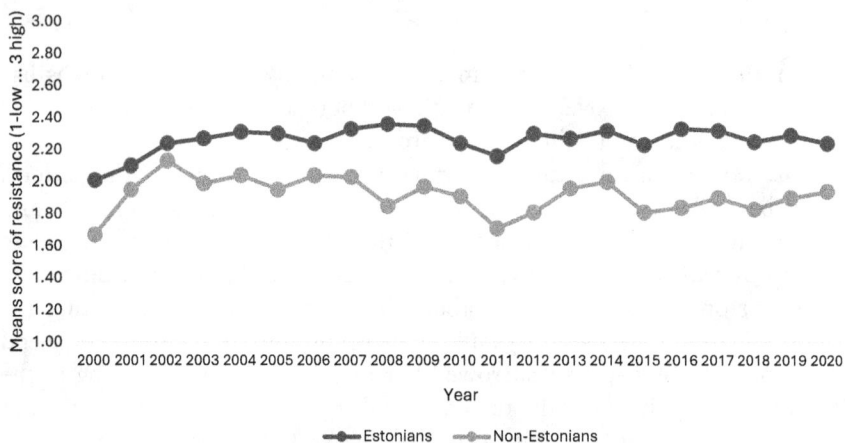

FIGURE 13.2   Mean score of resistance index, by ethnic background.

Source: created by the authors

Estonia in the event of an attack as components of the resistance index requires an aware counterparty in the face of the threat.

Resistance and threat perception can be associated in two ways. First, a stronger will to defend is accompanied by a sharper and more diverse threat perception and (considering the core features) an orientation towards military threats. Second, one could think the opposite: the multiplicity of threats makes the need for protection pointless. To explain this, we looked at each respondent's "threat number", considering the threats to be conditionally equivalent and counting the number of threats considered serious by each respondent from a given list in a given year. Refusing to answer and being unable to answer were not distinguished in the data, and we considered them unaware of the danger.

Looking at the data regardless of ethnicity, in most years, we found that a higher resistance rate is characterized by a higher "threat number" on average. Unfortunately, in all six years from 2015 to 2020, it turned out that Estonians brought up more threats on average than representatives of other nationalities (who also had more non-responses/not know other characteristics in the above analysis). The national groups' total number of threats differed only temporarily on average, according to the core features of the will to defend and the level of resistance. This means that the connection between the level of resistance and the highlighting of threats that appeared at first glance may be due to ethnic differences, not so much the level of resistance (recall that the core characteristics of the will to defend and the level of resistance in the sense considered here vary by nationality).

The threats to Estonia survey block covered the following: (1) the military conflict in Eastern Ukraine; (2) Russia's attempts to occupy its neighbors; (3) the arrival of war refugees in Europe; (4) the intervention of a foreign country to influence Estonian politics or economy; (5) an attack on the Estonian information systems; and (6) a large-scale military attack by some foreign countries. The first three are of course currently the most salient. We again distinguish Estonians and other ethnic groups and look for a relationship between the resistance rate and risk assessment (Figures 13.3 and 13.4).

In all years of the observed period, Estonians perceived the armed conflict in Eastern Ukraine as a threat statistically significantly more often than representatives of other nationalities. In contrast, the difference in threat assessments tends to decrease over time (the coefficient of the statistical relationship between ethnicity and threat assessment distributions based on the chi-square statistic was 0.28 in 2015 and 0.18 in 2020).

Among Estonians' answers, it is difficult to notice a clear connection between the resistance rate and the considered threat assessment (in two years—2017 and 2019—the variability of the threat assessment in the levels of the resistance rate turned out to be statistically significant, although the importance of the option "definitely" shown in Figure 13.3, square 1 does not show a clear uniform pattern from year to year). Among representatives of non-Estonians (Figure 13.3, square 2), the picture is more uniform in different years: the armed conflict in eastern Ukraine is perceived as a threat more often with a high resistance rate (3) than with a low resistance rate (1).

The immigration of war refugees and refugees to Europe (Figure 13.3, squares 5 and 6) is recognized as a threat in the observed period (choice "definitely"), but with a somewhat decreasing importance every year. It is impossible to confirm variability in estimates of this risk by nationality. There is also no correlation between the data and the resistance rate (with the exception of 2016).

Figure 13.4 presents analogous data on threats that could materialize in Estonia. The probability of the threat was assessed on a four-point scale, and the figure shows the proportion of those who tend to consider the corresponding threat as probable. As a statistically significant variation based on nationality emerged for all three threats considered here, the importance of persons considering the threat as probable is illustrated again separately among Estonians and representatives of other nationalities. It is essential to add that the variability by nationality decreases over time for all three threats.

A large-scale military attack is the least likely of the three threats, and no clear relationship emerged between resistance and threat perception (Figure 13.4, square 1 and 2). The threat of a cyberattack is probable in 80% of Estonians and 50–60% of the rest of the population (increasingly over time). Except for 2019, a high resistance rate among Estonians is accompanied by a higher threat perception. The intervention of a foreign country to

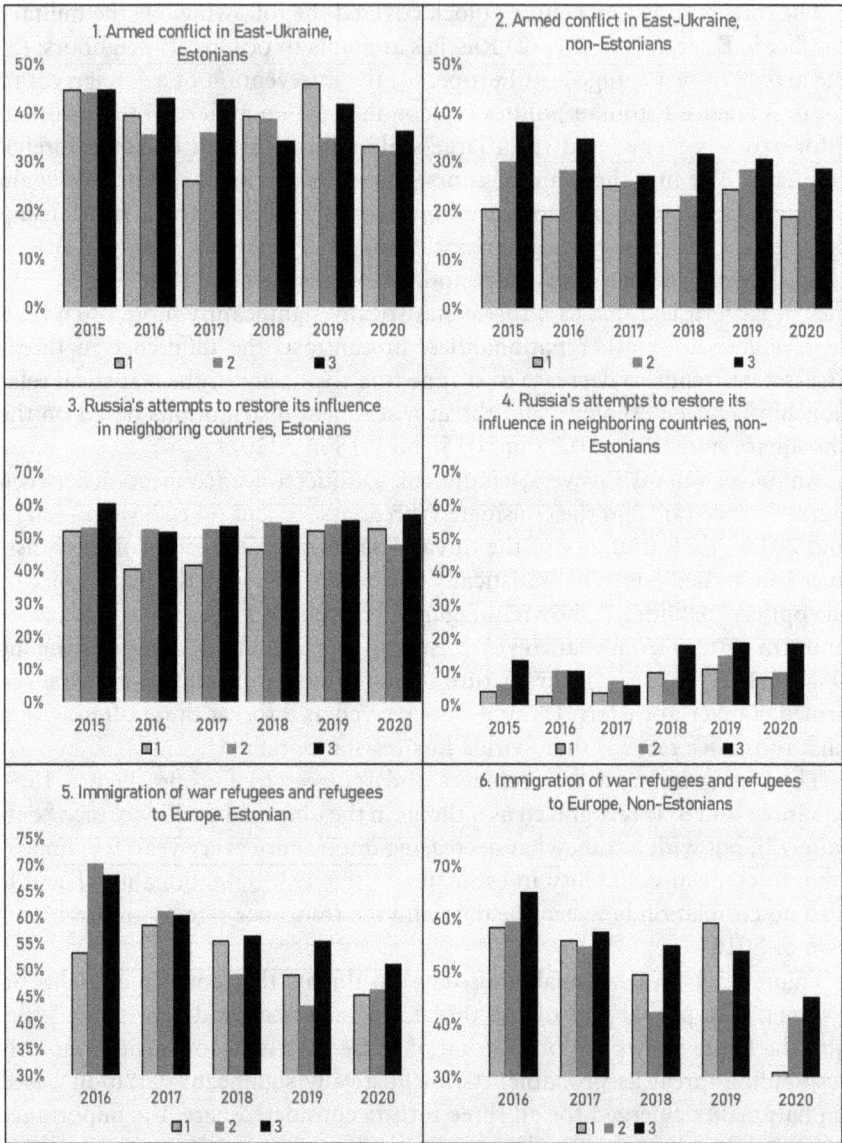

FIGURE 13.3   Awareness of threats to peace and security in the world depending on the level of resistance (1, 2, and 3) among Estonians and non-Estonians. The proportion of the answer "certainly", %

Source: created by the authors; data: "State defence and public opinion" https://datadoi.ee/handle/33/81

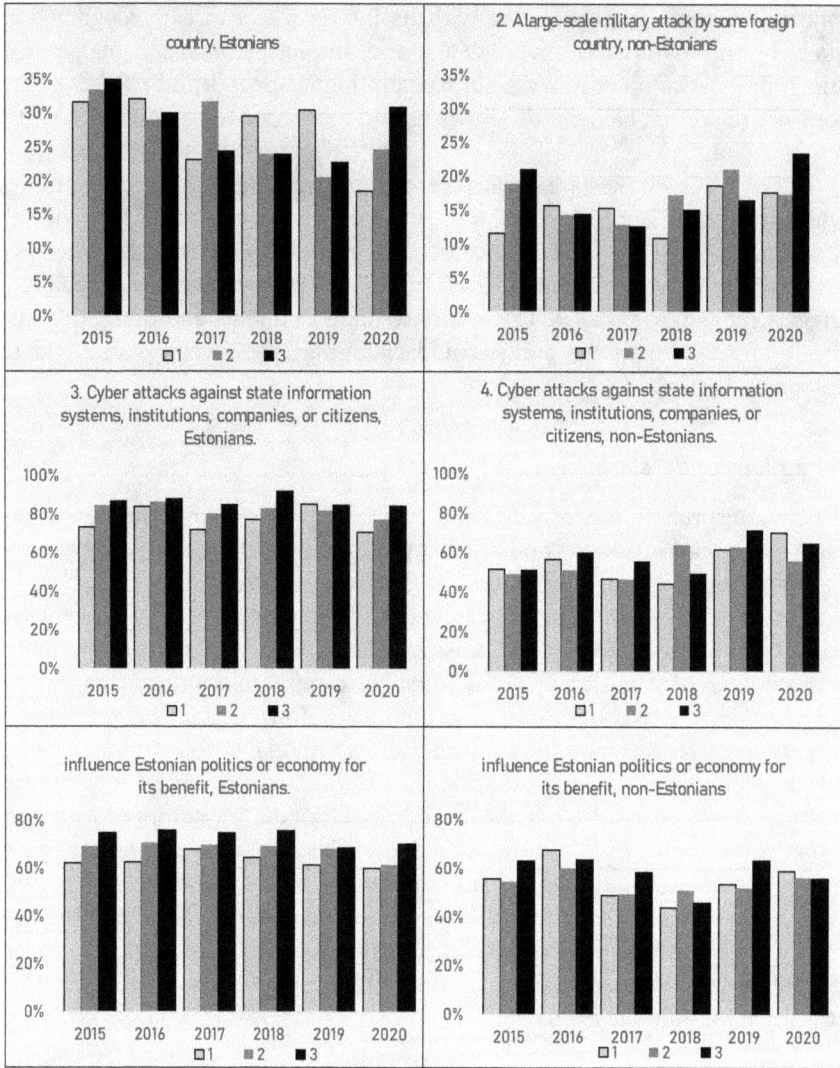

FIGURE 13.4 Awareness of threats that may materialize in Estonia among Estonians and representatives of other nationalities, depending on the level of resistance (1, 2, and 3). The proportion of "very likely" or "quite likely" responses, %.

Source: created by the authors; data: "State defence and public opinion" https://datadoi.ee/handle/33/81

influence Estonia in accordance with its interests is a threat recognized as likely by the majority of respondents, and among Estonians, a high resistance rate is accompanied by a statistically higher proportion of those who consider the threat likely in the entire period compared to persons with a low resistance rate.

In conclusion, in the case of the threats considered here, there are examples where the degree of resistance and the perception of danger are proportional. It did not appear for all risk factors, but it could be considered when we describe the will to resist in its broader sense. The awareness of the danger triggers the will to resist it. The ability to analyze threats and their adequate communication with the public could encourage the spread of the will to defend and resist.

Limitations of the study

When interpreting the results, one must keep two things in mind: the long-term trend in society and relativity. In the long term, the perception of risks and threats is directly related to the corresponding experience, that is, the threat background and the perception thereof at a specific moment. In the case of hybrid threats, time sensitivity must be considered, for example, a specific terrorist attack at the moment of time X in country Y raises the perceived threat to a higher level in country Y for a certain period. Therefore, for a better interpretation of trends, it is essential to look at longer-term trends. Relativity—when evaluating the cohesion of society, the degree of resistance, we proceed from the expected standard, we compare to something, either with previous years, other countries, or ethnic groups with each other. Therefore, judging whether something is high or low depends on the background or level taken as a basis for comparison. In this analysis, we compared how the institutional trust, threat perception, and resistance readiness of the country's main ethnic group and the population with a different ethnic background have changed.

Conclusion

In a situation where we are faced with ever-increasing hybrid threats, society's resilience becomes an increasingly important issue. How well can states and societies cope with various external and internal threats that threaten society's resilience and the country's independence? Front-line countries, relatively small in population and territory, face the big question of ensuring internal social cohesion in the event of an external threat. This chapter, therefore, analyzed what ethnic diversity means for general resistance in Estonia, and how this diversity relates to social trust.

Based on the Estonian case five general conclusions can be drawn: (1) there are differences of interpretation of hybrid threats among main ethnic group and minorities; (2) targeted communication is vital; (3) large minority groups of aggressive neighbors might not recognize the neighbor as a threat; (4) protracted crisis can bring threat perceptions of the main ethnic group and the large minority closer to each other; and (5) higher perception of threat also heightens the intention to resist.

The first is perhaps quite self-evident but also hardest to tackle; different hybrid threats and crises are interpreted differently by the main ethnic group versus ethnic minorities. Ethnic minorities tend not to interpret possible threats as deeply existential as members of the main ethnic group. However, as our data showed universal threats, transcending political or strategic motivators, such as the COVID pandemic can be interpreted in much the same way by most members of society regardless of their ethnic belonging. Second, just as Berndtsson in his chapter in this volume has stressed, communication and messaging are of vital importance within the comprehensive defence network. The messaging and targeting of the strategic narratives most probably need to differ considering the target group. The level of mediatization in contemporary societies also necessitates that the appropriate channels of mass media are exploited in order to reach the target audiences.

Third, countries with large minority groups from an unstable or aggressive neighbor state might be faced with the dilemma that the neighbor simply is not regarded as a threat by the large minority. Among the main ethnic group, high existential threat perception such as threat to independent statehood has a mobilizing effect, but not so much for the minorities. This can be explained by different interpretations and social construction of past, present, and future wars and their heroes (Kasearu et al., 2024). However, the data also shows that polarization in terms of the will to defend and to resist might stem rather from a non-alignment with the states' political and strategic choices on behalf of the minorities but is more similar with the majority when social and economic welfare is threatened. Fourth, regardless, as our data indicates, protracted, intense crises such as the war in Ukraine can mean that the large minority can experience a rise in threat perception, though not reaching the levels of the main ethnic group, the perception levels might become close to each other.

Fifth, in terms of the will to defend, we found higher rates of the intention to resist across all of the population if more threats to the state were considered as credible. Therefore, awareness of threats and adequate communication to different target groups could boost the spread of the will to defend and resist across ethnic groups. Social studies at times suffer from the tendency to the heterogenic nature of minorities. For example, half of non-Estonians have similar attitudes to Estonians. However, at the same time, it also means that we have to be aware that there is also a risk that minority groups are internally diverse and more vulnerable to hybrid threats.

In a situation saturated with hybrid risks, it must be taken into account that we cannot always distinguish, so to speak, more vulnerable, risk-susceptible, and more manipulable groups based on socio-demographic characteristics. Instead, creating a common information field and space is becoming essential.

Note

1 The Estonian Defence League is a voluntary national defence organization operating in the area of government of the Estonian Ministry of Defence, which is organized in accordance with military principles, possesses weapons, and holds exercises of a military nature. The purpose of the Defence League is to enhance, by relying on free will and self-initiative, the readiness of the nation to defend the independence of Estonia and its constitutional order (*Kaitseliit*, n.d.).

References

Berglund, C., Kasearu, K., & Kivirähk, J. (2022). Fighting for the (Step) Motherland? Predictors of defence willingness in Estonia's post-Soviet generation. *Journal of Political & Military Sociology*, 49(2), Article 2. https://doi.org/10.5744/jpms.2022.2002

Chandler, D. (2012). Resilience and human security: The post-interventionist paradigm. *Security Dialogue*, 43(3), 213–229. https://doi.org/10.1177/0967010612444151

Dinesen, P. T., Schaeffer, M., & Sønderskov, K. M. (2020). Ethnic diversity and social trust: A narrative and meta-analytical review. *Annual Review of Political Science*, 23(1), 441–465. https://doi.org/10.1146/annurev-polisci-052918-020708

Discussion Document—Resilient Social Contracts | United Nations Development Programme. (2018). https://www.undp.org/policy-centre/oslo/publications/discussion-document-resilient-social-contracts

Ehala, M. (2009). The bronze soldier: Identity threat and maintenance in Estonia. *Journal of Baltic Studies*, 40(1), 139–158. https://doi.org/10.1080/01629770902722294

Fiala, O. C. (2019). *Resistance operating concept* (O. C. Fiala, Ed.). Swedish Defence University.

Friedkin, N. E. (2004). Social cohesion. *Annual Review of Sociology*, 30(1), 409–425. https://doi.org/10.1146/annurev.soc.30.012703.110625

Joint Communication to the European Parliament and the Council. (2016). *Joint framework on countering hybrid threats a European Union response*. https://eur-lex.europa.eu/legal-content/EN/TXT/?uri=CELEX%3A52016JC0018

Juurvee, I., & Mattiisen, M. (2020). *The bronze soldier crisis of 2007: Revisiting an early case of hybrid conflict*. International Centre for Defence and Security (ICDS). https://www.jstor.org/stable/resrep54443

Kaitseliit. (n.d.). Retrieved June 13, 2019, from http://www.kaitseliit.ee/

Kasearu, K., Truusa, T.-T., & Lillemäe, E. (2024). From zero to hero: A conceptual framework of creating and recreating heroism based on the Estonian experience. In E. Ben-Ari, U. Ben-Shalom, N. Stern, & R. Moelker (Eds.), *Heroism in a post-heroic era: Polymorphic forms*. Routledge.

Larsen, C. A. (2013). *The rise and fall of social cohesion: The construction and de-construction of social trust in the US, UK, Sweden and Denmark*. Oxford University Press. https://doi.org/10.1093/acprof:oso/9780199681846.001.0001

Lillemäe, E., Kasearu, K., & Ben-Ari, E. (2023). Conscription and social transformations: Estonia between security needs and social expectations. *Journal of Baltic Studies*, *0*(0), 1–20. https://doi.org/10.1080/01629778.2023.2212914

Lowe, J. B., Barry, E. S., & Grunberg, N. E. (2020). Improving leader effectiveness across multi-generational workforces. *Journal of Leadership Studies*, *14*(1), 46–52. https://doi.org/10.1002/jls.21681

Mitzen, J., & Schweller, R. L. (2011). Knowing the unknown unknowns: Misplaced certainty and the onset of war. *Security Studies*, *20*(1), 2–35. https://doi.org/10.10 80/09636412.2011.549023

National Defence and Society | Kaitseministeerium. (n.d.). Retrieved November 29, 2023, from https://kaitseministeerium.ee/en/objectives-activities/national-defence-and-society

National Threat Assessment. (2020). *Second investigation department under the Ministry of National Defence and the State Security Department of the Republic of Lithuania*. https://www.vsd.lt/wp-content/uploads/2020/02/2020-Gresmes-En.pdf

Objectives for Swedish total defence 2021–2025—Government bill 'Totalförsvaret 2021–2025'—Government.se. (2020). https://www.government.se/government-policy/defence/objectives-for-swedish-total-defence-2021-2025-government-bill-totalforsvaret-20212025/

Putnam, R. D. (2007). And community in the twenty-first century the 2006 Johan Skytte prize lecture. *Scandinavian Political Studies*, *30*(2).

Rahvaarv | Statistikaamet. (n.d.). Retrieved November 29, 2023, from https://www.stat.ee/et/avasta-statistikat/valdkonnad/rahvastik/rahvaarv

Schiefer, D., & van der Noll, J. (2017). The essentials of social cohesion: A literature review. *Social Indicators Research*, *132*(2), 579–603. https://doi.org/10.1007/s11205-016-1314-5

Silm, K., Kasearu, K., Truusa, T.-T., & Torpan, S. (submitted). Are the concepts "will to fight" and "will to defend" different or the same? *Journal of Peace Research*.

Szymański, P. (2020). *New ideas for total defence: Comprehensive security in Finland and Estonia*. OSW Report 2020-03-31 [Other]. http://aei.pitt.edu/103309/

Truusa, T.-T. (2022). *The entangled gap: The male Estonian citizen and the interconnections between civilian and military spheres in society* [Thesis]. https://dspace.ut.ee/handle/10062/76076

Van der Meer, T., & Tolsma, J. (2014). Ethnic diversity and its effects on social cohesion. *Annual Review of Sociology*, *40*, 459–478. Available at SSRN: https://ssrn.com/abstract=2475574 or http://dx.doi.org/10.1146/annurev-soc-071913-043309

Veebel, V., & Ploom, I. (2018). Estonia's comprehensive approach to national defence: Origins and dilemmas. *Journal on Baltic Security*, *4*(2), Article 2. https://doi.org/10.2478/jobs-2018-0007

Wither, J. K. (2020). Back to the future? Nordic total defence concepts. *Defence Studies*, *20*(1), 61–81. https://doi.org/10.1080/14702436.2020.1718498

14

HYBRID THREATS AND THE POWER OF IDENTITY

Comparing Croatia and Ukraine

Gordan Akrap and Maksym Kamenetskyi

Introduction

This chapter explores how a range of hybrid threats may be used to undermine the identity of a nation. The development of a defense system for early recognition and warning against hybrid threats is one of the fundamental conditions that must be met for an individual state. Respond must involve institutional, organizational, and human knowledge, abilities, and skills. It is necessary to work on constant analysis of existing as well as recognition of new signs for early warning system of upcoming threats. Analysis of hybrid threats and risks must pay special attention to rarely occurring events that can have an extremely high level of negative consequences on the nation's identity.

Theoretical foundation

We describe hybrid threats as: *aggregation of planned, managed, directed, and coordinated actions that attack society/state, exploit existing, and create new vulnerabilities of the democratic system and institutions by using political, economic, civil, security, intelligence, energy, media, information and communication systems and assets, military, and other attack vectors, simultaneously or in phases.*

It is common to all these attack vectors that the aggressor uses them as a means by which he tries to impose his will on the target audience. Armed force has been replaced by information in all its derivatives regardless of the domain of offensive actions, content, meaning, and context come from. Information is a key mean by which hybrid attacker attempts to influence the

DOI: 10.4324/9781032617916-18

information environment of the target audience's cognitive processes and its decision-making process. The aggressor aims to impose some new divisions or to deepen the existing ones on the target audience. Attempts are made to take advantage of the vulnerabilities of democratic society in the form of achieved levels of freedom and rights of individuals, groups, and communities. Feelings of helplessness, restlessness, anger, fear, vagueness, powerlessness, and insecurity are aroused. Mistrust in state institutions is encouraged.

Different situations, expressed needs, goals, and conditions influence the diversity of the activities. In practical use, integrated action is encouraged by using multiple attack vectors in different time intervals due to the increase of malicious effectiveness through their cumulative, multiplicative action. For this reason, we cannot speak of a uniform model of hybrid threats that can be applied in all conditions and in accordance with expressed needs in the same way. This diversity and diffusion of implementation activities significantly complicates the processes of threat recognition, unmistakable identification of attacks, attackers, and goals to be achieved.

Several institutions and organizations may participate in the planning, implementation, supervision, control, and adaptation of hybrid threats on an individual or joint level. They can use secret, covert, or open methods and actions of civil and military institutions. One of the key features of hybrid threats is that military means can achieve the desired results in the civil and military domains—and vice versa. It is also important to state that with tactical means, an aggressor can achieve strategically important results. With special emphasis on operations conducted in, and coming from, cyberspace.

Methodology

This chapter is based on a comparison of hybrid threats against Ukraine and Croatia. It compares the violent and brutal disintegration of Yugoslavia (1990–1999) and the Russian aggressions against Ukraine (2014–2022).

The empirical background of the study was events in the two countries over a long period. The data analysis was based on:

- open-source information
- personal experience in public service
- research
- direct participation in the decision-making process at the strategic and tactical levels.

Each of the authors has acted as both a participant and an external observer in each of the cases. One author is a member of the NATO SAS-161 expert research task group directly connected with the topic of this paper.

Croatia and Ukraine—main similarities and differences

The violent and brutal disintegration of Yugoslavia (1990–1999) and the Russian aggressions against Ukraine (2014–2022) have numerous similarities but also differences.

The Croatian case

We observe the aggression against Croatia in a period that is wider than the kinetic, physical war (1990–1995). The kinetic aggression was preceded by numerous offensive actions (especially emphasized during the 1980s of the last century) using several different instruments related to influence operations. It is also important that with the cessation of kinetic aggression (in 1995) against Croatia, Serbia did not stop its offensive and malicious activities using influence operations. That part is also considered in this chapter because it is a process that Ukraine has yet to go through after achieving a military victory on the battlefield. Ukraine must be ready for those challenges to defend itself and not to repeat the mistakes that happened to Croatia.

The aggression against Croatia occurred at a time of serious and powerful political and social changes and challenges on the international scene. Global security architecture was changed due to the collapse of the USSR and the communist bloc. That change led to global tectonic consequences in the, between other areas, political, economic and security spheres, which in turn initiated the redefinition of international relations on a global level.

At the time of the beginning of the armed aggression against Croatia (August 1990), Croatia was *de iure* and *de facto* an integral part of the Yugoslavia and as such had no international legitimacy or legality. Croatia had no state institutions, no human resources at its disposal, no diplomacy, no armed forces, no defense sector, no intelligence community, and no support from the international community to create its own state. Croatia first organized its armed formations through the system of the Ministry of the Interior as police units. Croatia bought weapons on the international black market. In the economic and financial sense, Croatia had to rely on its own emigration, which financially supported the process of independence and defense of the state and its approach to Euro-Atlantic integration. Croatia was not supported by international community until, almost, the end of Homeland war.

The Ukrainian case

The Russian aggressions against Ukraine initiated (international, even global) processes that pointed to the necessity of redefining the existing security architecture, in accordance with significantly changed conditions and processes. The aggression against Croatia was a consequence of the changed global security architecture, while the Russian aggression against Ukraine indicates the necessity to redefine existing security architecture.

Ukraine was attacked by a global continental power. Russia is a permanent member of the UN Security Council. Russian foreign policy assumes that Russia is a global power that believes that they can act aggressively towards other subjects of international law. Russia reserves the right to interpret it and, if necessary, redefine it in accordance with its wishes. Russia is a force that directly influences numerous processes at the global level. Russian aggressions against Ukraine have negative consequences (of varying intensity) on numerous domains of human life and activity, almost on a global level (either directly or indirectly).

Ukraine gained its independence precisely with the collapse of the USSR at the time when the aggression against Croatia was being prepared and started to be carried out. Ukraine was going through a very complex political process of transformation and numerous political changes because of strong (pro)Russian actions directed against the key interests of Ukraine—access to Euro-Atlantic integration. However, Ukraine had time to develop its own state instruments, defense and security institutions and capabilities, and the economy before the war started.

TABLE 14.1 Differences and similarities between Croatian and Ukrainian cases

	CROATIA	*UKRAINE*
International recognition in time of aggression	No	Yes
Support from, and approach to, international organizations	No	Yes
State institutions existed in time of aggression	No	Yes
Arms embargo	Yes	No
Aggressors advantage in human capital at least 3:1	Yes	Yes
Aggressors Grand Strategy	Greater Serbia (Serbian World)	Greater Russia (Russian World)
Aggressor	Regional power	International/Global power
Nuclear potential of aggressor	No	YES
Consequences of the aggression	Local and regional	Global
Demonized by aggressor as neo-Nazi state?	Yes	Yes

Source: created by the authors

Influence operations

Identity

The identity of a group or community represents an important fact around which population bonds are built; establishes a common system of values, beliefs, and principles; recognizes and interprets the past; recognizes the present and defines the future (Tuđman, 2003). It acts as an integrating factor on the community. As long as it exists, if it is clear and indubitable, it is built as an indisputable fact in democratic processes by the joint action of the entire society without the introduction of politicking and malicious influences.

This is precisely why identity is the most common target of malicious activities from the spectrum of hybrid threats that attempt to influence the cognitive and decision-making processes of the target audience. If the hybrid attacker succeeds in his intention to shape and reduce the trust of the community in the existing paradigms on which identity is determined and on which it rests, the attacker will create favorable conditions for further malign activities to bring discord within the attacked audience. A hybrid attacker often uses influence operations in which facts related to identity issues change their real context and put them in a context that suits the attacker. In this way, there is a change in the interpretation of the affected identity issues, and a new information environment is created that can have a negative effect. The disintegration of the community and the growth of distrust in the institutions of the system built on the existing paradigm, because of the malicious actions of the adversary, potentially lead to the growth of discontent among the population and can lead to strong internal turmoil. These activities can also be violent, which further weakens the attacked audience.

Croatia and Ukraine have experienced similar challenges with shaping national identity during the time of communist regimes. National identity issues were treated as harmful if they were not in the function of supporting the communist authorities. The formation of national identities for Croats and Ukrainians had similar features—they were redeveloped in the shadow of a neighboring larger nation that had only partial commonalities with them.

The Croatian case

In numerous activities that preceded the armed aggression against Croatia, the aggressor tried to influence the Croatian identity. History was exposed to tendentious and inaccurate pro-communist revisionism; non-existing language (Serbo-Croatian) was imposed (example of Novosadski dogovor in 1954); Croatian historical and cultural heritage was attributed to Serbia (Kutleša, 2012); songs and historical features with national content were banned and their use was punished by imprisonment (Horvatić, 2020). The media were under the full control of the communist party. Journalists were treated as "ideological and political workers" (Akrap, 2010). The Catholic Church in Croatia, as one of the bearers of religious identity, was strongly informationally, politically, organizationally, and financially attacked by communists (Akmadža, 2012). However, Croatian identity was maintained, thanks to the activities of the Catholic Church, and through the activities of individuals and groups who were sentenced to prison terms because of their pro-democratic activities (Knežević & Mihaljević, 2018), and the activities of political and economic emigration. Thanks to these activities, Croatia, in an ideological and political sense, welcomed the collapse of the Soviet Union.

With the establishment of a new democratic government in Croatia (May 1990), activities were intensified to re-identify the population with the Croatian national identity and distance themselves from the artificially created Yugoslavian nation. Communist ideology was mainly removed from the interpretation of history and from social processes. The Catholic Church in Croatia, like other religious communities, gained absolute freedom. An additional integrating factor in the process of homogenization of the Croatian population and strengthening of national unity was the state policy of unity and general national reconciliation, as well as the threat to Croatian existence that came from Yugoslav and Greater Serbian circles and ideologies. That helped Croatia to develop societal resilience against many different influence and kinetic operations planed and conducted by aggressor.

The Ukrainian case

Like the case of Croatia, Russia used the issue of language similarities/differences as well as issues related to the targeted selection of historical topics and their tendentious interpretation for political purposes. Some topics were forbidden, some were ignored, and some were deliberately repeated to shape public knowledge in favor of Russian political, social, religious, economic, and security interests.

Identity and information strategy

In the period from 2010 to 2013, ethnic identity issues were not as important as those related to language. Many Ukrainians spoke Russian, followed Russian TV and radio stations, listened to Russian music, read Russian books, and watched Russian films. In this way, numerous disinformation was also distributed to Russian-speaking Ukrainians, especially from the domain of history and politics. This narrative shaped and formed the information environment in accordance with Russian plans. We will mention two important reasons: (a) In 1991, the Ukrainian media space was divided into private commercial media that were controlled by a few oligarchs. These media did not have a specific policy towards the necessity to protect and strengthen

Ukrainian identity issues, and (b) Russia acted to attract Russo linguistic Ukrainians and citizens of other countries with a Russian population in the interest of expanding Russian state policy. Therefore, the vacuum of the national information policy in Ukraine was replaced by Russian information operations of influence, which acted with the aim of replacing the Ukrainian national identity with the Russian one.

Identity and religion

The prolonged period of Soviet rule had a significant impact on the role of the church in the perception of the general population in the former USSR. Confessional matters in Ukraine were only utilized as a means for political forces to gain power. In contrast to Croatia, none of the churches in Ukraine held any traditional authority. The Orthodox Church, based in Moscow, historically had the most believers in Ukraine prior to the USSR's dissolution. However, after Ukraine gained independence, the autocephalous church of the Kyiv Patriarchate also came into existence, albeit with a small following. Nearly two decades later a political movement began to establish an Orthodox Church in Ukraine that was not affiliated with the Moscow Patriarchate. The efforts made by Petro Poroshenko, who was originally a believer of the Moscow Patriarchate, were crucial in the formation of the Ukrainian Orthodox Church. This process was accelerated after the Russian aggression against Ukraine. However, it is still slow and incomplete. The issues of property ownership of the Orthodox Church under the control of the Moscow Patriarchate and the transition to the Ukrainian Orthodox Church are still open and unresolved.

Identity and history interpretation

The history of Ukraine was and still is very often subject to recontextualization in accordance with the interests of Russian influence operations. Namely, history always contains topics that, placed in a different context, provide an information attacker with the opportunity to plan and execute malicious influence operations. The famine, as well as the history of the actions of the Ukrainian armed forces in WWII, is often the subject of debate and malicious interpretation. The Russians often use this in their actions, trying to create and shape information environment, at the national and international level, according to which Ukraine does not have its own history but has always been part of Russia. The Russian claims that the Ukrainians were closely associated with the Nazis, which is why even today they do not have the right to their own independent state. Russian television went so far in its manipulation that it illustrated the meeting between President Putin and the

President of the Constitutional Court of Russia with their conversation connected to a map from the 18th century, on which the name of Ukraine was not written. They used this as an argument that Ukraine has no right to exist, even though there is no name of the Russian Federation on that map also. Recontextualizing information and denying the existence of other important information related to it, is a well-known model that tries to mislead the target audience, leading them to wrong conclusions and decisions. These activities are an integral part of the spectrum of activities called influence operations (Silverman & Kao, 2022).

It was the return of the Ukrainian national identity to historical narratives through all its visible and spiritual forms that was one of the key activities that the leadership of Ukraine had to carry out to rebuild and strengthen the national identity. It was expected that the positive results of those activities would prevent all the negativity that characterized Ukraine's almost completely failed defense against the first Russian armed aggression in 2014.

Disinformation

Data and information are focal points of all influence operations, especially those that are trying to negatively influence the system of values, beliefs, and principles of the target audience. Until recently, following dictum was used quite often: Information is a power. Today, this is no longer true. Modern information and communication technologies have led to the need to integrate several different processes into a harmonized and coordinated activity. Possession of quality, accurate, true, complete, and usable information can no longer bring a decisive advantage in a conflict or war. There are two more essential conditions that must be met for information to truly become a powerful tool: it must be obtained in time and early enough so that it can become intelligence and used effectively, and there must be a reliable and secure communication channel, which will serve to distribute the same information/intelligence content on time, in an appropriate and clear manner, and in an integral form to the target audience.

The Croatian case

Before, during, and after the Croatian war for independence, a lot of disinformation was placed in the Croatian, Yugoslav/Serbian, and international information environment. Various topics were used that were supposed to fulfill their strategic goal: to remove the new democratically elected government in Croatia by force and install a pro-Serbian puppet regime. In this sense, and with the help of numerous agents of the Serbian and Yugoslav secret services in various communication channels (printed media, radio, and

TV stations, by distribution of leaflets), they spread numerous disinformation and created prejudices (Akrap, 2011):

- Demonizing and identifying Croatia with the Nazi creation during WWII and accusing it of committing numerous fictitious crimes during the Homeland War.
- Encouraging divisions within the society: by encouraging regionalism and strengthening intolerance towards employees of the former Yugoslavian administration on the one hand and toward returnees from emigration on the other. That was a direct attack against on President Tudjman's strategic concept of creating national unity and reconciliation.
- Encouraging the growth of distrust of the population in state institutions and accusing leading state representatives of corruption and crime.
- Equalization of guilt and division of responsibility between defender (Croatia) and aggressor (Yugoslavia and Serbia).
- Insistence on the "fraternal and very close Slavic identity, national and historical proximity of the two brotherhood nations."

Fighting disinformation

Most frequently used methods were information laundering (Akrap, 2011) and decontextualization and recontextualization. Typical examples are two articles (Hedl, 2000). Namely, disinformation planners, with the help of their network, first planted disinformation in some of the foreign media and then misused that foreign media as the source (whose credibility should not have been doubted) to share in the Serbian, Croatian, and foreign media space (Domović, 2019; Lončar, 2010). Croatia had to react quickly.

Disinformation began to shape the Croatian and international information space in a negative sense. There was the emergence of internal divisions, doubts about the correctness and efficiency of state decisions in the management of defense and the international recognition of Croatia, and the development of additional social crises. Croatia has initiated many different activities to inform national and international auditorium through the Ministry of Information, Foreign affairs, other state institutions, and the media with the aim of combating disinformation. Many were recognized as such and publicly debunked. At the same time, the state also launched its own repressive system (primarily the intelligence community) in identifying the planners and perpetrators of these malicious activities, and actively fighting against identified individuals, groups, and organizations. Intelligence community was very successful in those activities by identifying many planers of different disinformation and communication channels that they used. A huge number of disinformation were publicly debunked. As a result of analyses of numerous disinformation attacks, many important and valuable papers, books, and dissertations were written. Disinformation continued to be a weapon against Croatia and its interest of that time aggressor after the Homeland war and is still active. Fight against disinformation and malicious activities that are trying to shape information environment is 24/7 process.

The Ukrainian case

Russian disinformation activities

When Putin took the power, he initiated the process of gaining full control over the media in Russia. The Russian information environment almost completely falls under their control, supervision, and management. Media continued to be used as an instrument of power in the hands of the state (Павловский, 2015). After the stabilization of the state budget, the establishment of media houses that were primarily oriented towards the West (such as Sputnik, Russia Today) was initiated. This significantly differed from the former Soviet information policy, which was full of propaganda activities of lower quality, not so accepted in the West, and thus with reduced effectiveness. The information content that Russia created and distributed in the last decade aimed to achieve the following goals:

- Shaping of public opinion in Russian society aimed at supporting the Russian government, glorifying Putin, emphasizing the need that Russian must become superpower again, and developing super-potent weapons that no one (especially the United States and NATO) has anything efficient to defend against them.
- Creating necessary conditions and the use of various factors that should enable the achievement of Russia's goals in neighboring countries.
- Forming a positive image of Russia among the population of foreign countries, while the actions of Russia's opponents are interpreted as the result of actions that do not correspond to the interests of the entire humanity (Hinton, 2022; Павловский, 2015; Полякова, 2022).

In its information aggression against Ukraine, Russia used a situation that Ukraine did not have centralized information policy. This was a key point for fighting against Russian aggression, especially in information domain. Before and during the aggression on Ukraine, Russia developed and distributed a lot of disinformation tailored to different audiences (Mandić & Klarić, 2023; Baig, 2022).

Unlike the Croatian case where the aggressor used audio, video, and printed materials, in case of Ukraine, the Russian aggressor constantly used the web to distribute disinformation. They often used AI to create various disinformation messages (Euronews, 2022). As in Croatia where some (already debunked) disinformation continues to exist in the public space,

Russian disinformation remains to be influential in some Western countries as well as in countries outside the EU and the NATO alliance.

Targets and goals

With the aim of shaping the information environment in Ukraine, Russia directed its malign information activities toward two key target audiences: the pro-Ukrainian population (by spreading defeatism and mistrust against state institutions) and the pro-Russian population (by strengthening solidarity and connection with Russia). Therefore, Russia encouraged and through covert activities financed the "Ukrainian" media which were under the control of pro-Russian owners. Just like in a case in Croatia, Russia also used foreign channels for "information-laundry" activities. To hide real source of the (dis)information, it was often sent to public via different, mainly foreign media. Then those (dis)information appeared in Russian media environment and some other countries as an information without Russia origin. Russian attributed those disinformations as a relevant and reliable, and the source as unbiased (Idzelis, 2017).

The main purpose and the targets of Russian disinformation were:

- Discrediting the highest political and military leadership of Ukraine.
- Discrediting Ukraine's cooperation with its friends and partners at the international level.
- Justification and recontextualization for Russian military activities that will be the basis for charges of war crimes.
- Equalization of guilt between aggressor and a victim (Camut, 2023; Полякова, 2022; Hinton, 2022).

As a response, Ukraine controlled its information space very well, taking care of the information and operational security of political and military decisions and actions. Ukraine engaged countermeasures such as strategic and crisis communication with the aim of confusing the Russian aggressors and reducing the Russian ability of early forecasts of Ukrainian operations (Ekman & Nilsson, 2023).

Cyberspace

Hybrid threat aggressors exploit the many advantages that modern technologies provide (see Soldal, ch. 5). Analysis of conflicts and wars in recent several decades (especially these cases) shows that future conflicts and wars will use cyberspace as a central battlefield. Cyberspace can be defined as a space that exists in four different domains: (1) geographical regarding the regional distribution of the physical parts of the system, (2) system of networks communication and functions that enable its functioning, (3) artificial intelligence, and (4) the human factor that has both physical and cyber dimension.

Social networks, just as mobile communication applications, are very often used as a communication channel by malicious users. Hybrid attackers use them for rapid, intensive, anonymized, covert distribution of disinformation, and non-knowledge aimed at influencing target audience decision-making process. Social networks have also emerged as a productive platform for recruiting online supporters (agents and for networking) in the enemy's territory. It is mostly these agents who provide and continue to provide information to launch strikes on enemy targets.

Many different target audiences can be attacked at the same time regardless of their physical presence. Social networks also serve for collection of a huge amount of data and information necessary for effective planning of influence operations. Not every audience is the same and does not react equally to different stimuli. Therefore, it is necessary to hire own or another intelligence and security system, or specialized companies, with the aim of collecting required data, which in that case act as intermediaries.

The Croatian case

During the aggression of Serbia and Yugoslavia, Croatia was not faced with threats from the cyber space. At that time (1990–1996), the Internet was still in its early developing phase. Digital social media did not exist. Mobile phones were very expensive and had limited availability. Instead, printed media, as well as TV and radio, were used as channels to distribute disinformation.

The Ukrainian case

The digital space was intensively used by the Russian aggressor in times of increasing tension, targeted at strengthening existing and creating new divisions in Ukraine. Before, but also after the aggression carried out in 2014, Russian offensive malicious operations in the cyberspace came to the focus of experts and the public, especially before the second aggression in 2022 (Huntley, 2023; Masters, 2023; Duguin & Pavlova, 2023). The Russian intelligence community was identified as a mastermind that stands behind those cyberattacks. Units 26165 and 74455 from the Russian Military Intelligence Service (GRU), Unit 71330 from the Russian FSB, and an unnamed group from the Russian SVR were identified (Knapczyk, 2022).

In addition to offensive activities with the aim of disabling the operation of Ukrainian key critical infrastructures, the cyberspace was also used to share numerous disinformation. These instruments were targeted at reducing the defense potential through undermining Ukrainian identity, introducing divisions into society and the state, preventing the functioning of the state, and encouraging distrust of the population in the ability of Ukraine as a state to defend itself against Russian aggression.

The use of social networks as well as applications for mobile communication in the spread of disinformation has shown, in this Russian aggression against Ukraine, the range of possibilities that (totalitarian) societies can use in attacking the democratic order (Scott & Kern, 2022). By creating a (dis) informational environment that the attacked public would not recognize as such, it would lead to numerous harmful consequences for the attacked target in numerous domains of activity (social, informational, security, defense, economic, energy, financial, etc.).

It should also be noted that Ukraine learned numerous lessons after the Russian aggression in 2014 and initiated processes of educating, training, organizing, and institutionalizing its own defense capabilities in the period from 2014 to 2022. One of the capabilities that Ukraine has developed, both at the state level and at the level of private business entities, is the ability to recognize offensive malicious activities, identify a digital attacker, and improve their counterattack abilities. At the beginning of the Russian aggression against Ukraine, Ukraine showed that is capable of successfully defending itself against numerous complex digital attacks (in the digital space as well as threats coming from the digital space) and moving into offensive actions against the aggressor (Lambert, 2022; Badanjak, 2022; HINA, 2022).

Conclusion

Activities from the spectrum of hybrid threats may be in the center of future conflicts. They will, among others, be based on the planning and execution of influence operations. Kinetic operations may represent secondary means of imposing the attacker's will on the target audience. The cyberspace may represent the primary area of influence operations related to cohesive factors of society and community. This includes the identity issues, system of principles, beliefs, and values on which the strength and unity of a certain community rests. It may also include critical infrastructure (such as energy, cyber, information-communication, and water-food production and distribution). The goal of attacker is to acquire a state of information supremacy in the spectrum and domains of its activities.

In this study, we have shown how the attacked countries faced opponents who should have won by pure mathematical calculations. In both cases, the aggressor was stronger, more numerous, more technically equipped, and in a situation of controlling the information and media environment of the defenders. However, thanks to their own armed forces, unity of the population around key national identity questions, their readiness to fight for their families, homes, and country, and thanks to friends and partners, as well as the institutions of the state and society that were in the process of being developed or were already developed, Croatia and Ukraine were able to defend themselves. That enabled the defenders to change the initial information advantages of the opponent and gradually transfer it to a state of information supremacy. As it can be seen from this short overview, general themes and goals of the aggressor's disinformation policies were almost the same. The essential difference was in communication channels used. In both cases, defending societies had to work very hard to create, strengthen, and protect their own national identity; control and monitor communication channels; take intensive counter-disinformation actions; and link the capacities of society and the state with the leading media with the aim of providing accurate and objective information to both the domestic and foreign public.

Croatia successfully fought against numerous disinformation. However, we still occasionally see some of them in the media today. They are distributed by some circles for clear political purposes, especially during the pre-election period. Ukraine will continue to face a lot of disinformation at both national and international levels. Ukraine must not allow itself to constantly react passively and subsequently to Russian disinformation. An active approach should be taken together with its partners to timely and reliably inform the target audience about Russian and pro-Russian disinformation policies and goals so that the population can independently recognize them. On the other hand, it is necessary to develop trust between the population and state institutions, as this is one of the key prerequisites for an effective fight against disinformation.

The next challenge for states and the international community today is the definition of the rules of modern warfare. There are no rules for hybrid threats. Where there are no rules, everything is allowed. That means that nothing is prohibited, which opens space for unsuspected attack models and the choice of many different targets and use of many different attack instruments. Therefore, it is necessary to intensify efforts to reach a consensus on establishing the rules of hybrid threats.

References

Akmadža, M. (2012). Katolička crkva i Hrvatsko proljeće. Zagreb: *Časopis za suvremenu povijest, 44*(3), 603–630. https://hrcak.srce.hr/95080

Akrap, G. (2010). Mač i štit u rukama partije—Represivni sustav u funkciji oblikovanja korpusa javnoga znanja. *National Security and the Future, 11*(4). https://nsf-journal.hr/online-issues/case-studies/id/1196

Akrap, G. (2011). Informacijske strategije i operacije u oblikovanju javnog znanja, Zagreb: Filozofski fakultet, doktorska disertacija [*Information strategies and operations in public knowledge shaping*]. PhD thesis.

Badanjak, I. (2022). *Rusi krenuli na posao, a onda se s radija zaorilo: Uzbuna, hitno u skloništa, prijeti raketni napad!. Jutarnji list.* https://www.jutarnji.hr/vijesti/

svijet/rusi-krenuli-na-posao-a-onda-se-s-radija-zaorilo-uzbuna-hitno-u-sklonista-prijeti-raketni-napad-15308622

Baig, R. (2022). The deepfakes in the disinformation war. *DW-In Fokus*. https://www.dw.com/en/fact-check-the-deepfakes-in-the-disinformation-war-between-russia-and-ukraine/a-61166433

Camut, N. (2023). Berlusconi blames Zelensky for war in Ukraine. *Politico*, February 13, 2023. https://www.politico.eu/article/silvio-berlusconi-blame-volodymyr-zelenskyy-war-ukraine-russia/

Domović, R. (2019). 20 godina o dezinformacijama, krivotvorinama i medijskim manipulacijama. *National Security and the Future*, 20 (3). https://nsf-journal.hr/online-issues/focus/id/1273

Duguin, S., & Pavlova, P. (2023). *The role of cyber in the Russian war against Ukraine: Its impact and the consequences for the future of armed conflict; DG for external policies*. https://www.europarl.europa.eu/RegData/etudes/BRIE/2023/702594/EXPO_BRI(2023)702594_EN.pdf

Ekman, I., & Nilsson, P.-E. (2023). Ukraine's Information Front—Strategic Communication during Russia's Full-Scale Invasion on Ukraine. FOI-R-5451-SE.

Euronews (2022). *Deep fake Zelenskyy surrender video is the 'first intentionally used' in Ukraine war*. https://www.euronews.com/my-europe/2022/03/16/deepfake-zelenskyy-surrender-video-is-the-first-intentionally-used-in-ukraine-war

Hedl, D. (2000). Feral Tribune, number 791, from November 11, 2000, and number 792, from November 18.

HINA (2022). *Čaki Rusi priznaju: 'Stranice Kremlja, Vlade, Aeroflota i Sberbanke su pod neviđenim cyber napadima'. Jutarnji list*. https://www.jutarnji.hr/vijesti/svijet/cak-i-rusi-priznaju-stranice-kremlja-vlade-aeroflota-i-sberbanke-su-pod-nevidenim-cyber-napadima-15171455

Hinton, A. (2022). *Putin's claims that Ukraine is committing genocide are baseless, but not unprecedented*. https://theconversation.com/putins-claims-that-ukraine-is-committing-genocide-are-baseless-but-not-unprecedented-177511

Horvatić, P. (2020). *8. siječnja 1879. 'Još Hrvatska ni propala dok mi živimo'– komunisti zabranjivali domoljubne pjesme u Jugoslaviji i slali ljude na robiju! Narod.hr*. https://narod.hr/kultura/8-sijecnja-1879-ferdo-livadic-jos-horvatska-ni-propala

Huntley, S. (2023). Fog of war: How the Ukraine conflict transformed the cyber threat landscape. *Threat Analysis Group*. https://blog.google/threat-analysis-group/fog-of-war-how-the-ukraine-conflict-transformed-the-cyber-threat-landscape/

Idzelis, L. LtCol (2017). LTU LAF Strategic Communications Department—slide 15—example from February 23, 2015. Lecture held in Zagreb, Croatia, June 2017.

Knapczyk, P. (2022). *Overview of the cyber weapons used in the Ukraine—Russia war*. https://www.trustwave.com/en-us/resources/blogs/spiderlabs-blog/overview-of-the-cyber-weapons-used-in-the-ukraine-russia-war/

Knežević, D., & Mihaljević, J. (2018). Political trials against Franjo Tuđman in socialist Federal Republic of Yugoslavia. *Review of Croatian History*, XIV (1), 353–381. https://hrcak.srce.hr/214273

Kutleša, S. (2012). *Na Istoku ništa novo II—ili Kako znameniti Hrvati postaju Srbi. Matica Hrvatska: Vijenac, 478*. https://www.matica.hr/vijenac/478/Na%20Istoku%20ni%20C5%A1ta%20novo%20II.%20ili%20Kako%20znameniti%20Hrvati%20postaju%20Srbi/

Lambert, F. (2022). Hacked electric car charging stations in Russia display 'Putin is a d*ckhead' and 'glory to Ukraine'. *Electrek*. https://electrek.co/2022/02/28/hacked-electric-car-charging-stations-russia-displays-putin-dckhead-glory-to-ukraine/

Lončar, M. (2010). Činjenice o Optužnici"koju je pročitao Siniša Glavašević". *National Security and the Future*, 11(2–3). https://nsf-journal.hr/online-issues/case-studies/id/1198

Mandić, J., & Klarić, D. (2023). Case study of the Russian disinformation campaign during the war in Ukraine—propaganda narratives, goals and impacts. *National Security and the Future*, 24(2). https://www.nsf-journal.hr/online-issues/case-studies/id/1471; https://doi.org/10.37458/nstf.24.2.5

Masters, J. (2023). *Russia-Ukraine war: Cyberattack and kinetic warfare timeline.* https://www.msspalert.com/news/ukraine-russia-cyberattack-timeline-updates-amid-russia-invasion

Novosadski dogovor (1954). *Institut za hrvatski jezik i jezikoslovlje.* http://ihjj.hr/iz-povijesti/novosadski-dogovor/70/

Scott, M., & Kern, R. (2022). *Social media goes to war. US social media companies have been pressured to move off the sidelines to block Russian propaganda as the Ukraine war escalates.* https://www.politico.eu/article/social-media-goes-to-war/

Silverman, C., & Kao, J. (2022). *In the Ukraine conflict, fake fact-checks are being used to spread disinformation.* https://www.propublica.org/article/in-the-ukraine-conflict-fake-fact-checks-are-being-used-to-spread-disinformation

Tuđman, M. (2003). *Prikazalište znanja.* Zagreb: Hrvatska sveučilišna naklada.

Павловский, Г. (2015). Система РФ. Источники российского стратегического поведения. М.:Европа.

Полякова, Т. (2022). Путин перевернул все с ног на голову. https://www.forbes.ru/society/456925-putin-perevernul-vse-s-nog-na-golovu-mirovye-smi-o-voennoj-operacii-na-ukraine

15

BUILDING COMPETENCE AGAINST HYBRID THREATS

Training and exercising hybrid command organizations

Bjørn T. Bakken, Thorvald Hærem and Inger Lund-Kordahl

Introduction

Training and exercises play a fundamental role in developing situational awareness, a critical component of effective response to hybrid threats. Equally important is decision-making at all levels (strategic, operational, tactical) as well as communication, both within the command organization (whether military or civilian—policing, firefighting, medical emergency) and to the public. By providing individuals and organizations with realistic scenarios and hands-on experiences, training and exercise programs enable participants to understand the complexities of hybrid threats and develop the necessary skills to identify and respond effectively to such threats. Exercises, for example tabletop simulations, "wargaming," and full-scale (field) exercises, allow stakeholders to test their preparedness, evaluate response strategies, and identify areas for improvement. Through these activities, situation awareness is heightened, fostering a proactive and coordinated response to hybrid threats. By engaging in a broad spectrum of scenarios, in particular situations that can be called wicked (or VUCA—Volatile, Uncertain, Complex, and Ambiguous) learning environments (Bakken et al., 2023), it is possible for exercise participants to generate complex and versatile mental models that would not be attainable in a real-life wicked learning environment.

The research question we ask is: *Which parameters could be included as command organizations pursue training and exercise programs in a hybrid threat environment, and how are these parameters linked to optimize the outcome?*

To respond to the research question, we explore how training and exercises can be effectively used by a multi-disciplinary command organization

DOI: 10.4324/9781032617916-19

to improve situation awareness, decision-making, and communication skills in crisis management. We build on earlier studies and observations, discussing the characteristics and merits of different organizational models—such as centralized, distributed, and "hybrid," that is, a mix of hierarchy and network, when it comes to either facilitating or impeding crisis management. We are particularly interested in the functioning and effectiveness of what we may call multi-disciplinary, cross-functional, or inter-agency command organizations. We will also pay attention to the impact of internal team processes, as well as the contextual demands, the latter conceptualized as complexity and dynamics of the situation (Bjurström & Bakken, 2022). Based on the theoretical overview and data from a present case, we suggest effective strategies for building, sustaining, and innovating crisis management competence in the face of hybrid threats. We also propose how to use modern technology, such as AI and VR, to effectively support crisis management training for hybrid threats.

Theory

Hybrid threats are prevalent and may occur in the entire crisis spectrum from peace to crisis and armed conflict (Malerud & Hennum, 2023; Mattingsdal et al., 2023). Particularly challenging are complex threats that manifest in the "gray zone" between peace and crisis, and that can pose a threat to both societal and governmental security (see Ellingsen, ch. 3; Heier, ch. 8). Hybrid threats create uncertainty and ambiguity that challenge the ability to discover, attribute, and understand whether one is in fact subject to an attack—that may also be the start of a longer-lasting military campaign. The uncertainty imposed contributes to making it more difficult to identify, interpret, and understand an enemy actor's intentions and goals (Malerud & Hennum, 2023; see also Thiele, 2021; Cristiano et al., 2023). Overall, this vague and unclear "threat picture" challenges the ability of a command organization to achieve the necessary situational awareness as a basis for making good and timely decisions. In addition, the "hybrid" concept is itself vague and much debated regarding its usefulness and relevance for the academic study of conflict, crisis, and warfare (Libiseller, 2023).

Situational awareness

On a day-to-day basis, an enemy can exert long-term influence with, for example, the use of political/diplomatic, informational, economic, and legal instruments aimed at targets in various sectors of the society (Malerud & Hennum, 2023; see also Thiele, 2013, 2021; Cristiano et al., 2023; Pedersen, 2023). This type of threat often represents small deviations from the normal situation and may therefore be difficult to detect (Bakken & Hærem, 2020).

The challenge to detect and identify (perceive) small changes in the backdrop of an otherwise normal situation corresponds to attaining situational awareness (SA) at level 1 (Endsley, 1995; see also Valaker et al., 2018; Hærem et al., 2022). Comprehension of the threat—its nature and severity in relation to the specific context and our (relative) ability to respond—will constitute situational awareness at level 2. Finally, and probably, the most important and most demanding challenge will be to gain an understanding of the enemy's intentions, goals, and objectives in the longer run. This long-term comprehension of the future developments conforms to situational awareness at level 3. Thus, SA 3 encompasses building a dynamic mental map of the situation and projecting future events and actions as they may unfold subject to both own and enemy actions, as well as to external events (Endsley, 1995; Valaker et al., 2018; Sætren et al., 2023). Several chapters in this book discuss topics related to building and maintaining situational awareness; for example, Sandbakken and Karlsson (ch. 11) on information sharing; Borch on intelligence (ch. 12); and Magnussen et al. on strategic competence development (ch. 16).

Decision-making

Decision-making and the decision processes during a crisis are marked by significant time constraints, extensive uncertainty, and high stakes. Gary Klein has formulated a model, known as the Recognition Primed Decision (RPD) model, which effectively captures the intuitive and analytical elements of decision-making in real-world scenarios, including operational and emergency situations (Bakken & Hærem, 2011, 2020; Crichton & Flin, 2017; Flin & Arbuthnot, 2002; Steigenberger et al., 2017; Bakken et al., 2023). It is reasonable to anticipate that the utilization of novel technological tools, that is, Artificial Intelligence (AI), for decision support under hybrid threats will lead to enhancements in both the swiftness and precision of decision-making. In the longer run, an autonomous AI agent may even run entire operations without the direct involvement of human decision-makers (Bakken, Lund-Kordahl et al., 2023).

Dynamic decision-making (DDM) refers to the interconnected process of decision-making that occurs within an environment characterized by constant changes over time. These changes can arise from either the decision-maker's prior actions or external events beyond their control. In such situations, the decision-maker must be mindful of the evolving nature of the system. Berndt Brehmer introduced another crucial element by emphasizing that decisions in DDM scenarios need to be made in real time (Brehmer, 1992, 2000, 2009; Brehmer & Dörner, 1993; Bjurström & Bakken, 2022). The time dimension adds an additional layer of complexity to dynamic decision-making, as the decision-maker must explicitly consider not only what actions should be taken but also the optimal timing for those actions. It is insufficient to solely know *what* needs to be done; understanding *when* it should be done is equally essential.

When it comes to facing off hybrid threats, we believe that the DDM perspective is particularly useful as a foundation for training and exercises (Bjurström & Bakken, 2022; see also Bakken et al., 2017). According to Bjurström and Bakken (2022), the crucial point for a crisis manager in general (or a military commander in particular) facing a complex, dynamic, and uncertain situation (which is the hallmark of a wicked learning environment; Bakken et al., 2023) is that a decision-maker should be trained in dynamic decision-making (DDM) and sensemaking (e.g., Barton et al., 2015; Weick & Sutcliffe, 2015) to have acquired complex mental models (featuring *non-linear delayed feedback*) of similar situations (e.g., Brehmer, 1992; Sterman, 2000). We posit that having acquired such mental models is a prerequisite for successful proactive decision-making when facing a hybrid threat environment. The basic idea is, therefore, for a decision-maker to be able to "foresee" a threat unfolding, including any unwanted events or side-effects that may cascade across sectors and levels, so that the threat can be neutralized before it becomes too severe and complex to handle with available resources, capabilities, and tactics (Bjurström & Bakken, 2022).

Communication

Effective response to hybrid threats requires robust team processes, dynamics, and communication channels both within and between crisis management organizations and the public. Strong interagency coordination, collaboration, and information sharing enhance the overall situational awareness and response capabilities (Valaker et al., 2018; Sandbakken & Karlsson, ch. 11). Clear communication channels with the public, including timely dissemination of accurate information, help prevent the spread of disinformation and maintain public trust. Furthermore, fostering a culture of trust, psychological safety, and diversity within crisis management teams promotes effective communication and decision-making and enhances overall resilience (Larsson et al., 2021; Bakken & Larsson, 2023; Sætren et al., 2023).

Effective communication also involves considering the selection of appropriate communication channels (e.g., email, radio, face-to-face) and utilizing them efficiently and effectively (Valaker et al., 2018). This applies to internal communication within the command organization as well as communication directed towards the media and the public. An intriguing area worth exploring is whether emerging technologies like virtual reality (VR) and augmented reality (AR) together with AI can help prevent and minimize misunderstandings in tactical, operational, and strategic communication during crisis management (Cristiano et al., 2023). These novel technologies hold the potential to enhance communication clarity and reduce the impact of misinterpretations in critical situations (Valaker et al., 2018; Bakken et al., 2022; Bakken, Lund-Kordahl et al., 2023).

Organization structure

The structure of a command and control (C2) organization, along with the chain of command, pertains to the coordinated allocation of roles, tasks, and responsibilities among its members, whether they operate individually or in collaborative work groups and teams (e.g., Mintzberg, 1980). A crucial aspect is the degree of centralization and hierarchy versus distribution and networking within the organization (Alberts & Hayes, 2003; Bakken et al., 2006; Sandbakken and Nordhus, ch. 11). A distributed model promotes decentralized decision-making and empowers local entities to respond swiftly to threats in their areas. A centralized model consolidates decision-making authority, enabling a coordinated response at a higher level. Alternatively, a hybrid model combines elements of both centralized and distributed approaches, striking a balance between agility and coordination. The choice of organizational model—for example, how tasks, roles, and responsibilities are distributed between the three layers of command (tactical, operational, and strategic)—will depend on the specific context and threat landscape faced by the society (Flin, 1996; Flin & Arbuthnot, 2002; Atkinson & Moffat, 2005; Mattingsdal et al., 2023). In this book, Cullen (ch. 4), Lund (ch. 5), Heier (ch. 8), and Kasearu et al., ch. 16) present and discuss different contexts that in various ways present challenges that may call for a diverse range of organization models, depending on the nature of the conflict itself.

Another consideration is whether command decisions are issued as explicit orders or as conceptual directives based on intentions (e.g., Builder et al., 1999). Explicit orders provide an (illusion) of control and may be effective in stable, risk-less environments, while directives based on intentions imply a capacity for agility and may be effective in response to a changing, elusive threat such as the hybrid. Both augmented reality/virtual reality (AR/VR) and artificial intelligence (AI) technologies hold significant potential in supporting a decentralized, networked command organization, when it comes to taking measures against hybrid threats. These technologies can enable flexibility and agility in responding to unforeseen threats while at the same time ensuring efficiency in swift and resolute reactions to familiar forms of hostilities (Bakken et al., 2022; Bakken, Lund-Kordahl et al., 2023; Cristiano et al., 2023).

Training and exercises

Training and exercises have long been a fundamental concern in organizations that have roles and responsibilities within preparedness and crisis management, such as the police, the military, firefighters, and medical emergency units. These organizations rely on learning strategies, advanced technology, and development efforts to prepare their workforce for adverse scenarios. Crisis management training is intended to develop decision-makers' capacity and skills to respond to the new and atypical demands presented by an emergency or a disaster, as well as developing norms and routines of carrying out a task or exercising a specific skill (Bakken et al., 2017, 2020, 2022; Magnusson et al., 2019). Eduardo Salas and colleagues have developed a methodology for effective simulation-supported education, training, and exercises, based on a meta-analysis of training and development studies (Salas et al., 2009; see also Salas et al., 2002). In essence, this is a feedback process model that starts out with determining the learning needs of a particular training audience; then goes through stages of scenario development and the training session(s) itself, including debrief(s) and evaluation; then ends with an updated analysis of learning needs before the start of a new cycle. According to Sætren et al. (2023), effective practice-based learning requires a focus on three areas: the individual, the learning environment, and the participants' experience. What separates adult, experienced exercise participants from other, less experienced, is that they are probably more intrinsically rewarded; are more self-regulated in learning situations; and have the benefit of previous learning experiences that makes further learning more effective. Experienced participants will also have increased self-efficacy and motivation for learning due to their feeling of being competent and respected within the learning context.

Innovation is key to staying ahead of hybrid threats, and this is also true when it comes to innovations in training and exercises for hybrid threats. Organizations must actively explore emerging technologies, methodologies, and best practices to adapt and enhance their competence (e.g., Sætren et al., 2022). This includes leveraging advancements in artificial intelligence, machine learning, and data analytics to detect and analyze hybrid threats (Cristiano et al., 2023; Lund, ch. 5). Furthermore, fostering a culture of innovation encourages the development of novel response strategies and the integration of lessons learned from real-world incidents (Bakken et al., 2022; Venemyr, in press).

Other variables

Multiple variables can influence the perception of the severity of a crisis management situation, also including hybrid threats, which subsequently impacts the response to it. The contextual variables encompass temporal and spatial factors, the potential magnitude and likelihood of consequences, the physical or socio-psychological proximity, and the nature of the threat itself. Moderating variables that can either mitigate or intensify the perceived severity include psychological, social, and organizational resources as well as time pressure and the complexity of the situation (Hannah et al., 2009). In experimental designs, these contextual variables can be controlled by deliberate manipulation to examine the effects of systematic context variation, or by randomization to neutralize context effects.

Moreover, it is necessary to account for individual differences by controlling for variables such as domain experience, cognitive ability (IQ), personality traits (referred to as the "Big Five"), cognitive style (intuitive versus analytic), as well as leadership—whether this is oriented along the traditional pragmatic-charismatic distinction, or more in the direction of leadership for an adaptive organization (Bakken et al., 2023; Bakken & Larsson, 2023). "One of the biggest challenges facing leaders today is the need to position and enable organizations and people for adaptability in the face of increasingly dynamic and demanding environments" (Uhl-Bien & Arena, 2018, p. 89). In an experimental design, effective control can be achieved through matching or randomization (e.g., Bakken, 2013; Bakken et al., 2023); see also our analyzed data from previous simulation-based experiments below.

Findings from earlier experimental studies

We have together with colleagues over the last 25–30 years designed, conducted, and analyzed data from simulation experiments of crisis management decision-making to explore what parameters and factors may play a role in determining the effectiveness of organizational and individual outcomes from training and exercises. Our approach has been to select and operationalize key theoretical concepts (variables), such as organization structure, mode of communication, and other contextual factors, and implement these as controllable (independent) variables in a software simulation platform. In addition, we have measured individual differences in variables, such as cognitive style, professional and educational background, personality factors, as well as domain experience and expertise. As outcome variables, we have measured process variables related to situational awareness, decision-making, and communication process, as well as effectiveness and outcome variables (e.g. mission success rate). In the following, we will review some interesting findings from earlier experimental studies.

Network centric versus hierarchical command structure

In environments that demand a high degree of flexibility together with rapid and accurate decision-making, network centric command structures have been promoted as "the" organizational solution to meet these demands. Network centric command structures, arguably, enhance the situation awareness and the understanding of the situation. However, our results show that a network centric organization does not necessarily lead to higher perceived situation awareness or better understanding of the situation. In fact, we found evidence of the opposite (Bakken et al., 2006).

Our results indicate that operational and tactical command levels tended to perceive the success and effectiveness of the operation significantly different, as the structure shifted from a hierarchical structure to a network structure (Bakken et al., 2006). We speculate that this may be due to the abolishment of buffering and delegation principles that the hierarchical command structure entails. In addition, the self-synchronization processes required in the network structure seemed to place a heavy burden on the information processing capacities of the tactical level decision-makers. While our findings contrasted with contemporary writings on the organization of military operations (e.g., Alberts & Hayes, 2003; Atkinson & Moffat, 2005), they still make sense in light of the basic theories of information processing in organizations (e.g., Tversky & Kahneman, 1981; Simon, 1987; Newell, 1990; Morgan, 1998). A main impression from this set of experiments is that many aspects of human interaction must be managed before a network centric structure may give a full range of benefits in operations.

Co-located versus distributed command structure

In this study, we employed a military multiteam system that prosecuted time-sensitive targets to study how a co-located versus a distributed communications setting influenced the shared situation awareness and whether the shared situation awareness again influenced the outcome of the decision processes. We found that performance fell when the integration team shifted from a co-located to a distributed setting (Hærem et al., 2022). The fall in performance seemed to be mediated by a corresponding fall in situation awareness. Moreover, while the performance improved for each run in the co-located setting, we did not see such learning in the distributed setting. Qualitative observations revealed that misunderstandings lasted longer in a distributed configuration than in a co-located setting. We found that situation awareness at level 3 was the only level of situation awareness significant for predicting all dimensions of performance. In summary, our findings suggest that the proximal location of the team members is favorable compared to a more distal location. This pertains to both situational awareness and performance, as well as learning during the task runs (Hærem et al., 2022; see also Valaker et al., 2018).

Methodology

In the following, we present an original, experimental case study of a crisis management exercise with a hybrid threat scenario. The exercise was set at the operational level, with a tactical sub-command, involving inter-agency collaboration in a co-located setting. The participants were around 20 in total. Both sexes were represented, and for reasons of privacy, no demographic data on the participants are reported here. The participants were all in the final stage of a two-semester (15 ECTS) master's level course program in crisis management, communication, and collaboration, taking place at a Norwegian university. The exercise, which lasted around half a day, included both a short in-brief, and a concluding de-brief after the exercise. The participants were not pre-selected other than that that they had passed the basic requirements for master level studies, that is, a completed bachelor's degree or equivalent. In addition, the participants had to document a minimum of two year's relevant experience within a preparedness, emergency, or otherwise relevant sector or agency. For example, having the role as a preparedness coordinator at a municipality would qualify.

The practical arrangements were as follows: The participating students were divided into groups of three, each of which would play a role in the command organization—as a tasks force. Three students played fire chief, three played municipality representatives, and there were two police and health groups with three students in each group. Beyond this, the participants had to self-organize to solve the mission. The chosen solution was to appoint two representatives from the police groups as leaders of the operation. A time was then set for the first meeting, and the groups then went separately to solve the respective tasks for this meeting. Thereafter, a representative from each group met in "tactical command" to convey their group's work at the first meeting and subsequent status meetings. As a result, there was only a need for six to eight people to attend the meetings, rather than everyone in the class contributing. This probably led to more efficient use of time, and by splitting into smaller groups, everyone was able to contribute.

The hybrid aspect of the scenario was constituted by a suspected nuclear contamination incident—the source of radiation possibly carried by a civilian unmarked vehicle parked at a local camping area within the municipality. At the start of the exercise, there were observations of people in and around the surrounding forestry and lake, with symptoms consistent with exposure to nuclear substances. To simulate a search-and-rescue (SAR) capability, as well as a tool to assist in generating the situational picture, a "drone squad" was employed. The drone service team could be directed by the command group to survey areas for searching, and possibly come to the rescue. This rather novel concept for "Drone-as-a-service" (DAAS) was used to demonstrate the increased ability to generate and sustain a common operational picture (COP), which in turn strengthens shared situational awareness, communication, and team processes between the (simulated) tactical and operational level of the exercise.

Data analysis

In the following, we have summarized the student participants' own observations and reflections, edited from submitted, self-reflection reports ("narratives") written by the participants themselves short time after the exercise (i.e., the reports were retrospective). Handing in the individual report was a requirement for passing the course program and was completely anonymous, using the Inspera Assessment examination system. Here, we use no direct quotes from the reports, and all the reflections are translated (from Norwegian) and paraphrased by us, the chapter authors. The categories below are formed by a thematic grouping of the students' own reflections and are both concerned with the learning process itself, as well as factors that contribute to enhancing (or suppressing) learning in the wicked environment the exercise represented. In the analysis, we have attempted to follow the guidelines of thematic analysis, as described in, e.g., Braun and Clarke (2022). This means, among others, that the thematic categorization has not been restricted to established theoretical concepts. Rather, we have let the categories emerge, independently of predefined theoretical frameworks. During the categorization process, the following themes have emerged (not in any predetermined order):

Cross-Functional Training: With participation from several emergency services, municipalities and the armed forces, and the participants' relatively long experience from their respective disciplines, they made it possible to "mirror" the handling of a real unforeseen incident. It was a common opinion among the participants that the way in which the course program was designed, with both physical and digital teaching modules, including exercise-preparing group assignments preceding the exercise itself, constituted a fruitful approach to cross-functional interaction and learning. While it is not sufficient to interact only in a single exercise run where specific personnel are participating, the course program demonstrated that training and exercises should be repeated several times, over a timespan, and with the same people as participants throughout.

Awareness for the Unforeseen: Continuous and repeated exposure to the exercise environment and practice on unfamiliar situations (Herberg & Torgersen, 2021) will affect how we innovate and create alternative creative solutions to diverse threat challenges (Magnussen et al., ch. 13). The individuals participated in what they perceived as a safe arena for trial and error, as well as an arena to discuss and establish new points of view in relation to other participants' expertise, capacities, and capabilities. In addition, the awareness that the unforeseen is in fact unforeseen, and that one must face the individual incident both specifically as it evolves as a unique incident, while also recognizing that there may be more general and fundamental principles to apply, regardless of the incident involved. The participants pointed out that leadership, professional knowledge, and collaboration competence are redemptive. This, combined with acquired knowledge during the course program, will have made them better prepared for next time the unforeseen knocks on the door.

Information Sharing: That information is collected and dissipated throughout a command organization is essential to attain shared situational awareness, that is, ensuring that involved actors have identical perception, understanding, and projection of the situation (e.g., Hærem et al., 2022). It is common practice with incident and crisis management in Norway that the police are in lead of the operation (where "life and health" is at stake). As such, the participants in the role of police were in control of the "big picture," that is, they had the most recently updated situational awareness. Other participants had to be more patient, even though the demand for information was pressing. However, it is not always necessary—or desirable—that everyone gets every detailed bit of information at the same time. Nevertheless, to be successful at tasks that require cooperation, it is crucial that everyone has the same (shared) situational awareness. This requirement is demanding, and all participants had to make decisions within their area of responsibility during the exercise. It is, therefore, also essential that all participants contribute, by sharing the SA they possess, as seen from their point of view, so that the overall picture becomes as complete as possible.

Teams Familiarizing: It was considered a success factor for collaboration that everyone in the group got to know each other during the course program, preceding the exercise itself. It would have been a more difficult exercise if the group had been assembled without previous knowledge or familiarity with each other. This goes regardless of having (theoretical or practical) knowledge of the areas of experience that were represented as actors in the exercise. Discipline, respect, and humility for each other's profession are important factors, as well as being able to trust information that is given and received. Since each of the participants initially possessed only bits and pieces of the complete picture, interaction and learning along the way became some of the foremost factors for a successful exercise.

Distribution of Power: Even though the exercise scenario was unfamiliar to the participants, they benefited greatly from others' knowledge and willingness to share, based on their previous experience of real events and the daily professional practice from their own service. Participants who actively took lead in the exercise had previous experience of being in wicked situations. These participants made efforts to obtain an overview of the situation; they ensured that there was a good division of tasks; they were clear on the intention of the missions; they distributed power; they asked for help; they shared their own uncertainty; they involved actively other actors; they asked questions and learned in the process. This kind of behavior laid the groundwork for interaction and learning along the way. An experienced participant that assumes a leading position creates space to reflect and the capacity to solve other tasks such as communication and press management, safeguarding vulnerable individuals, and the initiating and follow-up of critical services.

Interacting with Specialists: The "learning along the way" from fellow participants made the solving of tasks in the exercise better than it probably would have been in a real situation. For example, participants who had not previously interacted with the municipality and therefore did not understand the municipality's role in the management of a critical incident, could learn along the way from several participants working in specialized roles, for example as leaders or coordinators in the municipality. At the same time, the gathering of information during the incident was essential to dealing with the incident itself. A discovery made by the participants was how closely interwoven the different roles and responsibilities of the actors were. Interaction and learning along the way in an incident will make us better prepared for handling the next similar event.

Individual Traits and Factors: As a collective category, we have the elements of trust, flexibility, perseverance, courage, mission-based leadership, involvement, and creativity. These were pointed out by several participants as the most important success factors, when it came to extract the potential and expertise of everyone involved in a fruitful way. In other words, it is the interaction between the people involved, the trust we have in each other, how we use the skills we have access to, and learn from each other along the way, that determines how the incident is handled (e.g., Bakken & Larsson, 2023).

Discussion

In our treatment of the narratives collected from the participating students, we have identified several factors or variables that contribute positively to situational awareness, communication, decision-making, and team dynamics (the performance variables). In turn, these variables influence an entity's capability to respond to hybrid threats, which is the main goal of such training and exercises. It became apparent that the factors—independent variables—we identified can be grouped in two: enabling variables and emerging variables. The former pertain to factors specified at design-time of an exercise, while the latter are factors that can be observed or inferred during the "run-time" of an exercise. Causally, the enabling variables precede the emerging variables. Figure 15.1 shows the independent and dependent

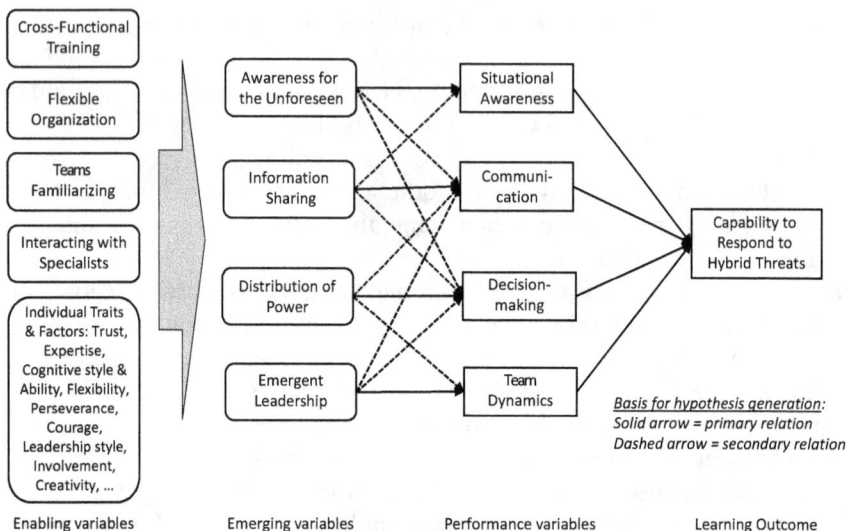

FIGURE 15.1   Model for effective training and exercise

Source: created by the authors

variables as a causal diagram, linked to the performance variables and the main goal (effect or outcome) variable.

While not being explicitly mentioned by the participants, it became evident that two additional factors or variables were crucial for the functioning of a "hybrid" command organization, namely flexible organization (an enabling variable) and emergent leadership (an emerging variable). These variables were also over-arching when it came to understanding the mechanisms underlying a well-functioning command organization that is capable of detecting, identifying, and managing hybrid threats.

An overall impression we as researchers and instructors got when observing the exercise, and later when reading the participants' own retrospective reports, was that the lines of communication were well designed "ad hoc" by the participants themselves, so that information could quickly and correctly be passed on to and shared with the right recipient(s). Various subgroups and individuals anticipated the information needs of others before they were requested, in a proactive manner, which was both time-saving and reduced confusion and uncertainty (Espevik, 2011; Barton et al., 2015; Bakken et al., 2017). A success factor appeared to be that participants shared their established tactical routines with surrounding participants, and working in a transparent manner overall made it easy to communicate and share questions and information. An example of this was the use of a whiteboard, where participants noted key points they deemed important for further work, so that everyone would have easy access and an opportunity to use this information in their work in between regular, joint briefing meetings. The importance of a flexible command organization when communicating in wicked contexts is emphasized by several researchers, e.g., Atkinson and Moffat (2005), Alberts and Hayes (2003), Flin (1996), Flin and Arbuthnot (2002), and Valaker et al. (2016). In a recent study, Mattingsdal et al. (2023) found that, in a hybrid warfare scenario, commanders' preferences for either unilateral or interagency forces depended on whether their decisions were made before (in peacetime), during, or after the conflict. They also found that commanders' preferences for interagency forces increased with their level of operational experience.

A phenomenon that is also enabled by flexible organization is emergent leadership. This is the notion that the best leader "in situ" is not necessarily a formally appointed leader within a strict hierarchical structure, but instead a leader that "rises to the occasion" and takes charge when the problem at hand falls within that person's area of expertise (Nesse, 2017, 2022; Nesse & Stensaker, 2021). For this to work, it is essential that the emergent, informal leader has the necessary trust and confidence within the organization (e.g., Bakken & Larsson, 2023). Notably, this informal, emergent approach to leadership enables mutual monitoring (Salas et al., 2005), a mechanism that could help "bringing order onto chaos."

Conclusion

The research question we asked was the following: *Which parameters could be included as command organizations pursue training and exercise programs in a hybrid threat environment, and how are these parameters linked to optimize the outcome?*

As hybrid threats continue to pose challenges to civic societies, building competence and enhancing societal resilience becomes increasingly important. This study has shown that training and exercises may contribute significantly to improving situational awareness, decision-making, communication, and other response capabilities. It will be vital to use innovative approaches and organizational models tailored to the very demanding contexts that hybrid threats pose to society. Importantly, we have found that a flexible organization structure, which enables emergent leadership, is crucial. Furthermore, fostering information sharing, distribution of power, and an awareness for the unforeseen are conducive to ensure cohesive and coordinated responses within and between crisis management organizations and the public. By collectively building competence against hybrid threats, societies can strengthen their resilience and be better prepared for an uncertain future. Advances in technology, such as AI, offer the potential for autonomy and adaptability in dynamic decision-making environments, but more development and rigorous testing are needed in this respect.

References

Alberts, D. S., & Hayes, R. E. (2003). *Power to the edge. Command and control in the information age.* DoD CCRP.

Atkinson, S. R., & Moffat, J. (2005). *The agile organization. From informal networks to complex effects and agility.* DoD CCRP.

Bakken, B. T. (2013). *Intuition and analysis in decision making. On the relationships between cognitive style, cognitive processing, decision behaviour, and task performance in a simulated crisis management context* [Doctoral dissertation]. BI Norwegian Business School.

Bakken, B. T., Hansson, M., & Hærem, T. (2023). Challenging the doctrine of "non-discerning" decision-making: Investigating the interaction effects of cognitive styles. *Journal of Occupational and Organizational Psychology, 97,* 209–232. https://doi.org/10.1111/joop.12467

Bakken, B. T., & Hærem, T. (2011). Intuition in crisis management: The secret weapon of successful decision makers. In Sinclair, M. (Ed.). *Handbook of Intuition Research* (pp. 122–132). Edward Elgar.

Bakken, B. T., & Hærem, T. (2020). Betydningen av enkeltindivider i krisehåndtering: Lederen som strategisk beslutningstaker. In Larssen, A. K., & Dyndal, G. L. (Red.). *Strategisk ledelse i krise og krig—det norske systemet* (ss. 101–121). Universitetsforlaget.

Bakken, B. T., Hærem, T., Ruud, M., & Frotjold, L. (2006). *The pros and cons of network centric organization.* Paper presented at the 2006 International Command and Control Research and Technology Symposium (ICCRTS).

Bakken, B. T., & Larsson, G. (2023). *Lederskap i langvarige kriser.* LUFTLED nr 2.

Bakken, B. T., Lund-Kordahl, I., & Bjurström, E. (2023). *AI in future C2—who's in command when AI takes control?* Proceedings of the 2023 International Command and Control Research and Technology Symposium (ICCRTS).

Bakken, B. T., Lund-Kordahl, I., & Sandberg, I. (2022). *Augmented and virtual reality (AR/VR) and artificial intelligence (AI) technology in systematic inter-professional crisis management training.* Proceedings of the 2022 International Command and Control Research and Technology Symposium (ICCRTS).

Bakken, B. T., Valaker, S., & Hærem, T. (2017). Trening og øving av krisehåndtering. En metodisk tilnærming. En Hafting, T. (Red.). *Krisehåndtering. Planlegging og handling* (s. 377–398). Fagbokforlaget.

Barton, M. A., Sutcliffe, K. M., Vogus, T. J., & DeWitt, T. (2015). Performing under uncertainty: Contextualized engagement in wildland firefighting. *Journal of Contingencies and Crisis Management, 23*(2), 74–83.

Bjurström, E., & Bakken, B. T. (2022). Dynamic decision making under uncertainty: A Brehmerian approach. *Journal of Behavioral Economics and Social Systems, 4*(2), 55–68. https://doi.org/10.54337/ojs.bess.v4i2.7749

Braun, V., & Clarke, V. (2022). *Thematic analysis. A practical guide.* Sage.

Brehmer, B. (1992). Dynamic decision making: Human control of complex systems. *Acta Psychologica, 81*(3), 211–241.

Brehmer, B. (2000). Dynamic decision making in command and control. In McCannn, C., & Pigeau, R. (Eds.). *The human in command: Exploring the modern military experience.* Kluwer.

Brehmer, B. (2009). *Command without commanders.* Paper presented at the 14th International Command and Control Research and Technology Symposium (ICCRTS).

Brehmer, B., & Dörner, D. (1993). Experiments with computer simulated microworlds: Escaping both the narrow straits of the laboratory and the deep blue sea of the field study. *Computers in Human Behavior, 9,* 171–184.

Builder, C. H., Bankes, S. C., & Nordin, R. (1999). *Command concepts: A theory derived from the practice of command and control*. Rand Publishing.

Crichton, M., & Flin, R. (2017). Command decision making. In Flin, R., & Arbuthnot, K. (Eds.). *Incident command: Tales from the hot seat* (ch. 11, ss. 201–238). Ashgate/Routledge.

Cristiano, F., Broeders, D., Delerue, F., Douzet, F., & Géry, A. (Eds.). (2023). *Artificial intelligence and international conflict in cyberspace*. Routledge.

Endsley, M. R. (1995). Toward a theory of situation awareness in dynamic systems. *Human Factors Journal, 37*(1), 32–64.

Espevik, R. (2011). *Expert teams: Do Shared mental Models make a difference*. PhD dissertation, University of Bergen.

Flin, R. (1996). *Sitting in the hot seat: Leaders and teams for critical incidents*. Wiley.

Flin, R., & Arbuthnot, K. (Eds.). (2002). *Incident command: Tales from the hot seat*. Routledge.

Hannah, S. T., Uhl-Bien, M., Avolio, B. J., & Cavarretta, F. L. (2009). A framework for examining leadership in extreme contexts. *The Leadership Quarterly, 20*(6), 897–919. https://doi.org/10.1016/j.leaqua.2009.09.006

Hærem, T., Valaker, S., Lofquist, E. A., & Bakken, B. T. (2022). Multiteam systems handling time-sensitive targets: Developing situation awareness in distributed and co-located settings. *Frontiers in Psychology, 13*, 864749–864749. https://doi.org/10.3389/fpsyg.2022.864749

Herberg, M., & Torgersen, G.-E. (2021). Resilience competence face framework for the unforeseen: Relations, emotions and cognition. A qualitative study. *Frontiers in Psychology, 12*, 669904. https://doi.org/10.3389/fpsyg.2021.669904

Larsson, G., Alvinius, A., Bakken, B., & Hærem, T. (2021). Social psychological aspects of inter-organizational collaboration in a total defense context: A literature review. *International Journal of Organizational Analysis*. https://doi.org/10.1108/IJOA-02-2021-2626

Libiseller, C. (2023). 'Hybrid warfare' as an academic fashion. *Journal of Strategic Studies, 46*(4), 858–880. https://doi.org/10.1080/01402390.2023.2177987

Magnusson, M., Venemyr, G. O., Bellström, P., & Bakken, B. T. (2019). *Digitalizing crisis management training [conference paper]*. 11th IFIP WG 8.5 International Conference, ePart 2019. https://doi.org/10.1007/978-3-030-27397-2_9

Malerud, S., & Hennum, A. C. (2023). *Situasjonsforståelse ved sammensatte trusler. Utfordringer og mulige løsninger* (pp. 20–23). LUFTLED nr 2.

Mattingsdal, J., Espevik, R., Johnsen, B. H., & Hystad, S. (2023). Exploring why police and military commanders do what they do: An empirical analysis of decision-making in hybrid warfare. *Armed Forces and Society*, 1–27. https://doi.org/10.1177/0095327X231160711

Mintzberg, H. (1980). *The structuring of organizations*. Prentice Hall.

Morgan, G. (1998). *Images of organization*. Sage Publications.

Nesse, S. (2017). *When leadership matters more than leaders: Developing a processual perspective on leadership during organizational crises*. PhD thesis, Norwegian School of Economics, Bergen, Norway.

Nesse, S. (2022). The emergence of collective leadership during a terrorist attack: Dynamic role boundary transgressions as central in aligning efforts. *Journal of Leadership & Organizational Studies, 29*(4). https://doi.org/10.1177/15480518221115036

Nesse, S., & Stensaker, I. G. (2021). Coping, not choking, under the pressure of a terrorist attack: A crisis leadership coping model. *Journal of Applied Behavioral Science, 58*(2). https://doi.org/10.1177/00218863211066569

Newell, A. (1990). *Unified theories of cognition*. Harvard University Press.

Pedersen, T. (2023). A small state's cyber posture: Deterrence by punishment and beyond. *Scandinavian Journal of Military Studies*, 6(1), 58–68. https://doi.org/10.31374/sjms.191

Sætren, G. B., Stenhammer, H. C., Andreassen, N., & Borch, O.-J. (2022). Computer-assisted management training for emergency response professionals in challenging environments. *Safety in Extreme Environments*. https://doi.org/10.1007/s42797-022-00066-0

Sætren, G. B., Vaag, J. R., Hansen, I. F., & Bjørnfeld, G. A. (2023). Situational awareness in a creeping crisis: How the initial phases of the COVID-19 pandemic were handled from a crisis management perspective in the Nursing Home Agency in Oslo. *Journal of Contingencies and Crisis Management*, 1–15. https://doi.org/10.1111/1468-5973.12458

Salas, E., Cannon-Bowers, J. A., & Weaver, J. (2002). Command and control teams: Principles for training and assessment. In Flin, R., & Arbuthnot, K. (Eds.). *Incident command: Tales from the hot seat* (ch. 12, pp. 239–257). Routledge.

Salas, E., Sims, D. E., & Burke, C. S. (2005). Is there a 'big five' in teamwork? *Small Group Research*, 36(5), 555–599.

Salas, E., Wildman, J., & Piccolo, R. F. (2009). Using simulation-based training to enhance management education. *Academy of Management Learning & Education*, 8(4), 559–573.

Simon, H. A. (1987). Making management decisions: The role of intuition and emotion. *Academy of Management Executive*, 1(1), 57–64.

Steigenberger, N., Lübcke, T., Fiala, H. M., & Riebschläger, A. (2017). *Decision modes in complex task environments*. CRC Press.

Sterman, J. D. (2000). *Business dynamics: Systems thinking and modeling for a complex world*. Irwin McGraw-Hill.

Thiele, R. (2013). *Towards comprehensive capabilities. ISPSW strategy series: Focus on defense and international security*, 244/2013. https://www.academia.edu/5920392/Towards_Comprehensive_Capabilities

Thiele, R. (2021). Competition and conflict. In Thiele, R. (Ed.). *Hybrid warfare. Future and technologies*. Springer Nature.

Tversky, A., & Kahneman, D. (1981). The framing of decisions and the psychology of choice. *Science*, 211, 453–458.

Uhl-Bien, M., & Arena, M. (2018). Leadership for organizational adaptability: A theoretical synthesis and integrative framework. *The Leadership Quarterly*, 29, 89–104. https://doi.org/10.1016/j.leaqua.2017.12.009

Valaker, S., Hærem, T., & Bakken, B. T. (2018). Connecting the dots in counterterrorism: The consequences of communication setting for shared situation awareness and team performance. *Journal of Contingencies and Crisis Management*, 26(4), 425–439. https://doi.org/10.1111/1468-5973.12217

Valaker, S., Lofquist, E. A., Yanakiev, Y., & Kost, D. (2016). The influence of predeployment training on coordination in multinational headquarters: The moderating role of organizational obstacles to information sharing. *Military Psychology*, 28(6), 390–405. https://doi.org/10.1037/mil0000123

Venemyr, G. O. (in press). Towards proactive crisis management innovation: A meta-narrative literature review of drivers and barriers. Forthcoming in *Nordic Journal of Innovation in the Public Sector*.

Weick, K. E., & Sutcliffe, K. M. (2015). *Managing the unexpected: sustained performance in a complex world* (Third edition.). John Wiley & Sons Inc.

16

COMPETENCE FOR HYBRID THREATS

A strategic competitive development model

Leif Inge Magnussen, Glenn-Egil Torgersen,
Ole Boe and Herner Saeverot

Introduction

Our key question in this research chapter: Is it possible to develop competencies to become better at dealing with hybrid threats (HT)—events that, due to their hybrid nature, are not always yet directly known or visible? And if the nature of HTs is shifty and multifaceted, there is a need to monitor the relevance of the detection tools. Hence, there is still a need for new research measuring the levels of HT across sectors and society, which may build on the foundations of the realms of the unforeseen (Torgersen et al., 2024) that relates to the known-unknown dimensions of meaning-making and learning processes. We ask whether traditional strategic (didactic) planning models for such training may be used, or is a new, more tailored one needed to embrace the characteristics of HT. We will discuss how existential threats and its relation to pedagogy can be analyzed and suggest prophylactic action based on education theory. The intent of hybrid threats, can be perceived as unforseen and unplanned, but in reality it can be orchestrated and planned in minute detail and demand new educational practices.

"Bricoleurs" in the total defense as a barrier

The effects of HT can be found not only at the individual level but also at all organization levels and globally (Rawat et al., 2020). The intent was to construct a practical model that can function as an aid to all actors that works with competence development related to HT and to build the relevant competence to be a horse's head so to speak in front of the attackers. It means that the educational model proposed will not only be of interest to leaders

DOI: 10.4324/9781032617916-20

of military units or emergency response organizations, but it may also be relevant to leaders of all organizations, plus the individual citizens.

With such an understanding of preparedness, where the entire society is included, so-called "Total defence," all actors will need a common insight and training related to what HT can be, and what should be done when attacks occur. To build a common resilience, it will also be necessary to have good and effective interaction, not only between emergency services and the Armed Forces but also between the agencies and the individual citizen. This will not happen by itself; training is needed, which in turn requires adapted pedagogical thinking and didactic models for planning and implementing training at all levels and parts of society, where there is a high degree of recognizability in both the pedagogy and the content, so that they do not diverge or work against each other. Similarly, because there is no set formula when it comes to HT, it will be important to develop competence to learn as you go, and to be able to innovate and improvise within a given framework. This kind of general total defense competence, whereas many people as possible become a "bricoleur" (Lévi-Strauss, 1966), who both individually and collectively can constitute the preparedness against HT. "Bricolage" refers to skills and mindsets that can put pieces together and solve unforeseen challenges in new ways with resources available at the time (Weick, 1993). Preparedness for HT will require such skills at all levels of society. Pedagogy must thus take the nature of HT and this competence requirement seriously.

Competence development models

A traditional model of didactic planning consists of questions like what's to be done, how to do it, when to do it, and why do it and is an educator's tool of implementing a pre-defined outcome (Försvarsmakten, 2000). Let's start with models of how skills development has traditionally been planned. Didactic models-practical application models related to teaching and learnings processes (Tobias & Fletcher, 2000), are used as strategic planning tools that show the overall process and what should be considered when planning education and training programs for the development of competence for specific tasks and challenges (Torgersen, 2015). One of the most used, especially in operational professions such as emergency preparedness and defense, is the instrumental so-called "goal-means" model (Figure 16.6, only the factors from competence needs to evaluation), or "Instructional Systems Development/Design (ISD) models" (Tennyson & Foshay, 2000).

Central to this model is identifying competence needs and then planning training on this basis. The different parts of the model are referred to as didactic phases. First, learning objectives are formulated, then teaching methods and scenarios to be practiced are determined, then this is carried out for the target group, and finally, learning outcomes and the learning process are evaluated. This – New formulation: Such educational practice works where there are clear and easily identifiable types of competence to be trained; the content of what is trained is well known; the context for both the learning and where the competence is to be exercised in practice is stable, clear, controllable, and predictable over time; and not least, solutions and procedures are known in advance. Such didactic models can also work if the topic or threat is reduced to something very specific, such as specific software sequences in cyber threats, or other specific technical systems that can be maintained and adjusted with simple measures if previous versions do not meet the need. The focus is on the content. For example, Steingartner et al. (2021) use such "goal-means" models in their development of what the content should be in educational programs for cyber deception provided in an HT context.

However, "goal-means models" will be dependent on the time aspect with regard to the validity of the knowledge over time. As soon as new types of specific threats are identified, the content of the training must change. In an educational context, an unforeseen event and HT as a phenomenon foster the need for alternatives to educationally recommended "smart goals" used in emergency preparedness training (Kristiansen et al., 2017). Such a "direct" approach is dependent on the ability to foresee what is necessary to train on beforehand and set learning objectives in measurable, realistic, specific, motivating, and time defined. Such training frameworks relate to past experiences and ensure that the planners are updated on the current status of HT. A model will also not work where the model is to cover all training for different types of HT, or HT in general, and in particular not from a strategic (long-term) perspective. In such cases, other types of models are needed. These are so-called relational didactic models, which are more general and contain selected general characteristics of the challenges to be solved. Here, these characteristics are seen in relation to various didactic phases given in the goal-means model.

In order to develop such a model, it is necessary to identify important characteristics of the phenomenon to be handled, and the development of these characteristics is both flexible and as time-independent as possible, while at the same time touching the core or depth of the nature of the phenomenon. We will do this by drawing on other relational training models that have been developed for similar phenomena, which are characterized by a high degree of unpredictability and require complex competence in order to be managed. Through these, we also derive the characteristics of HT that should be included in a strategic didactic model for HT in general, but which can also be applied to specific HT threats. Such a model will require a relational understanding of the factors involved, and that all choices made in connection with the planning, implementation, and evaluation of training measures must be matched against these characteristics to ensure that the competence developed is sufficiently adequate to deal with HT.

254 Leif Inge Magnussen, Glenn-Egil Torgersen, Ole Boe and Herner Saeverot

In these relational models, we see that the direct-indirect relationship will be of importance in regard to how learning processes should be understood and facilitated, including the planning and evaluation of competence measures, given the context of the nature of HT, which is characterized by a high degree of unpredictability. Thus, our main question will be guided by this relationship as well via the following questions: Can hybrid threats, as an indirect cause of unforeseen events in society, make the nature of the unforeseen itself indirect, as countermeasures to planned operations?

This opens for a counter-oriented Bow-tie analysis. We do not intend to cross the borders of counterintelligence; rather, we wish to argue that the public should recognize the options given by foundational theories related to basic research on *The unforeseen* and *Indirect pedagogy* (Torgersen, 2015; Saeverot, 2022). This will develop competencies needed to reduce the effects of hybrid threat, and by that become a countermeasure. The unforeseen can be defined as *a relatively unknown event or situation that occurs relatively unexpectedly and with relatively low probability or predictability to the individual, group, or community that experience and handle the event* (Translated from Torgersen, 2015, p. 30). A discussion of what is the dependent and independent variable in this chapter about the nature of the phenomena of HT and hybrid warfare (HW) can be problematic. We agree that this line of thinking can be fruitful in some methodologies like in measuring the effects of interventions. However, we perceive the phenomena of HT as more complex and a field where the orderliness of causation may be blurred and even chaotic. We thus wish to conceptualize basic strategic competences for formative and summative evaluation practices as a possible detection tool of HT. To counter HT with the competence needed, we need to recognize some of the patterns or nature of HT.

The challenges of hybrid threats to learning

To gain more insight into the nature of HT, more systematic research is needed, in particular regarding the characteristics of HT. This means not only to define the term more precisely and comprehensively, but also to be able to prevent and develop preparedness and measures to reduce damage, as well as being able to develop the legal systems related to HT. There is also need for developing adequate pedagogical learning designs to develop and convey such competence across society. Such needed set competencies are also discussed in Chapter 15 by Bakken et al. A pivotal point of this chapter is one of the basic models in military pedagogy, didactics for the demands of war, first developed by the Finnish military educationist Jarmio Toiskallio (2006). In the light of the nature of HT, the model will be further developed, including new research findings and perspectives from existential threats (Saeverot & Torgersen, 2022a) showing how pedagogy can pave the way for interorganizational learning at both a societal and personal levels.

Here it must be added that modern education, especially the kind of education developed by politicians with a basis in New Public Management thinking and the OECD's economically based education policy of human capital optimisation, is not suitable when it comes to the complexity of HT. The main reason is that this politicized education is too direct and linear, whereupon it will block that which is unforeseen and therefore overlook opportunities for new knowledge and learning. Therefore, we would like to propose a different form of education, which is free from political and ideological undertones. We are thinking of indirect pedagogy.

Direct versus indirect pedagogy

Indirect pedagogy opposes the smart goals strategy and the direct instruction that these goals presume. The reason being that indirect pedagogy does not follow a linear or teleological time pattern. While direct instruction is about someone pointing out directly, as clearly as possible, what the recipients should do and how they should do it, often with a clear goal in advance, indirect pedagogy is characterized by ethical communication and dissemination of thoughts, both verbally and non-verbally, in a way that to varying degrees deliberately obscures, hides or "wraps" the message through intermediaries, such as metaphors, symbols, allusions, silence, and irony (Saeverot, 2013, 2022). Such intermediaries can be used as pedagogical tools for teaching and learning, for example, in such a way that they make room for productive moments that may lead to new learning and knowledge. In direct instruction, the recipients must achieve a predetermined goal. In this way, time is perceived as homogeneous and chronological, like the arrangement of a clock or calendar.

In Greek antiquity, this kind of time was referred to as Khronos. Khronos, or the mechanical structure of homogeneous time, forces us to perceive reality in a predetermined way. Thus, we end up with a predetermined and fixed structure that lacks the conditions for opening to the unforeseen where the recipients are given opportunities to make their own choices and reflect and think independently. Instead, the recipients are trapped in a pattern characterized by habits and mechanical actions. Indirect pedagogy, on the other hand, creates space for a different time perspective, which the ancient Greeks referred to as Kairos. The Greeks saw Kairos as a God, more specifically the God of the fleeting moment. It is all about seizing time, the moment, which is coming toward us, in unbeknownst manners. We must so to speak catch Kairos as he comes at us, in order to seize time. If not, time will pass by, and we will miss time altogether. Once time has passed, it never comes back. One must therefore react quickly, so as not to miss a golden opportunity; namely, time itself. Unlike Khronos, Kairos cannot be predicted in advance, other than through certain preordained signals. Kairos is a metaphor for a temporal perception that can open the door to what is genuinely new.

We find traces of a linear thinking of time in the Bow-tie model (Figure 16.1) in which a linear timespan from threat (pre-event), occurrence (present), and recovery (post-event) (Torgersen, 2015). It belongs to the sphere of Chronos. The individual perception of time in the phase of occurrence can upheave the sense of chronology. There is a sensation of time stopping, and it can create situations of super-memory and distorted sense of time. It is the experience of Kairos where the situation dictates the response. The concept of Kairos represents the situation that dictates the learner or response but also the opportunity to step-up and grow. It represents a learning opportunity.

Bow-tie model thinking (Torgersen, 2015) is also traditionally linked to a pathway from threat, through incident and on to recovery. Unforeseen events materialize in UN-0 (the UNforeseen zero) and are relative. It can be unforeseen to some actors; to others, it can be something that has been foreseen. In addition, it can even be an event that is planned for. This is also the nature of HTs and events; some are planned like terrorism or military operations, but others can occur at random. Furthermore, the chapter conceptualizes how HT can be measured, interpreted, and communicated as training for HT. Multi-threats and their cascading effects may also represent warning signs in the Bow-tie model (Torgersen, 2015), which can be at least partly revealed through monitoring systems like UN-meth, but adjusted to different realms. An unforeseen event (Figure 16.1) is divided into five main categories—within a degree of: (1) relevance (to the target audience), (2) possibility (of occurrence), (3) how known (in advance by the target audience), (4) warning signs (scope/number), and (5) warning time (for given/identified warning signs and exercises, e.g. unannounced exercises).

UN-0 = The moment of occurrence of HT actions, which have a tangible impact and are experienced by individuals, groups, organizations, or society in general. In and just after UN-0, the event may have a high or low degree of unpredictability, depending on the degree of danger signals and whether they are identified or ignored, and how suddenly and clearly, they appear. Over time, the situation will become more familiar, and the degree of unpredictability will decrease. However, there may be unforeseen consequences of the events (in UN-0 or new ones along the way), which are then referred to as sidewinders. Sidewinders carry resemblances with "cascading effects" found in HT theories.

Jarmio Toiskallio launched an educational model at a strategic level, with the intent to make the demands of war more clearly in educational models at military academies at all levels (Toiskallio, 2006, p. 63). Traditional educational models for learning, with civilian origin, do not consider the demands of war in specific manners. Toiskallio (2006) called the model as the hermeneutical military-pedagogical planning model. The apex of the model is about the conceptions and the nature of war, battle, and crisis, connoted as "the demands of war." The demands of war can be seen as essential in

RISK EVENT CHAIN

PHASE 1   THREAT

PHASE 2   CRISIS

PHASE 3   DISRUPTION AND IMPACT

WARNING SIGNS   [INTERACTION DURING RISK   ]

THREAT

PREVENTION/BARRIERS   [INTERACTION FACTORS]   RECOVERY   CONSEQUENCES

THREAT   UN-0   UNFORESEEN PHENOMENON DECREASES   SIDEWINDERS   CONSEQUENCES

CONTROL MEASURES

THREAT   LOSS OF CONTROL   CONSEQUENCES

[BEHAVIOUR TO INTERACTION   ]

UN-INTERVAL

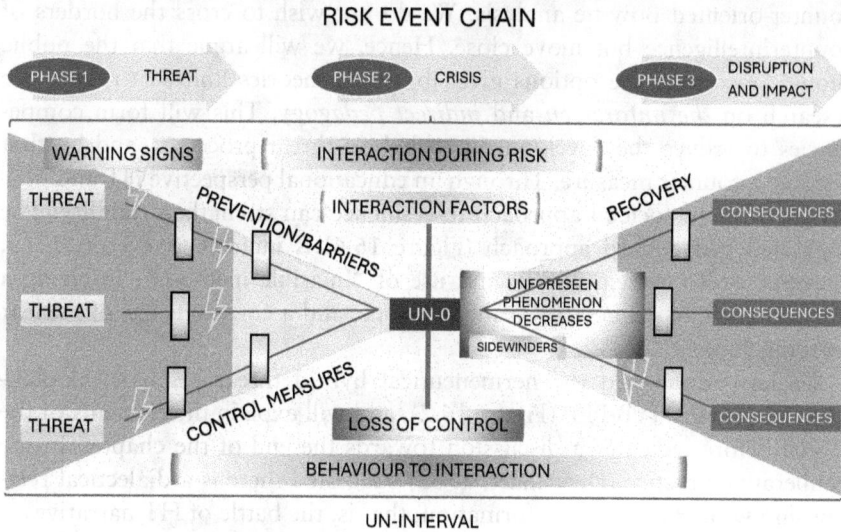

FIGURE 16.1   The UN Bow-tie model.

The unforeseen presented in a Bow-tie model, which also envisions that interaction are key in the coping with HT and unforeseen events at the three core phases.

Source: created by the authors

military training and of the learning processes, emphasizing the utility of the competence rather than the subject discipline. In our opinion, the model represents a form of pragmatism, meaning that it is useful related to training for real situations.

The Toiskallio-model and the thinking behind it also raise the question of who decides what is important to learn and how to learn it in the best possible manner. We will return to this question at the end of the paper when discussing how we know what sort of competences are needed to reduce the effects of HT and offer strengthened resilience. The unpredictability of the unforeseen and the complexity of HT contribute to a new understanding of the military educational foundations of "the demands of war." We will promote a new model of educational planning, which embraces HT and the unforeseen.

A revised model—the hermeneutical hybrid threat pedagogical planning model

Based on the foregoing discussion and Toiskallio (2006) educational model our new model represents a novel and practical tool of developing competences to handle HT. We will argue that the different conceptions inform educators at strategic, tactical and operational levels. With other words a counter-oriented Bow-tie analysis. We do not wish to cross the borders of counterintelligence but move close. Hence, we will argue that the public should re-cognize the options given by basic theories knitted to the basic research on *the unforeseen* and *indirect pedagogy*. This will form competencies to reduce the development of hybrid threat processes, and by that become a counter measure. Through an educational perspective HT and pedagogy, a broad societal approach to resilience can strengthen stability. The suggested pedagogical approach (Figure 16.6) is indirect (Saeverot, 2013; Saeverot & Torgersen, 2020), and use of "invisible methods", implying a minimum use of defined blueprint solutions, and a conscious use of unclear learning content.

We have developed the hermeneutical hybrid threat pedagogical planning model HE-HYB-PEP (Figure 16.2), and will explain the elements of the model before we enter a discussion towards the end of the chapter. In the deliberative practice (Degenhart, 1982) suggested there is a dialectical relationship with HT-sympathy formation, that is, the battle of HT narratives—HT expression and public opinion (national and international) sympathy.

The model (see Figure 16.2) "starts" with the actors' current conceptions of the HT. How manifest the threats are informs the educators on their conceptions of time and situation. Predictions of the futures by providing and hypothesis (and mental simulations) stimulate the thoughts and decisions on how to educate. Such processes are influenced by current objectives and the fostering of an experimental learning environment and mindset. Meta-evaluations of the objectives and methodological choices adapts to the conceptions at hand, influenced by the views of education and humans.

In Toiskallio (2006) didactic model there is a perspective of growth learning environment. We put weight on the learning environment but understand it more as situations where the teacher tries new ways of teaching, and the student are allowed to make errors (Försvarsmakten, 2001). If learning goals due to the unforeseen (Torgersen, 2015) and HT and its cascading effects are of less importance, there is a need of being dynamic in relation to the didactic relevance of training.

Learning and the unforeseen

Learning related to the unforeseen in organizations and society, together with strategic competence management, form the baseline in the development of curricula and competence development schemes at all levels and in all types of organizations in society. Furthermore, it can contribute to the formation of authorities and governments, not only nationally but also globally. A key feature is the collaboration between different actors, where an international perspective is added as a basis. Such collaboration efforts were one of the

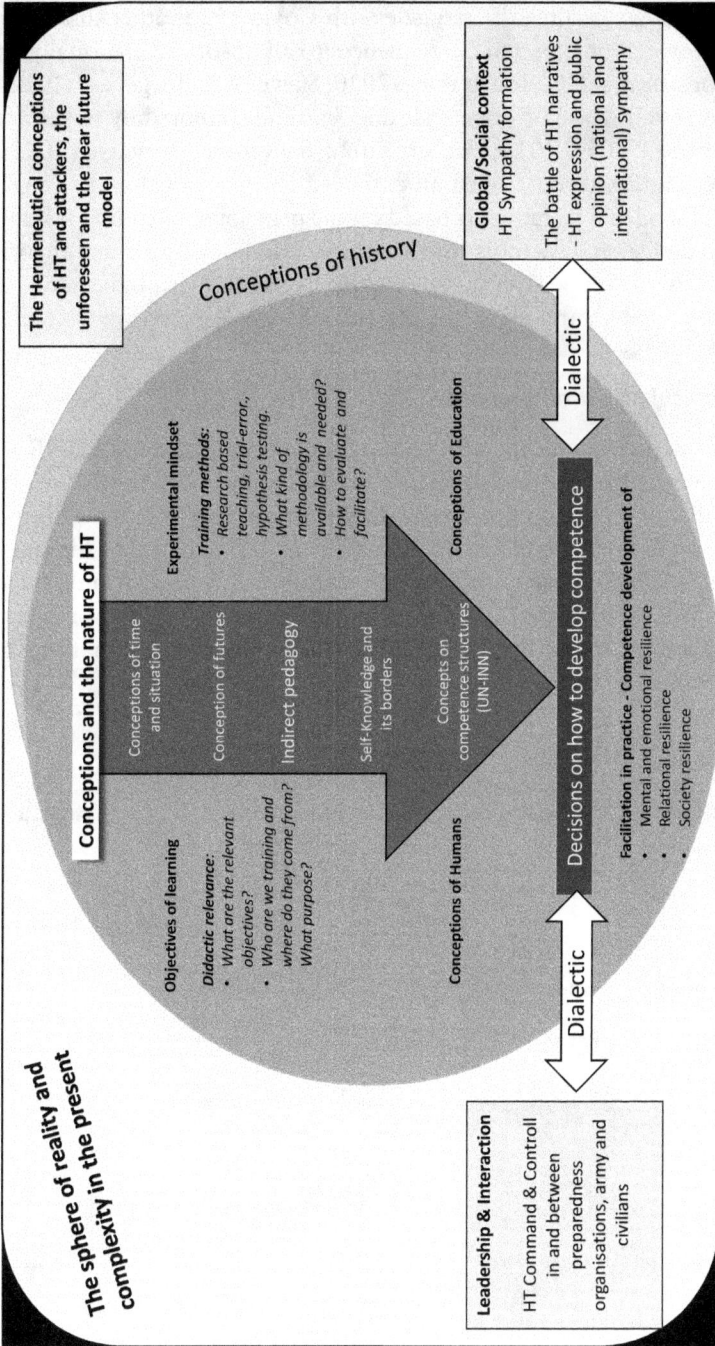

FIGURE 16.2 The hybrid threat pedagogical planning model.

Source: created by the authors

learning aspects given after the struggle with Covid-19, another unforeseen event that provided unexpected consequences to all of society, nationally and globally (Torgersen, 2018; Rawat et al., 2020; Saeverot & Torgersen, 2022a).

From the basic research competence and learning connoted by unforeseen events (Torgersen, 2015, 2018; Herberg, 2022; Saeverot & Torgersen 2022a), Torgersen et al. (2020) developed an extended hermeneutical and strengthened societal model. The model is based on the principles given by Toiskallio (2006), together with new tools for the construction of educational theories (Kvernbekk, 2005; Suppe, 1989), that include experiences from the Covid-19 pandemic and civilian and military perspectives.

The pedagogical preparation model

Figure 16.3 visualizes a pedagogical preparation didactic-model (PED-PREP), where the new central aspects in relation to Toiskallios (2006) basic models are essentially the evaluation processes, that is, learning from events, both in prevention while they happen and in the aftermath. In particular, the model emphasizes *underveislæring* (concurrent learning and evaluation), which both contains formative and summative principles of evaluation. Managerial or leadership competence during complex and time-sensitive situations are core objectives of learning processes.

FIGURE 16.3   PED-PREP model

Source: Torgersen et al. (2020, p. 849)

The PED-PREP model represents a concrete didactic tool for all organizations and societies through competence development in the challenges found in meeting with larger unforeseen events. In relation to pandemics, Torgersen et al. (2020) identified 22 competence structures or descriptors, which ought to be developed to be prepared for such types of challenges and divided them into three main categories of competence: mental and emotional resilience, relational resilience, and society resilience. "Resilience" in this context means resilience and endurance—that a system, group, or individual has the skills, routines and leadership to withstand the impact and preferably beat or manage the situation (Herberg, 2022). On a principal level, the apex of the PED-PREP-model (conceptions of the nature of virus, handling, and crisis) is similar to the demands of war in Toiskallios (2006) model and can be used as a didactic tool to develop identified competence needs, given the PED-PREP-model's left column (CS-PAN) (Torgersen et al., 2020).

The conception, nature, and basic competence structures on hybrid threats

We will now sketch out the competence structures that educators should consider when dealing with the demands of war, which originate from HT. We do this with perspectives that challenges the relevance of the perceived competence demands and the evaluative: are we asking the right questions? Further, thus, to handle HT and prevent effects of HT by pro-active use of evaluation theory (Stufflebeam, 2000), investigating the relevance of the measurement tools used to "detect" HT is of importance.

HT and competence structures

Based on pandemic research (Torgersen et al., 2020), we suggest that there is a need of understanding HT cascading effects through an educational perspective on resilience addressing factors of mental and emotional resilience, relational resilience, and societal resilience in the terms the competence to meet a pandemic situation (CS-PAN as seen in Figure 16.3).

The CS-PAN and the 22 resilience factors can all be targeted by HT-actors and can again be met by competence structures. To facilitate the learning and understanding of HT as different fields of competence, we can look to how different subjects can be organized to best inform learning processes and communicated. We will outline how CS-PAN, as a further tool of communication, can be learned (also indirect) through the principles laid out by Phenix (1964). Through *analytic simplification*—targeted communication of phenomena throughout the life-cycle, *dynamism*—the content is presented at increasing complexity at different stages and *synthetic coordination*—how the subjects relate to other subjects. The principles laid out aims to facilitate learning and communicate findings in a way that fosters understanding and clarity. Synthetic coordination is about the organization of disciplinary concepts and "logic," and the relationships between such constructs and taxonomies. The cascading effects of HT can be understood through the different realms or fields of science to use Phenix, that is, HT can be seen or experienced differently in different sectors and levels of society, that is, educational systems, politics, computer sciences, emergency agencies, electricity supply and at the individual sphere affecting beliefs and disbeliefs. To shape up the fuzzy field of HT the different realms can be seen as multiple and with some overlaps. Three levels/spheres of resilience can be addressed: mental and emotional, relational, and societal resilience, and discussed in the light of competence levels, formation, and action levels (Torgersen et al., 2020).

Pro-evaluations—basis for educational action

Educational theory may enhance action competence to meet the demands of HT and suggest prophylactic action based on education theory. We make the connotation that if the nature of HT is shifty and multifaceted, there is a need to monitor the relevance of the competence detection tools. As for the Toiskallio (2006) model, this can help to determine the individual, organizational, and societal needs of competence based on the demands from HW. Hence, there is a need for meta-tools measuring the levels of HT across sectors, in which learning needs competence levels and the HT action level can form learning processes proactively, concurrently, or reactively.

Scriven (1969) launched the perspectives on meta-evaluations, evaluations of evaluations, evaluation system, or evaluation devices. The partly tacit nature of HT strengthens a societal need of multi-faceted evaluative tools and educational learning and/or communicative systems. Stufflebeam (2000) developed a toolkit for developing a systematic meta-evaluation, addressing the realms of utility, feasibility, propriety, and accuracy, on which we build our methodologies on. This resonates with Evans (2020) and his work on hypothetical thinking—how can we be asserted that our perceptions of the future, (present and past) are the informed ones.

If HT carries four basic structures in its nature, it (i) is progressively complex, (ii) is grey zone oriented, (iii) is existential, and (iv) carries a high degree of the unforeseen. This forms four realms in the PHL-model apex—the demands of war or HT. At the same time, there is also a need to bring in the attacker perspective: Who is behind the threats? What is the motive? Are the actions planned and a collaborative project, or is another unforeseen or spontaneous event, such as rapidly occurring conflicts, terrorism, natural disasters, or accidents, supplemented by deliberate HT? The attacker perspective is necessary to include in competence development for HT. This should also be seen in the context of social context and socially oriented sympathy assessments, that is, how the actions of the attackers are perceived in public opinion, together with countermeasures from those exposed to HT. This will always influence national and international political assessments, and thus affect civil society to a large extent with unforeseen consequences.

Didactic relevance and applied training methods are key. To approach HT conceptions of humans and education will form the deliberation backdrop.

The link between conceptions of HT and educational decisions

We suggest in the core of educator's decision-making lies an awareness of the current conceptions of time past (historicity), present (Kairos), and the future. Further, the understanding of the in-direct education might foster alternatives ways of being a "teacher." Self-knowledge is about the process of constantly questioning what we know, how we know it, and what we still don't know (Figure 16.5).

Objectives of learning	Experimental mindset
Didactic relevance:	*Training methods:*
• *What is the overall purpose of the learning and formation measures and plans – based on competence challenges that the nature of HT provides?* • *Who are we training, their prerequisites and where do they come from?* • *What are the relevant objectives? (Or, not having specific learning goals – unforeseen, imply generic competence structures)*	• *What kind of educational principles will fit best? I.e. research based teaching, trial-error, hypothesis testing, case-based, discourse/dialogue-based, discovery-based, problem-based, unforeseen, and innovative based)* • *What kind of educational methodology is available and needed? I.e. traditional lectures, simulation (ICT/AI), full-low scale (including notified – unannounced) exercises, knowledge/skills methods (mastery of procedures/routines/ equipment/technology, societal mindset, interaction, existential and ethical, innovative thinking/bricolage)* • *How to evaluate and facilitate? I.e. observations, forms, video, solution-focused, formative (during the learning process), summative (final assessment), metacognitive (reflection), lessons learned-focused)*
Conceptions of humans	Conceptions of education

FIGURE 16.4   Didactical relationships on competence challenges

Source: created by the authors

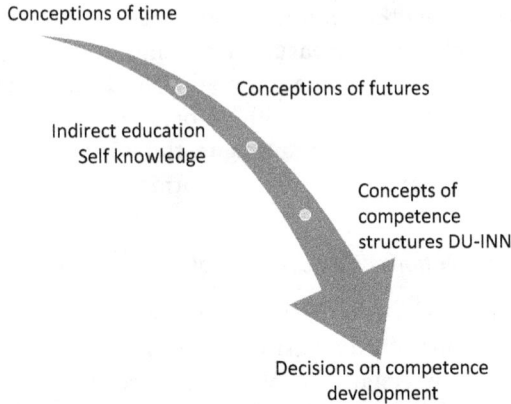

FIGURE 16.5 Conceptions on HT and education.

Source: created by the authors

The inner logic of how, what, and why of competence development, in the unclear and unpredictable of world of HT/HW, can require questioning about the learning contents (Figure 16.5). Questioning what one knows and how one knows it, understanding history, and the current learning culture are then of importance.

The core of a new competence development model

First, conceptual training will be required in the model itself, and often also the background to it, as we have derived it in this chapter. The target group for such training will be military education, civilian emergency preparedness environments, and higher education subjects that deal with emergency preparedness and international cooperation. It will also be necessary to introduce HT as a topic in schools, especially at upper secondary level.

In Norway, there are currently no guidelines in the official curricula for HT as a teaching topic. This will be a political issue, which our model also includes, especially in relation to the factors "Leadership & Interaction" and HT Sympathy formation (in A Global/Social context). If the understanding of HT is not built into society's educational systems, measures that will have to be taken to prevent HT and its management will not be a matter solely for military units and emergency response organizations. In addition, any invasive measures, such as the internet, strict ICT security, exercises, and general vigilance, will be something that civil society will not accept or will be seen as unnecessary.

Through a combination of unforeseen methodology (UN-meth) (Torgersen, 2018) and globCuret (Saeverot & Torgersen, 2022b)—a global curriculum aimed at existential threats, we conceptualize how HT can be measured and communicated. Bow-tie model thinking is traditionally connected to threats, through incidents and recovery from the impact of threats. While certain incidents can be experienced as unforeseen to some people, others may experience the same incident as foreseen. Furthermore, certain incidents can even be planned beforehand. Predictability can also be in the nature of HT and events, some events are planned, like terrorism or military operations, but others can occur at random.

Meeting HT through educational planning may represent educational chaos, if everyone formulates their own learning paths. However, by using the validation methodologies by Stufflebeam (2000), we can ask us are we providing the right education to demands of war or our conceptions of HT and by that establish a more unified understanding of the how's, what's, and why's of teaching. The methodologies at hand can be direct or indirect. A growth learning environment with the belief that we can be a cause in the matter is of importance together with the understanding that objectives of learning is also a set hypothesis, which we need to constantly monitor and ask if they are the right ones. Generic competence structures to meet the unforeseen and innovation play a key part in the decisions on how to educate and train (NIFU/USN, 2021).

In the professional education of personnel in defense and emergency response agencies, interaction between the various agencies will be a key competence component in general, but in light of the nature of HT, there will be an additional need for interaction in a wide range of infrastructural areas, which will ultimately also apply to private individuals in their own homes. This is also covered in the HE-HYB-PEP model (Figure 16.2), via "Leadership & Interaction."

The dynamism of time, objectives, learning environments, competence forms the why's, how's, and what's of the educational decisions. The complexity of training for the HT demands new ways of practicing educational planning. When decisions on what to teach are clear, educators can be in a dialectical relationship with the classic models of teaching (Figure 16.6).

The core of the *holistic HT didactic planning model for competence development to handle HT* model (Figure 16.6) is that educational planning for HT must include specific learning objectives that focus on both concrete threats that one already knows (history), unforeseen situations (the unthinkable and flexibility), and the whole, including the formation of society. This combination is not easy but will emerge as the only and best solution for building holistic resilience against HT. The holistic HT didactic planning model for competence (Figure 16.6) with the learning topics established within the HE-HYB-PEP model (Figure 16.2) can be straightforward and a matter of educational execution based on the ever-ongoing process of establishing the learning content based on the "demands of HT". One solution and topic to

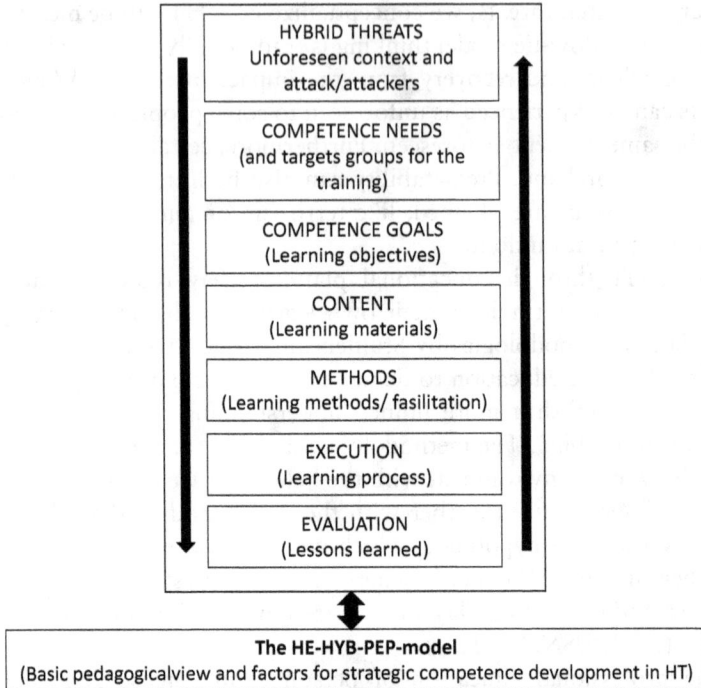

```
┌─────────────────────────────────────┐
│  ┌───────────────────────────┐  ↑    │
│ ↓│     HYBRID THREATS         │       │
│  │   Unforeseen context and   │       │
│  │      attack/attackers      │       │
│  ├───────────────────────────┤       │
│  │    COMPETENCE NEEDS        │       │
│  │ (and targets groups for the│       │
│  │        training)           │       │
│  ├───────────────────────────┤       │
│  │    COMPETENCE GOALS        │       │
│  │   (Learning objectives)    │       │
│  ├───────────────────────────┤       │
│  │       CONTENT              │       │
│  │   (Learning materials)     │       │
│  ├───────────────────────────┤       │
│  │       METHODS              │       │
│  │(Learning methods/ fasilitation)│   │
│  ├───────────────────────────┤       │
│  │      EXECUTION             │       │
│  │   (Learning process)       │       │
│  ├───────────────────────────┤       │
│  │      EVALUATION            │       │
│  │   (Lessons learned)        │       │
│  └───────────────────────────┘       │
└─────────────────────────────────────┘
                ↕
┌─────────────────────────────────────┐
│          The HE-HYB-PEP-model        │
│(Basic pedagogicalview and factors for│
│ strategic competence development in HT)│
└─────────────────────────────────────┘
```

FIGURE 16.6 A holistic HT didactic planning model for competence development to handle HT, where all didactic phases must be matched against the HE-HYB-PEP model.

Source: created by the authors

more research would be to introduce templates for the development of learning objectives for increased resilience, which can be used in the development of exercises and scenarios for emergency response agencies, the armed forces, and civilian educational institutions.

Conclusion

In this chapter, we have asked whether traditional strategic (didactic) planning models for HT/HW training may be used to embrace the characteristics of HT. We have analyzed how existential threats and its relation to pedagogy can be met through a hermeneutical approach and suggest prophylactic action based on education theory. If a hermeneutical perspective on learning is the way forward in the encounter with HT and HW, the question is how such a complicated competence development model can be used in practice. For that more research is needed, conveying the dynamism of piloting through testing and training.

In this chapter, we have also shown that the general planning models for training and competence development will be too rough and unvarnished when developing competence for HT. It is, therefore, necessary for the model to contain an adapted pedagogical philosophy (views) and factors that show the specific and distinctive characteristics of HT/unforeseen contexts and attackers. We have shown that all planning, implementation, and evaluation (didactic phases) of such training must be sparred, challenged, and prepared against the factors given in the HE-HYB-PEP model (Figure 16.2), in order to ensure precise and relevant competence development for handling HT in the present and future. Organizations and learning cultures that align with new public management, will probably oppose the suggested advanced didactical model, when they constrained by accountability and time, becomes an HT themselves. An empirical testing of the suggested model is therefore of high importance.

References

Degenhart, M. A. B. (1982). *Evaluation and the value of knowledge*. George Allen and Unwin.

Evans, J. S. B. T. (2020). *Reasoning, rationality and dual processes*. Psychology Press.

Försvarsmakten. (2000). *Pedagogiska grunder*. Försvarsmakten.

Försvarsmakten. (2001). *Pedagogiska grunder*. Försvarsmakten.

Herberg, M. (2022). *Competence for the unforeseen: The importance of individual, social and organizational factors* [Dissertation—PhD]. NTNU.

Kristiansen, E., Magnussen, L. I. & Carlstrøm, E. (Eds.). (2017). *Samvirke. En lærebok i beredskap*. Universitetsforlaget.

Kvernbekk, T. (2005). *Pedagogisk teoridannelse [Pedagogical theory construction]*. Fagbokforlaget.

Lévi-Strauss, C. (1966). *The savage mind*. Weidenfeld and Nicolson.

NIFU/USN. (2021). *Education for the unforeseen*. https://www.theunforeseen.no/

Phenix, P. H. (1964). *Realms of meaning*. Mc-Graw-Hill.

Rawat, S., Boe, O., & Piotrowski, A. (2020). *Military psychology response to post-pandemic reconstruction* (Vol 2). Rawat Publications.

Saeverot, H. (2013). *Indirect pedagogy. Some lessons in existential education*. Springer. https://doi.org/10.1007/978-94-6209-194-8

Saeverot, H. (2022). *Indirect education. Exploring indirectness in teaching and research*. Routledge. https://doi.org/10.4324/9781003193463

Saeverot, H., & Torgersen, G. E. (2020). Time, individuality, and interaction: A case study. In P. Howard, T. Saevi, A. Foran, & G. Biesta (Eds.). *Phenomenology and educational theory in conversation. Back to education itself* (pp. 128–139). Routledge.

Saeverot, H., & Torgersen, G.-E. (2022a). Basic constructs in the science of sustainability education. In H. Saeverot (Ed.). *Meeting the challenges of existential threats through educational innovation* (pp. 11–26). Taylor and Francis. https://doi.org/10.4324/9781003019480

Saeverot, H., & Torgersen, G.-E. (2022b). SSE-based frame of reference. Outlines for a global curriculum—turning existential threats into resources. In H. Saeverot (Ed.). *Meeting the challenges of existential threats through educational innovation* (pp. 27–41). Taylor and Francis. https://doi.org/10.4324/9781003019480

Scriven, M. S. (1969). An introduction to meta-evaluation. *Educational Products Report, 2*(5), 36–38.

Steingartner, W., Galanen, D., & Kozina, A. (2021). Threat defense: Cyber deception approach and education for resilience. Hybrid threats model. *Symmetry, 13*, 597. https://doi.org/10.3390/sym13040597

Stufflebeam, D. L. (2000). The methodology of metaevaluation as reflected in metaevaluations by the Western Michigan University Evaluation Center. *Journal of Personnel Evaluation in Education, 14*(1), 95–125.

Suppe, F. (1989). *The semantic conception of theories and scientific realism.* University of Illinois Press.

Tennyson, R. D., & Foshay, W. R. (2000). Instructional systems development. In S. Tobias, & J. D. Fletcher (Eds.). *Training and retraining: A handbook for business, industry, government and the military* (pp. 111–147). Macmillan.

Tobias, S., & Fletcher, J. D. (Eds.). (2000). *Training and retraining: A handbook for business, industry, government and the military.* Macmillan.

Toiskallio, J. (2006). *Ethics, military pedagogy and action competence.* Finnish National Defence College.

Torgersen, G.-E. (2015). *Pedagogikk for det uforutsette.* Fagbokforlaget.

Torgersen, G. E. (Ed.). (2018). *Interaction: Samhandling under risk (SUR) – A step ahead of the unforeseen.* Cappelen Damm Akademisk.

Torgersen, G.-E., Rawat, S., Boe, O., & Piotrowski, A. (2020). Integrative conclusions to post-pandemic reconstruction. In S. Rawat, O. Boe, & A. Piotrowski (Eds.). *Military psychology response to post-pandemic reconstruction* (Vol 2, pp. 847–852). Rawat Publications.

Torgersen, G.-E., Boe, O., Magnussen, L. I., Olsen, D. S., & Scordato, L. (2024). Innovation in the realm of the unforeseen: A review of competence needed. *Frontiers in Psychology, 15*, 1166878. https://doi.org/10.3389/fpsyg.2024.1166878

Torgersen, G.-E., Boe, O., Rawat, S., & Piotrowski, A. (2022). A pedagogical and psychological model for post-traumatic growth. In S. Rawat, O. Boe, & A. Piotrowski (Eds.). *International perspectives on PTSD and PTG* (pp. 518–523). Rawat Publications.

Weick, K. E. (1993). The collapse of sensemaking in organizations: The Mann Gulch disaster. *Administrative Science Quarterly, 38*(4), 628–652. https://doi.org/10.2307/2393339

PART V
Conclusions

17

TOWARD A HYBRID THREAT RESPONSE MODEL

Odd Jarl Borch and Tormod Heier

Introduction

Throughout this volume, our team of researchers has pursued one overarching aim: to describe hybrid threats and the tools needed to mitigate them. The purpose has been to provide research-based knowledge to a new generation of students, researchers, and civil servants; practical skills and insights that incentivizes public and private stakeholders to take bolder steps for collaborative action. Not least to thwart the countless avenues of malign activities that exploit liberal communities' transparent nature.

This chapter compiles the knowledge presented throughout the chapters. The aim is to deduce and suggest a model of explanation: a *Hybrid Threat Response Model*. This is a model that seeks to explain the contextual relationship between aggressors and their targets, and how a victimized community may organize a robust response. Effective response revolves around three mutually reinforcing lines of operations: *Strategic Planning*, *Operational Execution*, and *Knowledge Development*. To explain the model, the chapter is organized as follows: First, the generic interplay between aggressor states and their target audience are described. Thereafter, we discuss how targeted states may protect themselves by aligning material and non-material resources along the three aforementioned lines of operations. The chapter concludes by deducing the *Hybrid Threat Response Model* and synthesize the book's key findings.

Inter-connectivity between aggressors and target states

In the previous chapters, we have claimed that in a globalized world of mutual dependency, preparedness against hybrid threats does not evolve in

DOI: 10.4324/9781032617916-22

a vacuum. Robust welfare state rests on a broad array of interconnectivities, politically, economically, and technologically. Incidents far away may easily have local implications and cause popular anxiety, just as local activities easily may escalate, spiral out of control, and cause grave international repercussions.

Responses to hybrid threats thereby evolve within a dynamic context. Within a world of inter-connectivity and mutual dependencies, reactions and counterreactions are constantly pitted against each other. Perpetrators and responders are both energized by a broad register of digital, financial, judicial, and physical instruments; these are operational arenas where a diverse conglomerate of state and non-state actors is weaponized (Galeotti, 2022). A dialectic tit-for-tat logic, spurred by mutually opposing wills, characterizes the operational context between aggressors and targeted states.

The aggressor's perspective

As the first part of the book points out, aggressors have few limitations in their code of conduct. Except for the perpetrator, no one knows for sure what the aggressor's avenue of least resistance will look like. The employment of a neatly orchestrated course of action, therefore, allows the aggressor to take the initiative, define the tempo, display agility, and thus impact the calculus of targeted communities. Operating in a global, transparent, and interconnected environment, it is only the perpetrator's cynicism, creativity, resources, and executive skills that limit its' course of action.

One thing, though, is constant: the deliberate search for the responder's critical vulnerabilities and the avoidance of its strongest points. The aggressor will always seek to exploit critical vulnerabilities within the targeted society. Simultaneously, the perpetrator will meticulously protect their own shortcomings. To ensure that critical capabilities are in place is also a precondition for success. Seeking avenues of least resistance is not only logical but also the least risky strategy to avoid an uncontrollable escalation of violence; a situation that easily may spiral out of control (Heier, chapter 8). Attribution may otherwise lead to retaliation or outright war, an outcome that dramatically alters the aggressor's cost-benefit calculus.

Staying below the threshold of war may thereby coincide with the targeted state; a victim that often seeks to avoid uncontrollable escalation and thus wants to hide their cards of retaliation. If hybrid threats intend to operate below the threshold of war, aggressors and their target audiences are bound by a common interest: to avoid full-scale war. Finding the right arena for testing out their actions and counteractions is therefore important. One defensive approach may be to target one sector, one ethnic group, or one specific geographical area, preferably so by means that are difficult to detect and attribute, or unpleasant for the target to criticize. In this way, aggressors may quietly test the targeted state's responses and evaluate their effectiveness. The aggressor may in this respect achieve its goal with minimal efforts. If the response is feeble, hesitant, or inadequate, in accordance with a risk-avoidance strategy, the anomaly will easily be accepted and become the "new normal" inside the targeted community (see Borch, Chapter 12).

Aggressors and their targeted states may also pursue a more offensive strategy of escalation. The aggressor may, as in the case of Russia's 2022 invasion of Ukraine, use hybrid threats as a prelude to war: an instrument that finally culminates into organized violence in full scale. The subtle role of hybrid threats thereby changes. From being supported by the other instruments of power, hybrid threats end up as *the* supporting activity in a broader and more violent confrontation between opposing wills (Mahda et al., Chapter 6; Hordiichuk et al., Chapter 7). Whether the threat is below or above the war threshold, the political objective remains the same: to impose the aggressor's will upon the target audience. Coercion may be used by deliberately targeting the social contract between the citizens and the state—the social fabric that keeps people, governments, and their Armed Forces together as one cohesive unit. This may represent the aggressor's primary target to hit the responder's centre of gravity; the hub from where all resistance and response measures originates from (Clausewitz, 1831/1976, p. 89). Any rupture or crack within the social contract is likely to be exploited ruthlessly by the perpetrator. Success is thereby defined by the systematic degradation of a responder's material capacities, like its Armed Forces. Success is measured along qualitative criteria such as the undermining of social cohesion, confidence, trust, and commitments to a specific set of beliefs and value systems (Ellingsen, chapter 3).

The targeted state's perspective

How may a targeted state or its civic community respond to such threats? The key concern relates to how own vulnerabilities can be safeguarded. The most critical area of protection is the public's confidence; the trust between governments, majors, police chiefs, and generals, on the one hand—and their benevolent, dedicated, and loyal citizens, on the other. Without public confidence, decisive and comprehensive responses are unattainable. Unless responders address hybrid threats effectively and professionally, and with a sense of strategic and public urgency, the populace may instead start questioning the responders' legitimacy. Not necessarily from one day to another, but gradually over time—as weeks, months, and years goes by. The responders must thereby address not only the short-term threat but also long-term implications deeply rooted beneath the surface.

As endemic shortages on basic deliveries wither or do not fulfil the expectations in areas such as fundamental health care, education, transportation, or even proper food, warming, and housing, the social contract becomes strained. The economic and social fragility may easily undermine the political cohesion and the national resilience needed for any government aiming to sustain legitimacy vis-à-vis their fellow citizens. Unless these root causes are addressed, the social fabric—the confidence-based contract between the citizen and the state—will be a key vulnerability. And thereby also a "high value target" for any aggressor (Akrap and Kamenetskyi, Chapter 15). As confidence withers, so may also the responder's first line of defence: the collective, supportive, and dedicated effort from much of the populace. One may lose the motivation of social groups to "stand up and be counted," and thus support their ministers, mayors, police chiefs, and generals as unexpected crisis suddenly arise (Truusa et al, Chapter 16). Popular support, therefore, is a strategic asset that needs to be constantly cultivated and nurtured by authorities at the local, regional, and national level in the chain-of-command.

The targeted state's response system

The contextual interconnectivity between aggressors and target states implies that preparation and responding to hybrid threats rests on a whole-of-government approach; a *modus operandi* where collaborative planning and execution take place along three lines of operations; within an operational context characterized by a myriad of public and private agencies that each and every one tries to align their resources within local municipals and in conjunction with efforts at the regional and national level. First, requirements for strategic planning are explored; thereafter, operational execution and knowledge development.

Strategic planning

The term "strategic planning" involves the organization's leaders and their vision for the future. Not least when it comes to *how* goals and objectives are reached. This is because planning and execution to mitigate hybrid threats take place in a highly politicized landscape. Any wrongdoings or failures may easily accelerate misperceptions, anxiety, or create more rather than less frustrations within a targeted community. Strategic planning therefore involves long- and mid-term planning, but also instant crisis management throughout the organizations' chain-of-command "here and now."

As responders seek to mitigate vulnerabilities, an almost endless range of contingencies must be considered. The rationale is encapsulated by Patrick Cullen (Chapter 4), where states like China exploit its "cultural, economic, political, technological, and private sector engagement with other nations as vectors for ambiguous and plausibly deniable malign hybrid threat activities." Mitigation thereby needs to include a threefold approach: aggressors' economic-financial control over industry and infrastructure; the aggressors' control and pressure over national citizens that may have residential family members in the own country; and the aggressor's co-optation of significant leaders in your own community. Variations in the aggressor's instrument allocation and possible infiltration thereby complicate efficient response planning, that is, because different ethnic or social groups may be targeted—which again may stir more rather than less domestic tension.

Strategic planning must also be a proactive and permanent process, a constant activity that seeks to coordinate ends, ways, and means with a myriad of public and private stakeholders. This again calls for numerous policies, strategies, and doctrines; a hierarchy of planning guidance that serves one purpose: To provide operational executors at the national, regional, and municipal levels with necessary political guidance, support, and resources. This may even include allied resources from other countries if the hybrid threat is assessed as being part of a broader joint international response.

A coherent alignment of political perspectives is necessary, even before any threat has materialized. On the one hand, planning needs to vitalize governmental vigilance and resolve, not least to strengthen public confidence and expectation in times of peril (Berndtsson, Chapter 10). On the other hand, planning should at the same time avoid creating more tension or anxiety: Protecting democratic values, as well as marginalized social groups, gender and equality perspectives, immigrants, and indigenous people's rights are but a few politically sensitive pitfalls that will arise as governmental and municipal bodies perform crisis management.

Cross-sectoral collaboration among different political arenas, government institutions, and sectors are therefore a precondition for success. Of particular concern is the civil-military interaction. The military apparatus is often the state's largest and most professional readiness organization against hostile action. The unrelented quest for an even closer and more efficient exploitation of military skills and competence may proactively energize civilian-led planning and execution. A confidence-based relationship with military subordinates is particularly important in districts. As remotely and highly dispersed municipals often suffer from resource scarcity, informal and personal interactions over years, that is, between regional councilors, mayors, municipal directors, police chiefs and military personnel, are key remedies for inducing pragmatism and reducing inter-agency rivalry (Borch, Chapter 12). Closer civil-military cooperation, on a day-to-day or weekly basis, will thus energize a modern integrated civil-military framework; a whole-of-government approach or a whole-of-society approach that is locally tailored to the individual municipal's unique characteristics. A the same time it may serve as a beacon for a well-integrated civilian and military doctrine (Fiskvik and Heier, Chapter 9). This integrated approach may provide the necessary deterrence in the gray zones between peace and war (Leimbach & Levine, 2021).

Bolder and more pragmatic steps toward integration are, however, not a straightforward process. Aggressors will often seek to put strategic planners into dilemmas. One of the most precarious problems arises when perpetrators prefer to stay below the threshold of war. For the planner, a whole-of-government approach would still require a vivid collaboration between civilian and military resources. But if hybrid threats are combined with military coercion, should strategic planners change their priorities? Should military resources rather be reallocated toward the perpetrator's most dangerous course of action of outright war? If so, the civil-military framework would not only lose key resources but, as the military component slides toward conventional deterrence and kinetic operations, the roles may also change. As war looms and sovereignty is threatened, the civilian component may rapidly end up as the *supporting* rather than the *supported* element (Fiskvik and Heier, Chapter 9).

The ambiguity of strategic planning, below or above the threshold of war, therefore complicates an integrated approach. Resources, such as strategic communication, offensive cyber operations, and military protection of critical civic infrastructure, are particularly exposed. A key remedy is collaborative planning across governmental levels—an intimate and continuous coordination that allows planners to maintain flexibility. This is not least due to rapidly shifting roles between being the supporting versus the supported component. As adaptive perpetrators constantly change their hybrid courses of action, in accordance with changing circumstances, responders have no choice but to follow suit. Collaborative planning, therefore, requires frequent risk assessments and frequent revisions of existing planning documents. This flexibility challenges the traditional long-term planning processes that governments tend to follow, leaving sector-oriented strategies cemented for a four- or even five-year period, without revision. This rigidity may also hamper the comprehensive involvement of external stakeholders, that is, the commercial companies and industrial pillars that must be engaged to alleviate public resource scarcity. Stove piping in accordance with a ministerial sector-organization impedes these efforts.

Operational execution

Operations that are neatly tuned and calibrated to the threat is key to crisis management. Whether operations are executed from a national governmental body, a regional police station, or a mayor's remotely located office, the principle remains the same: calibration and tuning rest on a common situation awareness and coordinated joint action from several response institutions.

A common frame of reference is nevertheless difficult to achieve. Civilian preparedness systems often tend to be fragmented, partly so due to the political sensitivity that accompanies hybrid threat mitigations. Any misguided execution may easily lead to unintended consequences. Not least inside increasingly diverse ethnic, cultural, or even value-based communities (Kasearu et al., Chapter16; Borch, Chapter 12). To reduce risks, therefore, ministers and their civil servants, as well as regional directorates and local agencies, may be inclined to pursue micro-management within their own sectors of responsibility. This tendency is accelerated by the hybrid threat's character: As malign efforts unfold within many sectors simultaneously, responses are likely to be fragmented. For scarcely resourced municipals, a coherent information management regime becomes crucial not only for maintaining an updated threat assessment but also for measuring the public's perceptions of governmental operations.

Intelligence may nevertheless hamper an effective execution: As sources and methods are highly classified and must be protected, and as classified information also is a potential source of internal power, inter-agency rivalry and skepticism may occur. This is partly due to a wider problem of communication and information sharing, often rooted in dysfunctional or problematic inter-organizational relationships and differing mandate. But it may nevertheless motivate decision-makers to pursue a doctrine of "dare to share and compare" rather than a more conservative "need to know" doctrine. This is a principle that sets the tone for a broader and more comprehensive situation awareness (Karlsson and Sandbakken, Chapter 11; Borch, Chapter 12).

A neatly targeted and tailor-made narrative toward own citizens is therefore important; partly so to preclude misperceptions and distrust within local communities, but also to energize the social cohesion that generates a will to defend key values within own communities (Berndtsson, Chapter 10; Truusa et al., Chapter 16).

Such a narrative needs to build on a common situation awareness; a unifying perception that serves as a building block for an imminent orchestration of a coherent response—firmly coordinated across sectors and levels in the chain-of-command. This may include a broad range of "strange bedfellows" that usually do not cooperate, that is, various interest organizations, business companies, media, police and military forces, voluntary organizations, and locally recruited NGOs from a diverse civic community. A common situation awareness nevertheless needs to be made on (i) how hybrid threats evolve, (ii) how these threats impact popular perception of governmental countermeasures, and finally (iii) how overall response efforts underscore political objectives defined by strategic planners.

These parameters are key to continuously adapting and adjusting governmental efforts to new operational realities—circumstances that rapidly may change and need to be substantiated by political guidance inside governmental and municipal agencies. Therefore, within executive bodies, cross-boundary communication channels, common professional standards, a bricolage of different perspectives, and a network of continuous assessments are needed.

But this structure also needs regular trimming and exercise. Not least to foster a culture of creativity, flexibility, cross-boundary interaction, and coordination (Bakken et al., Chapter 14). These activities are operationally crucial to ensure efficient coordination between various instruments of power that are located at the international, national, regional, or local levels. The operative utility also stems from the perennial effort to energize private and public cooperation, such as joint efforts between police and military forces on the one hand, and the myriad of private entrepreneurs and volunteer NGOs tending to mobilize during a crisis. On the other hand, these requirements are particularly evident in the cyber community. In this domain, a multi-national network of cybersecurity experts may on short notice be mobilized and be tasked to protect, attack, or mitigate grave challenges to a target's digital infrastructure (Soldal, Chapter 5).

Refining a whole-of-society response thereby rests on a broad range of governance tools: formal-legal governance through legislation and instructions; cross-boundary organization structures; as well as cementing confidence through common values and norms keeping civic democracies together. An honest, direct, and proactive dialogue with all groups in society is therefore crucial. Public debates and open meetings that seek to identify and legitimize the right balance between the different governance mechanisms are needed. Of particular importance are reasoned laws and regulations; infringements that are easily interpreted as curtailment of democratic citizens' basic freedoms. Such infringements may hamper cross-border collaboration; they may also induce new systems for national surveillance that undermine public privacy. This approach calls for hands-on politicians; executives that have a broad understanding of the threat picture but also possess the communicative skills needed to invite citizens into an honest, realistic, and nuanced dialogue; a trust-based relationship that precludes unnecessary anxiety, frustration, or fear inside communities.

Knowledge development

To achieve the goals set in the discussion above, knowledge development is crucial and a continuous process. This includes the chain-of-command that seeks to mitigate hybrid threats in a highly politicized and sensitive civic society. But also, in the consecutive dialogue that authorities will have with their fellow citizens as the dynamic relationship between aggressors and responders unfolds in unpredictable ways. As pointed out by Ellingsen in Chapter 3, "there is a need to strengthen public vigilance on hybrid threats. A critical and educated population that can distinguish fact from fiction is an extremely important protection mechanism."

Responders therefore need to acknowledge that knowledge development is a strategic imperative; a precondition for neatly anchoring and calibrating civilian and military efforts in ways that mitigate rather than exacerbate the threat. In a hybrid threat environment, therefore, the role of strategic planners and executive agents is much wider than that of first responders alone.

To fulfill political objectives, the response needs to be competent. And competence, if it is to be professional and thus legitimate in the eyes of the public, rests on a knowledge-based performance. It entails deeply institutionalized insights into planning and execution; skills, values, and knowledge that throughout the years have been accumulated, refined, and re-evaluated for continuous improvement. The constantly changing character of hybrid threats therefore requires responders to identify themselves as first-line responders. As important is the self-perception of being an agent for unique and professional expertise. The long-term investment in continuous learning processes, such as embedded procedures for how to improve own planning and execution, has strategic significance. Not least in an operational environment where popular moods and perceptions are volatile and thus easily influenced by malign information campaigns (Soldal, Chapter 5). Such incidents may easily exacerbate an already anxious or polarized community, which again may cause panic and unpredictable movements among citizens. A culture of institutional "open-mindedness" that seeks to exploit new experiences, evaluate, and assess them and eventually translate or exploit the new knowledge into improved routines and procedures is therefore important.

Knowledge development has both a long-term and short-term dimension. In the long term, you must build and maintain an understanding of the range of potential and real threats, and how to respond and create resilience. In the short term, knowledge development is about refining and improving an intelligence system for strategic warning and informed decision-making processes (Sandbakken & Karlsson, Chapter 11). Included are capabilities for multi-source surveillance in a complex security environment with global reach. The knowledge is also gained in close collaboration with international partners and allies. Together, the knowledge serves as a backbone for any responder seeking to differentiate between normal and abnormal patterns in the security environment. For example, what should be regarded as a legitimate political discourse related to policymaking from opposing community groups, and what should be regarded as an emerging and clandestine political warfare campaign directed by a hostile foreign power?

These nuances may be important for governmental legitimacy, but they are also complex. New models for learning and dissemination, based on an indirect pedagogy concept, are needed (Magnussen et al., Chapter 13). Of particular importance is the formal and informal dialogue between agencies and people who, in their daily lives, work in different buildings, in different cities, and in different parts of the country, often with different roles and responsibilities tied to their respective sectors in the state apparatus. By means of more dialogue, that is, through conferences, seminars, table-top exercises, or informal networking, new thoughts, ideas, and experiences are disseminated and reflected upon, verbally and non-verbally. By communicating new experiences and insights, learning is communicated through links to issues of inter-personal and inter-organizational trust. More intimate and direct interaction between the various agencies in a wide range of areas, including private actors, must therefore be stimulated. This interplay may provide a process of interpretation, communication, and consensus building that creates "unity of purpose" and "unity of command" in a context that is often perceived as fragmented, hesitant, and chaotic.

Conclusion

In this chapter, we have described how aggressors and targeted states may act in the context of interconnectivity and mutual dependency. We have explored how this context can be mitigated by combining *Strategic planning*, *Operational execution*, and *Knowledge development*.

We highlight the circular and continuous interplay between strategic planning, operational execution, and knowledge development. The continuous interplay between these three lines of action unfolds within a context of continuous threats and countermeasures. A special focus is put on governmental policies directed toward its citizens in the planning phase, an executive system emphasizing a common situation awareness and response adjustments, and a knowledge development culture embracing systematic response evaluation

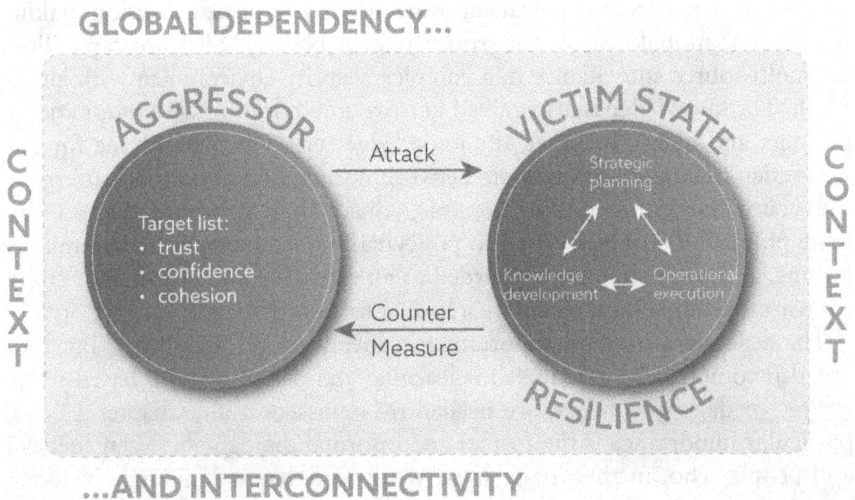

FIGURE 17.1 The hybrid threat response model

Source: created by the authors

and a broad-scale competence building. We believe these lines of operations will strengthen a more resilient and robust society; a community where public confidence is secured in times of peril.

The model's emphasis on a fast-paced development of tailor-made responses and countermeasures resembles with Weissmann et al.'s *Hybridity Blizzard Model* (2021). This is particularly so when it comes to two variables: first, the accelerating range of potential targets that perpetrators easily may identify within transparent liberal communities; and second, the potential leverage that aggressors may achieve by neatly orchestrating a mutually reinforcing specter of means available. A certain external validity is thereby visible.

The model allows us to deduce the following six findings on the book's key question: How may liberal states respond to hybrid threats?

First, there is a need for a continuous revision of policies, strategies, and doctrines based on up-to-date knowledge of the threat picture and bold, forthright decision-makers. The ability to respond effectively below the threshold of war is particularly important as efforts may escalate, and as a deterrent component may be regarded as a force multiplier in territorial disputes.

Second, there is a need to organize a flexible broad-spectrum operational response system. Military capabilities must be well-integrated into seamless collaborative relations with civilian authorities most often in charge of the total preparedness system under the threshold of war.

Third, a knowledge development system must be tailor-made for identifying and interpreting early indications of hybrid threats with an open and creative mindset, with a fine-grained, interactive information network across sectors and decision-levels.

Fourth, the process of learning and dissemination knowledge requires a new and more inclusive pedagogy, more inter-agency dialogue, extensive debate with the public, as well as creative minds engrossed by experimenting and learning from laboratories and real life.

Fifth, a comprehensive approach is needed; this is, in a Nordic parlor, labelled total defense concept: a strategy emphasizing the neatly orchestration of all operational capabilities within the entire society. Crisis management agencies must be flexible enough to change roles as hybrid threats fluctuate above and below the threshold of war. Hybrid threats with covert and unconventional methods may represent a stage in conventional war preparations to confuse responders' perception of imminent war.

Sixth, the sharing of responsibility between civilian and military resources may represent a challenging gray zone where adequate response measures may not be available or implemented too late. Alignment of organizations, knowledge, and culture is important to provide a seamless total defense system.

The Ukrainian experiences explored in this volume underline the importance of taking hybrid threats seriously from day one. There is a need to look into both short- and long-term security implications for the society and the nation as a whole, taking into consideration the innovative efforts of new aggressor instruments, their forces, areas of attacks, and elements of "maskirovka." Worst-case scenarios, where hybrid threats are seen as imminent steps toward broad-range, territorial war, should be at hand, as well as agile, vigilant, and firm decision-makers.

References

Clausewitz, C. (1831/1976). *On war*. Ed. and trans. by P. Paret & M. Howard. Princeton University Press.

Galeotti, M. (2022). *The weaponisation of everything: A field guide to the new way of war*. Yale University Press.

Leimbach, W. B., Jr., & Levine, S. D. (2021). Winning the gray zone: The importance of intermediate force capabilities in implementing the National Defense Strategy. *Comparative Strategy*, *40*(3), 223–234. https://doi.org/10.1080/01495933.2021. 1912490

Weissmann, M., Nilsson, N., Palmertz, B., & Thunholm, P. (Eds.). (2021). *Hybrid warfare. Security and asymmetric conflict in international relations*. I.B. Taurus.

INDEX

Note: Page numbers in *italics* indicate a figure and page numbers in **bold** indicate a table on the corresponding page.

For Product Safety Concerns and Information please contact our EU
representative GPSR@taylorandfrancis.com
Taylor & Francis Verlag GmbH, Kaufingerstraße 24, 80331 München, Germany

www.ingramcontent.com/pod-product-compliance
Lightning Source LLC
Chambersburg PA
CBHW060447240326
41598CB00088B/3897

9 781032 627014